Stefan Behme
**Manufacturing of
Pharmaceutical Proteins**

Related Titles

Gad, S. C. (ed.)

Pharmaceutical Manufacturing Handbook

Production and Processes

2008

ISBN: 978-0-470-25958-0

Gellissen, G. (ed.)

Production of Recombinant Proteins

Novel Microbial and Eukaryotic Expression Systems

2005

ISBN: 978-3-527-31036-4

Gruber, A. C.

Biotech Funding Trends

Insights from Entrepreneurs and Investors

2009

ISBN: 978-3-527-32435-4

Borbye, L. et al.

Industry Immersion Learning

Real-Life Industry Case-Studies in Biotechnology and Business

570 pages

2003

ISBN: 978-0-470-84327-7

Stefan Behme

Manufacturing of Pharmaceutical Proteins

From Technology to Economy

WILEY-VCH

WILEY-VCH Verlag GmbH & Co. KGaA

The Author

Dr.-Ing. Stefan Behme
Gardeschützenweg 64
12203 Berlin

Cover
Cover Picture with kind permission of
Rentschler Biotechnologie GmbH,
Laupheim, Germany.

■ All books published by Wiley-VCH are carefully produced. Nevertheless, authors, editors, and publisher do not warrant the information contained in these books, including this book, to be free of errors. Readers are advised to keep in mind that statements, data, illustrations, procedural details or other items may inadvertently be inaccurate.

Library of Congress Card No.: applied for

British Library Cataloguing-in-Publication Data
A catalogue record for this book is available from the British Library.

Bibliographic information published by the Deutsche Nationalbibliothek
The Deutsche Nationalbibliothek lists this publication in the Deutsche Nationalbibliografie; detailed bibliographic data are available on the Internet at http://dnb.d-nb.de.

© 2009 WILEY-VCH Verlag GmbH & Co. KGaA, Weinheim

All rights reserved (including those of translation into other languages). No part of this book may be reproduced in any form – by photoprinting, microfilm, or any other means – nor transmitted or translated into a machine language without written permission from the publishers. Registered names, trademarks, etc. used in this book, even when not specifically marked as such, are not to be considered unprotected by law.

Cover design   Adam Design, Weinheim
Typesetting   Thomson Digital, Noida, India
Printing   Strauss GmbH, Mörlenbach
Binding   Litges & Dopf GmbH, Heppenheim

Printed in the Federal Republic of Germany
Printed on acid-free paper

ISBN: 978-3-527-32444-6

Contents

Preface *XIII*

Part One Introduction *1*
1 **Biopharmaceutical Production: Value Creation, Product Types and Biological Basics** *3*
1.1 Role of Production in Pharmaceutical Biotechnology *3*
1.1.1 Relationship between Production and Development *6*
1.1.2 Relationship between Production and Marketing *8*
1.2 Product Groups *10*
1.2.1 Vaccines *11*
1.2.2 Pharmaceuticals from Blood and Organs *13*
1.2.3 Recombinant Therapeutic Proteins *13*
1.2.4 Cell and Gene Therapeutics *13*
1.2.5 Antibiotics *16*
1.3 Basics of Biology *17*
1.3.1 Cells and Microorganisms *17*
1.3.1.1 Structure and Types of Cells *18*
1.3.1.2 Metabolism *21*
1.3.1.3 Reproduction and Aging *22*
1.3.1.4 Viruses and Bacteriophages *23*
1.3.1.5 Protein Biosynthesis *24*
1.3.2 The Four Molecular Building Blocks of Biochemistry *26*
1.3.2.1 Proteins *26*
1.3.2.2 Nucleic Acids *31*
1.3.2.3 Polysaccharides *31*
1.3.2.4 Lipids *32*

Part Two Technology *33*
2 **Manufacturing Process** *35*
2.1 Role of the Manufacturing Process in Biotechnology *35*
2.2 Process Schematic and Evaluation *37*

Manufacturing of Pharmaceutical Proteins: From Technology to Economy. Stefan Behme
Copyright © 2009 WILEY-VCH Verlag GmbH & Co. KGaA, Weinheim
ISBN: 978-3-527-32444-6

2.2.1	Drug Substance Manufacturing	38
2.2.2	Drug Product Manufacturing	40
2.2.3	Key Factors for Process Evaluation	41
2.3	Cell Bank	43
2.3.1	Expression Systems	43
2.3.1.1	Microbial Systems	43
2.3.1.2	Metazoan Systems	45
2.3.1.3	Transgenic Systems	46
2.3.2	Manufacturing and Storage of the Cell Bank	46
2.4	Fermentation	48
2.4.1	Basic Principles	48
2.4.1.1	Cell Growth and Product Expression	48
2.4.1.2	Comparison of Batch and Continuous Processes	50
2.4.1.3	Sterility and Sterile Technology	52
2.4.1.4	Comparison of Fermentation with Mammalian Cells and Microorganisms	54
2.4.2	Technologies and Equipment	55
2.4.2.1	Fermentation in Suspension Culture	55
2.4.2.2	Adherent Cell Cultures	56
2.4.2.3	Transgenic Systems	59
2.4.3	Raw Materials and Processing Aids	60
2.4.3.1	Nutrient Media	60
2.4.3.2	Water, Gases and Other Processing Aids	61
2.4.4	Overview of Fermentation	62
2.5	Purification	63
2.5.1	Basic Principles	64
2.5.1.1	Basic Pattern of Purification	64
2.5.1.2	Types of Impurities	68
2.5.1.3	Principles of Separation Technologies	70
2.5.2	Technologies for Cell Separation and Product Isolation	72
2.5.2.1	Cell Separation	72
2.5.2.2	Cell Disruption, Solubilization and Refolding	73
2.5.2.3	Concentration and Stabilization	75
2.5.3	Technologies for Final Purification	79
2.5.3.1	Chromatographic Processes	80
2.5.3.2	Precipitation and Extraction	88
2.5.3.3	Sterile Filtration and Virus Removal	89
2.5.4	Raw Materials and Processing Aids	90
2.5.4.1	Gels for Chromatography	90
2.5.4.2	Membranes for TFF	92
2.5.5	Overview of Purification	93
2.6	Formulation and Filling	95
2.6.1	Basic Principles	95
2.6.2	Freeze-Drying	97
2.7	Labeling and Packaging	98

3	**Analytics** *101*	
3.1	Role of Analytics in Biotechnology *101*	
3.2	Product Analytics *103*	
3.2.1	Identity *104*	
3.2.2	Content *106*	
3.2.3	Purity *106*	
3.2.4	Activity *107*	
3.2.5	Appearance *109*	
3.2.6	Stability *110*	
3.2.7	Quality Criteria of Analytical Methods *111*	
3.2.8	Analytical Methods *112*	
3.2.8.1	Amino Acid Analysis *113*	
3.2.8.2	Protein Sequencing *113*	
3.2.8.3	Peptide Mapping *114*	
3.2.8.4	Protein Content *114*	
3.2.8.5	Electrophoresis *115*	
3.2.8.6	Western Blot *117*	
3.2.8.7	HCP enzyme-linked immunosorbent assay (ELISA) *119*	
3.2.8.8	Analytical Chromatography *120*	
3.2.8.9	Infrared (IR) Spectroscopy *122*	
3.2.8.10	UV/Vis Spectroscopy *122*	
3.2.8.11	Mass Spectrometry *123*	
3.2.8.12	Glycoanalytics *124*	
3.2.8.13	PCR *124*	
3.2.8.14	DNA/RNA Sequencing *125*	
3.2.8.15	Endotoxins and Pyrogen Testing *126*	
3.2.8.16	Bioburden Test *126*	
3.2.8.17	Virus Testing *127*	
3.2.8.18	TEM *127*	
3.2.8.19	Circular Dichroism *127*	
3.2.8.20	Differential Scanning Calorimetry *128*	
3.3	Process Analytics *128*	
3.3.1	Fermentation *129*	
3.3.2	Purification *129*	
3.3.3	Formulation and Packaging *131*	
3.4	Environmental Monitoring *132*	
3.5	Raw Material Testing *133*	
3.6	Product Comparability *134*	

Part Three Pharmacy *137*

4	**Pharmacology and Drug Safety** *139*	
4.1	Action of Drugs in Humans *140*	
4.1.1	Pharmacokinetics *141*	
4.1.2	Pharmacodynamics *144*	

4.1.2.1	Principles of Phenomenological Effects *145*
4.1.2.2	Parameters of Drug Effects *146*
4.2	Routes and Forms of Administration *148*
4.3	Drug Study *149*
4.3.1	Pre-Clinical Study *151*
4.3.2	Clinical Study *152*
4.3.2.1	Phases of Clinical Studies *153*
4.3.2.2	Design and Conduct of Clinical Trials *155*
4.4	Path of the Drug from the Manufacturer to Patients *158*
4.5	Drug Safety *159*
4.5.1	Causes and Classification of Side-Effects *160*
4.5.2	Methods of Supervising Drug Safety (Pharmacovigilance) *162*
4.5.3	Measures at Incidence of Adverse Reactions *162*

Part Four Quality Assurance *165*

5 Fundamentals of Quality Assurance *167*

5.1	Basic Principles *167*
5.2	Benefit of Quality Assurance Activities *167*
5.3	Quality Management According to ISO 9000 *169*
5.3.1	Fields of Activity *169*
5.4	Structure of Quality Management Systems *171*
5.5	Quality Management System Components in the Pharmaceutical Area *173*
5.5.1	Documentation *173*
5.5.2	Failure Prevention and Correction *174*
5.5.3	Responsibility of Management and Training of Personnel *178*
5.5.4	Audits *179*
5.5.5	External Suppliers *180*
5.5.6	Contract Review *181*
5.6	Quality Assurance in Development *181*

6 Quality Assurance in Manufacturing *183*

6.1	GMP *183*
6.1.1	Personnel *188*
6.1.2	Premises and Equipment *189*
6.1.2.1	Measures to Avoid External Contamination *189*
6.1.2.2	Measures to Avoid Cross-Contamination and Product Confusion *192*
6.1.3	Equipment Qualification *195*
6.1.4	Process Validation *197*
6.1.5	Computer Validation *199*
6.1.6	Documentation *200*

6.2	Operative Workflows under GMP Conditions	201
6.2.1	Product Release and Deviation Management	201
6.2.2	Changes in the Manufacturing Process	203
6.3	Production of Investigational Drugs	207

Part Five Pharmaceutical Law 209

7 Pharmaceutical Law and Regulatory Authorities 211
7.1   Fields of Pharmaceutical Law   211
7.2   Bindingness of Regulations   212
7.3   Authorities, Institutions and their Regulations   213
7.3.1   FDA   214
7.3.2   EMEA   216
7.3.3   German Authorities   218
7.3.4   Japanese Authorities   220
7.3.5   Other Important Institutions   221
7.4   Official Enforcement of Regulations   224
7.5   Drug Approval   225

Part Six Production Facilities 227

8 Facility Design 229
8.1   Basic Principles   229
8.2   GMP-Compliant Plant Design   233
8.2.1   Production Flow Diagram   234
8.2.2   Conceptual Plant Layout   236
8.2.3   GMP Flow Analysis   239
8.2.4   Zoning Concept   243
8.3   Basic Concepts for Production Plants   246
8.3.1   Single- and Multi-Product Plants   248
8.3.2   Fractal and Integrated Configuration   250
8.3.3   Flexible and Fixed Piping   251
8.3.4   Steel Tanks and Disposable Equipment   253
8.4   Clean and Plant Utilities   254
8.4.1   Clean Utilities   254
8.4.1.1   Water   254
8.4.1.2   Clean Steam   260
8.4.1.3   Gases and Process Air   261
8.4.2   Plant Utilities   261
8.4.3   Waste Management   263
8.5   Equipment Cleaning   265
8.6   Clean-Rooms   266
8.6.1   Separation of Zones by Clean-Room Design   267
8.6.2   Finishing of Floors, Walls and Ceilings   269
8.6.3   HVAC Installations   269
8.6.4   Qualification   270
8.7   Automation   271

| 8.8 | Quality Control Laboratories 273 |
| 8.9 | Location Factors 273 |

9 Planning, Construction and Commissioning of a Manufacturing Plant 277
9.1 Steps of the Engineering Project 277
9.1.1 Planning 278
9.1.2 Construction 279
9.1.3 Commissioning, Qualification, Validation 279
9.2 Project Schedules 283
9.3 Cost Estimates 285
9.4 Organization of an Engineering Project 286
9.4.1 Expert Groups Involved 286
9.4.2 Role and Selection of Contractors 287
9.4.3 Contracts and Scope Changes 288
9.5 Successful Execution of an Engineering Project 292
9.6 Legal Aspects of Facility Engineering 292
9.6.1 Health, Safety and Environmental Law 293
9.6.2 Building Law 294

Part Seven Economy 297

10 Product Sales and Manufacturing Costs 299
10.1 Lifecycle of a Drug 299
10.2 Position of the Manufacturing Costs in the Overall Cost Framework 303
10.2.1 Basic Principles of Cost Calculation 305
10.2.1.1 Nominal Accounting – Actual Accounting 305
10.2.1.2 Cost Accounting – Profit and Loss Accounting 306
10.2.1.3 Direct Costs – Overhead Costs 306
10.2.1.4 Fixed Costs – Variable Costs 306
10.2.1.5 Relevant and Irrelevant Costs 308
10.2.1.6 Cost Type, Cost Center and Cost Unit 308
10.3 Manufacturing Costs 309
10.3.1 Cost Types 310
10.3.1.1 Depreciation 311
10.3.1.2 Interest 311
10.3.2 Typical Costs of Biotechnological Manufacturing Processes 312
10.3.3 Methods of Calculation 313
10.3.3.1 Cost Calculations 319
10.3.3.2 Profit and Loss Calculation 323

11 Investments 325
11.1 Basic Principles 326
11.1.1 Investment Targets 326

11.1.2	Types of Investments	*326*
11.1.2.1	Classification According to the Object of Investment	*327*
11.1.2.2	Classification According to the Effect of Investment	*328*
11.1.2.3	Classification According to Other Criteria	*328*
11.1.3	Decision Processes	*329*
11.2	Value–Benefit Analysis	*332*
11.3	Investment Appraisal	*334*
11.3.1	Static Methods	*337*
11.3.1.1	Cost Comparison	*338*
11.3.1.2	Profit Comparison	*338*
11.3.1.3	Profitability Comparison	*338*
11.3.1.4	Static Payback Time	*339*
11.3.2	Dynamic Methods	*339*
11.3.2.1	Capital Value	*339*
11.3.2.2	Internal Rate of Return	*340*
11.3.2.3	Annuity	*340*
11.3.2.4	Dynamic Payback Time	*341*
12	**Production Concept**	*343*
12.1	Capacity Planning	*343*
12.2	Dilemma of In-House Manufacturing	*346*
12.3	Aspects of Manufacturing Out-Sourcing	*349*
12.3.1	Types of Cooperation	*350*
12.3.2	Contractual Agreements	*351*
12.3.3	Technology Transfer	*352*
12.3.4	Time Schedules	*354*
12.4	Make-or-Buy Analysis	*355*
12.5	Process Optimization after Market Launch	*357*
12.6	Supply-Chain Management	*359*

References *363*
Index of Abbreviations *369*

Index *375*

Preface

The present book introduces the basic knowledge of industrial manufacturing of biopharmaceuticals. It is written for those wanting to understand the landscape, interfaces and interactions between the different disciplines relevant for production; as such, aspects of technology and analytics, pharmacy, quality assurance, regulatory affairs, facility technology, and economic efficiency are illustrated. The work shall serve as textbook and reference at the same time, and is directed towards students as well as industry-experienced engineers, pharmacists, scientists or economists wanting to acquire a basic knowledge of biotechnological production.

My daily industrial practice has inspired this book. Manufacturing advanced drugs under Good Manufacturing Practice conditions can indeed be a critical factor for drug development and marketing. Being part of multidisciplinary teams, it became obvious to me that the technological and economic challenges of biopharmaceutical manufacturing and its interdependencies with adjacent disciplines are not understood everywhere. Decision making in interdisciplinary teams requires communication and appreciation of the constraints on the various counterparts in order to address them efficiently in the overall program. In contrast to this, particular disciplines become more and more specialized, using their language on a level difficult to understand for the counterparts foreign to the field, sometimes flavoring modern project work with a taste of the tale of the Tower of Babel.

Facilitating communication about manufacturing issues is the goal of this book. It does so by using numerous illustrations and simplifications, making the book easy to read. Correlations between disciplines are highlighted by cross-references and a detailed keyword index facilitates the search for special topics. After having read this book, the reader should have a high-level understanding of the roles, correlations between and terminology of the different disciplines engaged in the production of biopharmaceutical proteins. For those wanting to dig deeper into the topics, literature recommendations and weblinks are provided for further reading.

Manufacturing of Pharmaceutical Proteins: From Technology to Economy. Stefan Behme
Copyright © 2009 WILEY-VCH Verlag GmbH & Co. KGaA, Weinheim
ISBN: 978-3-527-32444-6

I would like to thank Andrea Rothmaler and Andreas Janssen for the valuable input into the manuscript, my students at the Technical University of Dortmund for their instructive questions, and my company Bayer Schering Pharma AG for the opportunity to participate in exciting biotechnological projects.

I hope that my readers will enjoy reading this book as much as I have enjoyed writing it.

Berlin, October 2008 *Stefan Behme*

Part One
Introduction

1
Biopharmaceutical Production: Value Creation, Product Types and Biological Basics

1.1
Role of Production in Pharmaceutical Biotechnology

Over recent years pharmaceutical biotechnology has developed very dynamically. An important driver for this success has been the enormous increase of scientific know-how in the areas of genetics and immunology that has created huge expectations for the development of innovative medicinal treatments.

The scientific pioneer spirit has been fueled by public and private sponsorship, resulting in a biotechnological landscape that has long been dominated by highly innovative, venture capital-based, small- and mid-size companies. However, before patients can benefit from scientific achievements it is necessary that the identified molecule is transformed into a medicine – fit for achieving the therapeutic target – and tested in comprehensive trials in the field. The production of such a medicine has to be carried out in officially licensed, often tailor-made technical manufacturing facilities.

This path from project to product usually lasts several years, and is associated with enormous costs and risks. On average, the development costs of a new compound are in the region of US$500–1000 million and only 10% of all projects that enter clinical trials find their way into the market.   *From project to product*

Facing these immense investments into drug development, the costs of drug manufacturing often seem acceptable, particularly as the costs are absorbed by sales of the marketed drug in the same accounting period; however, safe and efficient product supply is a cornerstone of the companies' success. In biotechnology, the overlap between development and market launch is particularly intensive, motivating companies to take care of manufacturing early on:

- Many targets of process development result from requirements of large-scale manufacturing.

Manufacturing of Pharmaceutical Proteins: From Technology to Economy. Stefan Behme
Copyright © 2009 WILEY-VCH Verlag GmbH & Co. KGaA, Weinheim
ISBN: 978-3-527-32444-6

- The classical separation of development (pre-marketing) and production (post-marketing) does not work for biologics, since both the manufacturing process and plant are factors that determine quality of the final medicinal product.

- Production is the basis for long-term market supply. Decisions about capital investment or outsourcing of manufacturing mostly have to be taken long before the market launch of the product.

- Biotechnological processes are much more difficult to control than small-molecule preparations. The limited ability to monitor and characterize the product results in increased manufacturing risks.

Significance of production in the value chain

The main target of production is to supply the product safely and cost-efficiently. It is positioned between development and marketing of a product. Figure 1.1 illustrates its significance in the value chain:

The chain starts with research that has a clear focus on the identification of targets by analyzing the interaction between the biochemical molecule and its potential therapeutic functionality. In the subsequent development phase, a process for the scale-up and more consistent manufacturing of the molecule is designed. Here, the target structure is developed into a pharmaceutical form, and tested in animals and humans as to its safety and efficacy. Once this is achieved, production

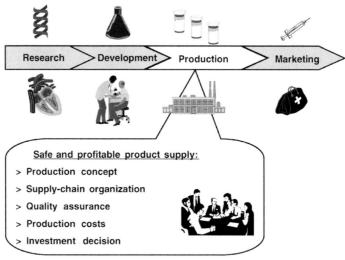

Figure 1.1 Role and tasks of production.

kicks in, taking care of a high-quality and profitable product supply, addressing the following main tasks:

- When, where and in which quantities should the drug be produced? *(Production concept.)*
- How should market supply be organized? *(Supply-chain organization.)*
- How should the quality of the product and Good Manufacturing Practice (GMP) compliance be assured? *(Quality assurance.)*
- What are costs of manufacturing and how can these costs be controlled? *(Manufacturing costs.)*
- How attractive is an investment in one's own facilities? *(Investment decision.)*

The marketing of the product stands at the end of the value chain; from this position, essential goals are formulated for production: supply safety and cost efficiency.

The integrated position of production in the value chain results in interdisciplinary tasks that are best treated in multilateral teams attended by experts in different disciplines like biology, engineering, chemistry, economics, law, pharmacy or medicine.

Production is interdisciplinary

Figure 1.2 shows the subject areas that are important for the understanding and control of production processes and workflows. This volume provides an overview over these subject areas, while special emphasis is given to the interaction between these areas.

Following this introductory Part, Part Two, 'Technology', focuses on process and analytics. This section illustrates why the manufacturing process plays such a large role in biotechnology, and to what extent product quality is determined by processes and analytics. Moreover, essential technologies for industrial manufacturing are described as well as methods and areas of application of analytical testing.

Part Three, 'Pharmacy', briefly elaborates on the basic principles of drug effects in humans and the essential steps of pre-clinical and clinical drug studies. The successful end of the clinical test marks the starting point of commercialization.

Product quality plays a crucial role in pharmaceutical manufacturing. Part Four, 'Quality Assurance', elucidates the organizational and operative workflows for quality assurance, including the rules of GMP.

Almost all activities of commercial production happen in the framework of legal regulations. Part Five, 'Pharmaceutical Law', describes drug regulations and laws as well as institutions and enforcing official authorities.

The translation of process technology into large-scale manufacturing capacities is described in Part Six, 'Production Facilities'. Basic principles of the design of GMP-compliant manufacturing facilities are given and different building concepts compared. The planning process that leads to industrial plants is illustrated. Here, we include a brief look

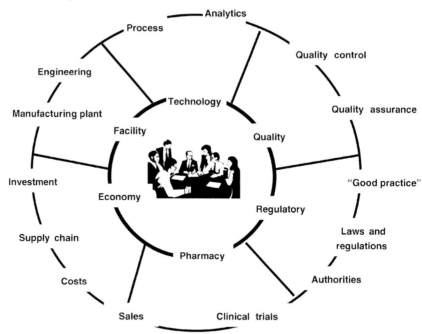

Figure 1.2 Subject areas in production. Inner circle = sections of this book; outer circle = subject areas treated in the sections.

at the regulations regarding health, safety, environment and construction that form the legal framework of industrial production facilities.

Commercial thinking is the spine of efficient production. Part Seven, 'Economy', introduces essential principles around product sales and cost of goods accounting. It compares concepts of in-house manufacturing with outsourcing strategies and elucidates the decision factors leading to capital investments into biotechnological plants.

The book closes with a Bibliography giving literature and web references, and an Appendix providing a list of abbreviations and an alphabetical index of key words.

1.1.1
Relationship between Production and Development

It is widely understood that production starts when development provides a marketable product and a commercially feasible manufacturing process. Ongoing market supply is secured by process optimization or the provision of additional manufacturing capacities, depending on how market demand develops. For biotechnological pharmaceuticals the flexibility to react to demand changes is reduced due to the following reasons.

Drug application as well as the manufacturing process are described and fixed in the regulatory license. Since the biotechnological manufacturing process is a quality-determining factor it has to be finally defined at the time point of regulatory submission and can thereafter only be changed with relatively high effort. The market application contains a proof of the safety and efficacy of the drug; it adds to the complexity that in biotechnology this proof has to be made – at least partly – with material from the commercial process and manufacturing site. Changes to the process or site require comparability exercises that can be more or less complex depending on the risk associated with the change. All of this means that the manufacturing process is fixed at a relatively early time point during development and can only be changed with quite some effort.

This coherence is illustrated in Figure 1.3. Product development consists of clinical development, on the one hand, and development of the manufacturing process and the analytical methods, on the other. The clinical development renders the proof of safe and efficacious use of the drug in humans. Ideally this proof is generated with material from the process and site designated to commercial supply. There is a challenge with this ideal approach: if the process would be finally established and only after that clinical development be initiated, the timelines of development would add up unacceptably. Therefore, the different branches in the development workflow occur in parallel;

Clinical and process development

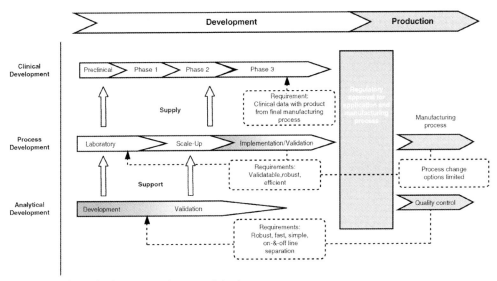

Figure 1.3 Relationship between production and development.

different stages of the clinical development are supplied with different development stages of the manufacturing process.

From lab to large-scale process

Coming from the laboratory scale, the process is evolved step by step into the final and mature manufacturing process. The scale and maturation a technical process achieves until first used for commercial supply depends on the characteristics of the process, the requested product demand and, often, on the available time for development. At the endpoint of development, the process is implemented in the designated commercial supply facility. Product generated in these so-called full-scale runs must be used in representative amounts in the clinical trial.

Validation and critical parameters

Process validation shows that the drug can be manufactured reproducibly and in good quality under consideration of applicable operating procedures. Product generated in these so-called 'validation runs' must usually be used in representative amounts in the clinical trial.

An 'easy-to-validate' process means that product quality is essentially independent of fluctuations of the critical process and equipment parameters. Critical parameters as well as measures to control them should be identified in the lab-scale process. These links result in interactions between production and development long before the actual supply of the commercial market.

Role of analytics

Due to the heterogeneous composition of biological pharmaceuticals, analytical methods play a special role. Just like the process, the developed methods find their way into the regulatory license documentation. Concurrently – while being optimized itself – analytics has to support the process development from very early on. Production requirements such as speed, robustness and simplicity of testing methods have to be taken into account. Moreover, it has to be decided which method should support processing, which method is necessary for product characterization and process validation, and which method should only be used in the development phase.

This short outline illustrates how deeply the aspects of production reach into the development phase. An early recognition of production aspects can help to avoid detours and project delays.

Production facilities

An important interface that is not shown in Figure 1.3 is the one to the facility in which the manufacturing process is carried out. The capital investment into a manufacturing plant, but also the alternative contractual obligation with an outside source, means an additional financial risk of considerable size. This issue is further discussed in the 'Economy' section.

1.1.2
Relationship between Production and Marketing

Production makes the final and packed product available for marketing (Figure 1.4). The packaging provides product protection and a possi-

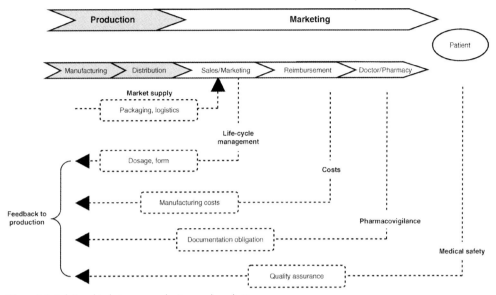

Figure 1.4 Relationship between production and marketing.

bility for attracting customers; especially in the pharmaceutical arena, the packaging contains a considerable amount of user information. The coordination and distribution of the country-specific final products is done by production logistics, which has to react flexibly to requirements from sales and marketing.

Life-cycle projects support the development of the project in the marketplace, and usually concern the pharmaceutical dosage and form, but also the indication of the product. In these cases production has to adapt to changes in demand, packaging materials or formulation processes.

The acceptable manufacturing costs are determined by the achievable price in the marketplace, which is often regulated by country-specific reimbursement systems. The construction or maintenance of manufacturing plants has to be justified by adequate profitability calculations that are based on estimates and expectations of the market situation, and the desired profit margin, on one hand, and the operating and capital expenses, on the other hand. These projections often reach far into the future (more than 10 years) and leave large room for variations.

Cost aspects

While the aforementioned characteristics also apply to other goods, there are indeed pharma-specific features, e.g. the governmental

Specialties of the pharma market

monitoring system, the exceptionally high ethical responsibility of pharmaceutical companies and the official regulation of drug reimbursement. Safety of patients is guaranteed by instruments for pharmacovigilance and intensive product quality assurance. Pharmacovigilance systems serve to register unforeseen adverse effects of drugs and route them to the supervisory body. To achieve this goal the pharmaceutical company collects and evaluates blinded patient data; in case of an unforeseen adverse event, a root-cause analysis has to be performed. To perform this analysis, it is necessary that the specific medicament used by the patient can be traced back to the manufacturing site and batch. It is the specific batch documentation that then provides insight into whether deviations have occurred during the operation that might have influenced the quality of the product. If yes, it needs to be clarified in a second step whether such a quality variation could have triggered the adverse reaction. Thus, the requirements of pharmacovigilance lead to a comprehensive documentation obligation of the whole manufacturing process.

The target of pharmacovigilance is to recognize risks retrospectively. As a complement in the framework of drug safety, there are intensive measures for prospective quality assurance. This has a significant impact on the operational workflows as will be shown in the 'Quality Assurance' section.

1.2
Product Groups

Pharmaceutical biotechnological products can be classified into:

- Vaccines derived from non-genetically modified organisms or blood.

- Therapeutics from blood or animal organs (e.g. Factor VIII, insulin, etc.).

- Antibiotics manufactured traditionally in biological processes. Usually this is done with non-genetically modified organisms.

- Recombinant proteins (i.e. active ingredients) derived from cultivation of genetically modified cells. Including monoclonal antibodies, these represent the biggest sector of current pharmaceutical biotechnology.

- A new branch of therapy opens up with the possibilities of cell and gene therapeutics. These complex interventions into the human body require the reassessment of the pharmaceutical safety concept, and demand special precautions from production technology and engineering.

The focus of the present work lies in the production of therapeutic recombinant proteins; however, the principles described can be applied to the other product groups as well. A closer look at the groups reveals interesting therapeutic and technological overlaps. For example, innovative gene therapy can learn from experiences in virus production gathered in the conventional vaccine field. Also, vaccines will face a new era due to the possibility to produce monoclonal antibodies (Section 1.3.2.1). In the following, the product groups – with the exception of the antibiotics – will be covered in more depth.

Manufacturing technologies of different product groups can be similar

Figure 1.5 schematically shows the production workflows for different product groups. There are differences regarding the genetic modification of the starting material. Genetically modified organisms are mainly deployed for recombinant proteins and gene therapeutics, but cell therapy can also use this technology. The products can be proteins, virus or bacterial fragments, cells, or intact viruses for gene therapy.

1.2.1
Vaccines

There are two principles of vaccination:

- *Passive vaccination*: antibodies against the pathogen are administered.
- *Active vaccination*: the immune system is confronted with alleviated pathogens and builds up its own immune defense against the causative organism.

Antibodies for passive vaccination are prepared by injecting the pathogen into animals. The immune system of the animals pours out so-called polyclonal antibodies into the blood system. Blood is collected from the animals, and the antibodies isolated and purified, so that they can be administered to humans.

Active vaccination uses inactivated germs that are no longer pathogenic, but are still immunogenic. The activation allows the immune system to recognize the real pathogenic germs much faster, and therefore fight them before they can spread out and cause the illness. It suffices to present only a moiety instead of the whole pathogen to enable the immune system to recognize the substance foreign to the body. This moiety can be the hull protein of a virus, whole inactivated cells or pathogen-specific deoxyribonucleic acid (DNA). The general term for these immune response-inducing agents is an 'antigen'. Active vaccines like the influenza vaccine can be proliferated in chicken eggs and reworked to vaccines.

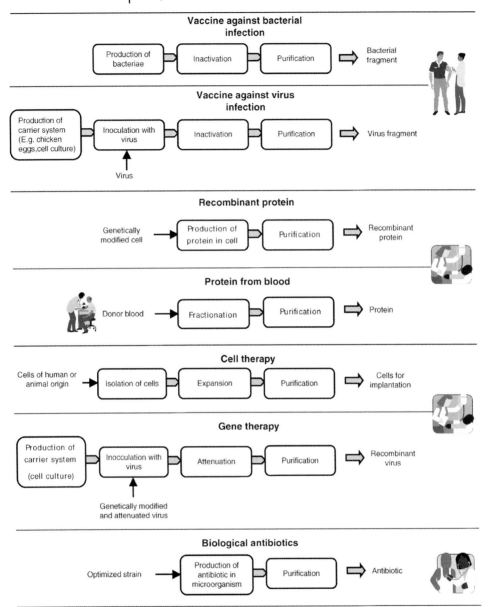

Figure 1.5 Schematic production workflows of important product groups. Product groups are shown on the right. Attenuation = elimination of reproducibility, but retention of infectivity; inactivation = elimination of reproducibility and infectivity.

1.2.2
Pharmaceuticals from Blood and Organs

Many diseases can be attributed to the lack of certain proteins in the blood. In part, these proteins can be extracted from animal or human blood or organs, such as insulin against diabetes or Factor VIII against bleeding disorders. Biotechnology has made it possible for these proteins to be obtained without being tied to these expensive and – under aspects of safety – questionable raw materials from natural sources. In some cases blood-derived products still play a role as it has not yet been possible to successfully replace them completely by recombinant proteins.

To isolate the proteins from the blood, it is first separated into its two main components: plasma and cells. The plasma is further fractionated to obtain the proteins. It is associated with considerable analytical and organizational effort to guarantee the safety of the raw material blood, especially the absence of viral contamination and transmissible spongiform encephalopathy (TSE)-inducing components. Despite the intensive surveillance of blood donors, the danger of safety-relevant incidents persists. It can be expected that the production of proteins will be more and more shifted to recombinant technologies, while whole-blood donations will remain irreplaceable for patient treatment in hospitals.

Plasma fractionation

Risks of protein extraction from blood

1.2.3
Recombinant Therapeutic Proteins

Recombinant proteins, including monoclonal antibodies, make up by far the largest group of biotechnological pharmaceutical products. Table 1.1 shows some examples; in addition to the medical indication and the functionality in the human organism, it provides details regarding the size and type of the molecule. Section 1.3 gives further insight into the structure of the proteins and the terms of amino acids and glycosylation.

It is expected that there will be a huge growth potential for monoclonal antibodies and antibody fragments.

The starting point for all protein production is the genetic modification of the host cell in which the protein should be expressed. The endpoint usually is a parenterally (per injection) administered product in liquid or solid form.

1.2.4
Cell and Gene Therapeutics

Therapeutic proteins are administered to compensate for the lack of the respective natural protein in the organism. Since the molecule is

Table 1.1 Examples for recombinant proteins.

Name	Indication	Functional group	Number of amino acids; glycosylation and fraction of sugars of molecular weight; molecular weight
Insulin	diabetes	hormone	AA 51; Gly no
Human growth hormone	dwarfism	hormone	AA 191; Gly no
Factor VIII	bleeding disorder	clotting factor	AA 2332; Gly to 35%; 300 kDa
Lepirudin	thrombosis	anticoagulant	AA 64; Gly no
Tissue plasminogen activator	thrombosis	thrombolytic agent	AA 72; Gly to 25%; 72 kDa
Interferons (IFNs)	diverse: multiple sclerosis, hepatitis, arthritis, etc.	immune modulator	IFN-β: AA 166; Gly yes and no
Interleukins (ILs) (13 different types)	diverse: asthma, HIV, cancer, mucositis, etc.	immune modulator, signal agent between immune cells	IL-2: AA 133; Gly yes; 15.5 kDa
Erythropoietin	anemia	growth factor	AA 165; Gly to 40%; 34 kDa
Granulocyte colony-stimulating factor (G-CSF), granulocyte macrophage colony-stimulating factor (GM-CSF)	infections, cancer	growth factor	G-CSF: AA 174–180; Gly yes; 19.6 kDa; GM-CSF: AA 127; Gly yes, 15.5, 16.8 and 19.5 kDa
Monoclonal Antibodies	cancer, transplantation, etc.	antibodies	IgG: AA about 1300; Gly yes; about 150 kDa

Gly = glycosylation; AA = amino acids; Gly to 30% = molecular weight fraction of glycosylation can reach up to 30%.

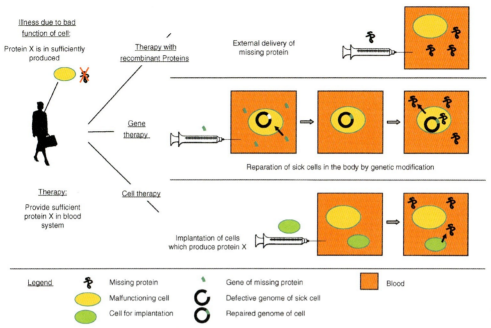

Figure 1.6 Schematic comparison of traditional therapy and cell and gene therapy.

eliminated either by degradation or excretion, the administration has to be repeated to achieve a constant active agent level. In contrast, cell and gene therapy is based on the idea to fight the disease at its source and to enable body cells to express the missing proteins by themselves. Figure 1.6 schematically illustrates the differences between the philosophies of protein versus cell and gene therapy treatment. The starting point is a disease caused by a lack of the example protein X. In conventional therapy, the protein is produced *ex vivo* and injected into the patient. Due to elimination processes the protein disappears after a while. In gene therapy, genetic information is injected together with the ability to infect suitable target cells. This combined capability is generated by means of biotechnological methods: the protein-encoding gene is linked to a molecular 'ferry' (or vector) that carries the gene into the designated target cells. This ferry is a virus that has been modified in such a way such that it retains its infectivity, but lacks its ability to replicate. After having been infected the cells start to produce the desired protein. In the ideal case – if the construct is genetically very stable and the expression rate high – this process has to be carried out only once. If the modified viruses are injected directly into the body, it is called *in vivo* gene therapy, but it is also possible that the cells are extracted from the patient and re-implanted after being infected in the lab (*ex vivo* gene therapy). The latter is basically a crossover between cell and gene therapy.

Cell therapy:
implantation of intact cells

The basic principle of cell therapy is to convey cells to the body that have the desired functionality, thus cell therapy does not aim at repairing dysfunctional cells, but rather replacing them. These cells can originate from animals (xenogenic) or humans, either patient proprietary (autologous) or non-patient proprietary (allogeneic). The allogeneic and xenogenic approaches raise questions regarding immunogenic responses, comparable to tissue rejection in organ transplantation. If the cells are genetically modified, they belong to transformed cell lines; cells which have not experienced any genetic modifications are called primary cells. Thus, allogeneic cell therapy with primary cells denotes a therapy in which cells of a foreign donor are implanted without genetic modification into the patient. The functionality of cells is not restricted to protein production – other molecules can also be expressed. A prominent area of cell therapy is the replacement of tissue or organs (tissue engineering). Consequently, the widely discussed therapy with human stem cells is only one form of cell therapy, which is characterized by the adoption of non-differentiated stem cells.

Cell production for therapy

Cell therapy production starts with isolating the cells intended for implantation. This is clearly a crucial issue of cell therapy, since the starting material is limited and the subsequent expansion of the cells is restricted by the natural limit of generation numbers. The implantable cells are available for surgical implantation after an additional manipulation step like washing or buffer exchange.

Gene production for
therapy

The genetic construct for gene therapy is produced by proliferating the DNA in a host organism which can be first expanded and subsequently transfected with a modified virus. This infected cell produces the desired virus. In order to control the infectivity, the virus can be weakened (attenuated); at the end of this production process the product consists of the attenuated, modified virus.

Safety questions around
cell and gene therapy

Cell as well as gene therapy are at the advent of their development. Despite the fact that the approaches seem plausible, complex questions around drug safety arise. The administration of generally replicable and propagatable substances is very different compared to conventional protein therapy. The scarce source and the handling of the living 'cell' system imposes new challenges on production as well as distribution processes.

1.2.5
Antibiotics

Antibiotics have been produced biologically for many decades. A penicillin-producing yeast strain is cultivated in a biological fermentation step and the expressed penicillin is further purified into a pharmaceutical product. In particular, the fermentation resembles that with recombinant microbial expression systems (Section 2.3.1). In contrast, the purification is different since penicillin is a relatively small and robust molecule compared to proteins.

1.3
Basics of Biology

This section is dedicated to some basic principles of biology and biochemistry that are relevant for the understanding of this book. After a short outline of cell biology and microbiology, the four basic molecular entities of biochemistry are introduced: proteins, nucleic acids, polysaccharides and lipids.

1.3.1
Cells and Microorganisms

Each form of life – plants, animals or microorganisms – consists of biological cells. While plants and animals constitute themselves as enormous networks of different cell types, microorganisms are predominantly single celled.

Microorganisms can survive in their selected habitat independently, while cells of higher organisms depend on their united cell structure. Cells are characterized by some general features:

- Cells contain a carrier of genetic information (double-helical DNA) and a single-strand ribonucleic acid (RNA) derived thereof. During replication – in the process of propagation – its genetic information is prone to erratic variations (mutation).

- Cells exchange nutrients and waste products with the environment (metabolism) for the purpose of energy recovery (catabolism) or substance construction (anabolism).

- They are confined by a membrane which allows for controlled substance exchange with the environment.

- They communicate via so-called receptors with the environment and can react to changes in external conditions.

- They are capable of replicating themselves and a number of higher cells differentiate into other cell types.

Figure 1.7 shows the phylogenetic (i.e. derived from genetic information) tree of living creatures. Bacteria and archaebacteria are prokaryotes; algae, molds, and animal and plant cells are eukaryotes (Section 1.3.1.1).

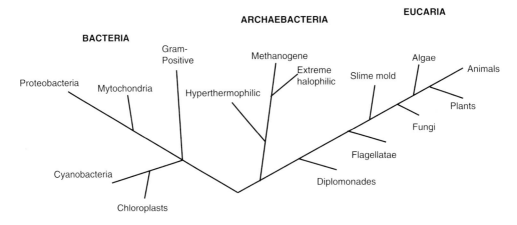

Important examples (genus) of microorganisms

Proteobacteria: *Escherichia, Pseudomonas, Salmonella*

Gram-positive: *Staphylococus, Streptococcus, Lactobacillus, Bacillus, Mycoplasma, Streptomyces*

Molds: *Penicillium, Aspergillus, Saccharomyces, Candida*

Figure 1.7 Phylogenetic tree of different organisms with important examples. The tree is restricted to selected important representatives. Modified from Madigan et al. (2000).

In addition to this genotypic characterization, which has been possible only in recent years thanks to the progress in gene technology, there is a phenotypic characterization based on the differences in shape, movement pattern, staining (Gram staining), metabolic pattern and preferred habitat.

Mycoplasmas are special bacteria. They do not possess a cell wall, and they are exceptionally small and resistant against types of antibiotics that attack the cell wall. As with viruses, they are not retained by filters of a pore size of 0.22 µm. Some members of this family are pathogenic (e.g. *Mycoplasma pneumoniae, Mycoplasma genitalium*).

Viruses are not cells, but consist of encapsulated genetic information in the form of DNA or RNA. They do not have their own metabolism and depend on other biological cells for their replication. Many representatives of this group are pathogenic [e.g. human immunodeficiency virus (HIV), hepatitis, herpes, etc.].

1.3.1.1 Structure and Types of Cells

There are two types of cells: the simple prokaryotic cells and the more complicated eukaryotic cells.

Structure of Prokaryotic Cells (Figure 1.8) Prokaryotic cells consist of a cytoplasm surrounded by a membrane, which is itself surrounded by a stabilizing cell wall. The most important functional units embedded in the cytoplasm are the ribosomes, the chromosome and the plasmids.

Chromosome and plasmids

The chromosome is a single-stranded DNA double helix, and contains the genetic information for the construction and the replication of the cell. Protein biosynthesis from the DNA happens at the ribosomes after the DNA information has been transcribed into RNA. In addition to the chromosome, prokaryotes often carry further genetic information in the so-called plasmids. These are ring-shaped DNA molecules located in the cytoplasm. They usually encode secondary functional proteins like the substances enabling penicillin resistance. The inclusion bodies shown in Figure 1.8 are storage locations for substances that for the time being are not required.

Inclusion bodies and secretion

The contemplated functional units have a high relevance in biotechnological production. Genetic modification can either be made in the plasmids (episomal or extrachromosomal) or in the actual chromosome (integrative or intrachromosomal). The modification consists of the insertion of the protein-encoding DNA sequence into the plasmid or chromosome, which ultimately leads to the production of the protein at the ribosomes (Section 1.3.1.5). This so-called gene expression can be controlled by external factors like temperature or supply of a specific agent (induced promoter), or happen spontaneously as a natural

Gene expression

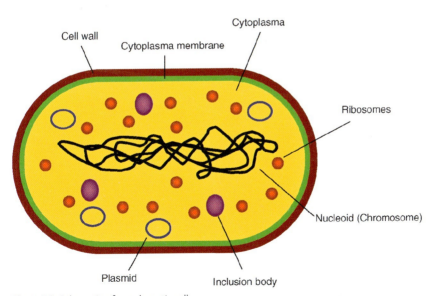

Figure 1.8 Schematic of a prokaryotic cell.

Inclusion bodies

cellular activity (constitutive promoter). If the proteins are delivered through the cell membrane into the environment, this is called 'secretion', but proteins are often aggregated in inclusion bodies (Section 2.3.1.1).

Structure of Eukaryotic Cells (Figure 1.9) Compared to prokaryotic cells, eukaryotic cells are strongly compartmentalized into functional units (organelles). Since they do not exist alone, but in a network of other cells, they have – unlike prokaryotic cells – no cell wall, but a structurally relatively weak cytoplasmic membrane. Unlike prokaryotic cells, the chromosome is contained in a cell nucleus. The genetic information of eukaryotic cells is much more complex than that of prokaryotic cells.

The ribosomes, which are responsible for the synthesis of the proteins, are located outside the nucleus in the cytoplasm. Proteins are transferred from there into the endoplasmic reticulum for further modifications. The mitochondria are the power plants of the cells; they generate the chemical energy to support all cell activities, like molecule synthesis or directed transport of substances. The dictiosomes are the 'glands' of the cell. Their role is to prepare the secretion of agents from the cell. Therapeutic proteins are expressed either via the intrachromosomal integration of a desired DNA sequence or the implantation of plasmids into the cytoplasm. Eukaryotic cells are approximately 10 times larger than prokaryotic cells.

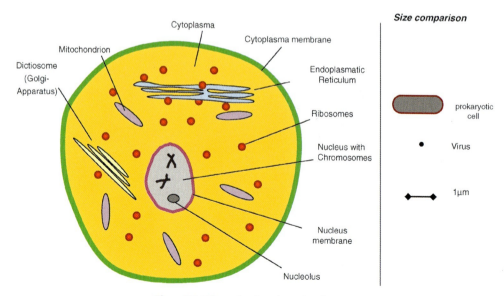

Figure 1.9 Schematic of a eukaryotic cell.

1.3.1.2 Metabolism

Cells constantly exchange substances with their environment. This activity is called metabolism; it serves to generate energy to support the living system (catabolism) and to absorb substances which are needed for creating new cell material during, for example, cell separation (anabolism).

Energy is generated by degrading carbohydrates, transferring them from a high to a low chemical energy level (Figure 1.10). Under addition of oxygen the carbohydrate molecule converts its hydrogen atoms into water and is finally decomposed to carbon dioxide. This oxidation reaction resembles an incineration with the exception that the cell controls the released energy and – along the reaction cascade – stores it in small usable portions, namely the molecule adenosine triphosphate (ATP). *Catabolism*

In the absence of oxygen the carbohydrate cannot be decomposed completely. Unfinished oxidation is called 'glycolysis'; in this case the cell lives under anaerobic conditions (anaerobic = without oxygen). However, most biopharmaceutically relevant organisms depend on aerobic respiration, which means on the supply of elementary oxygen being dissolved in the ambient water. In case of reduced oxygen supply the organisms can reduce their metabolism and produce other substances than if they had sufficient oxygen. That is the main reason why the homogeneous aeration of bioreactors is of great importance.

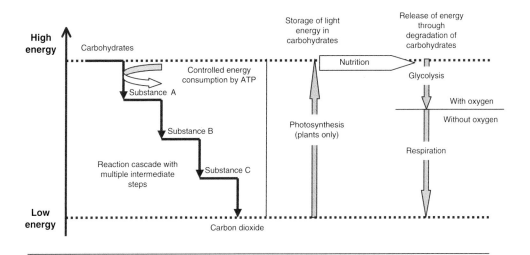

Gross chemical equation of energy generation $(CH_2O)_n + (O_2)_n \rightarrow (CO_2)_n + (H_2O)_n$

Figure 1.10 Schematic of energy metabolism.

Synthesis metabolism

During anabolism (synthesis metabolism) the energy recovered in catabolism is used to create the molecular building blocks of the cell (amino acids, purine and pyrimidine bases, sugar phosphates, organic acids, etc.) from the available nutrients (e.g. glucose, fatty acids, trace elements). The building blocks are utilized to assemble the biological macromolecules constituting the chemical backbone of biological life (nucleic acids, proteins, lipids, polysaccharides; Section 1.3.2). The metabolic pathways consist of highly complex chemical reactions including protein synthesis. The construction plans for the proteins are laid down in the DNA. After being activated this information is transferred to the ribosomes where the actual synthesis takes place.

Nutrient media

In industrial fermentation the nutrients have to be fed with the culture medium. Here, microorganisms are much less demanding than animal cells. While the former just require a menu of fundamental nutrients (carbon source, trace elements), the latter – in order to grow – depend on the provision of 'blood-like' conditions (vitamins, hormones, amino acids).

1.3.1.3 Reproduction and Aging

Cells proliferate by division, resulting in an identical copy of the original cell. The time that a cell needs to divide is determined by the complexity of the chemical reactions needed to synthesize the elements of the new cell. That is why the doubling time of simple prokaryotic cells is much faster than that of the much bigger eukaryotic cells. Under optimal growth conditions, the bacterium *Escherichia coli* needs 15–20 min for division, the yeast *Saccharomyces cerevisiae* 2 h, while mammalian cells have doubling times of up to 24 h.

Time for cell division

Cell growth

Provided that sufficient nutrients are available, microorganisms divide spontaneously and permanently as illustrated in Figure 1.11. Mammalian cells usually need an external trigger for growth. They often form adherent monolayers (i.e. they exclusively grow in single layers bound to surfaces), are contact inhibited and cease to grow once the surface is covered entirely. The total number of divisions is limited for most cell types. In order to achieve an unlimited growth potential, so-called continuous cell lines have been developed. This was achieved by

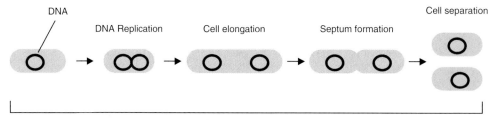

Figure 1.11 Schematic of cell division.

hybridizing the genetic information from a tumor cell and a different desired cell line.

Cells die after reaching their species-specific lifespan (apoptosis) and are prone to mutation. This spontaneous variation of genetic information can be caused by transmission errors during DNA replication as well as other factors. The genetic stability (i.e. the transfer performance of genetic information from generation to generation) is an important criterion for the evaluation of a modified cell line. *Genetic stability*

Section 2.4 (Fermentation) will provide further details on culture doubling times and Section 2.3.1 (Expression Systems) will elaborate on characteristics of different cell types.

1.3.1.4 Viruses and Bacteriophages

Viruses and phages are not constructed as cells, but consist of genetic information in the form of DNA or RNA that is encapsulated in a simple coat. They do not have their own metabolism and depend on cells for proliferation. This goal is achieved by transferring their genetic code into the cells (infection). While viruses affect animal cells, phages affect bacteria. *Viruses and phages are not cells*

Many viruses are famous for being pathogenic. The principle of active vaccination against virus-related diseases (smallpox, influenza) is that inactivated (non-augmentable) viruses are injected in order to evoke an immune response.

The ability of viruses and phages to transfer genetic information to cells has a great significance in biotechnology. A negative aspect is that the risk of viral contamination jeopardizes the safety of the drug; on the positive side, gene technology uses the ability of viruses and phages to introduce genetically modified DNA strands into cells. Both aspects will be further highlighted. *Viruses and phages can transfer genetic information*

The risk of viral contamination generally exists for bacterial and mammalian cells. Due to the inability of bacteriophages to infect human cells, and the inability of human viruses to infect bacterial cells, the virus threat for the patient is higher in mammalian cell culture than in microbial production processes. Figure 1.12 illustrates the main pathways of virus access to the product. There are several processing aids, recovered from human or animal blood, which have to be tested very diligently for the absence of viruses (HIV, hepatitis, etc.). Improper handling (e.g. in the lab when generating the cellular production system; Section 2.3.2) can lead to infection of the cell material. Particularly in this early stage of production, viral contamination can remain undiscovered as it may not lead to a detectable change of the cells. *Viral contamination*

In the simplest case the viruses proliferate in the cells outside of the host genome, destroy the cell and spread out large quantities of viral particles to the environment. Unfortunately, some viruses and phages integrate into the cell genome, are passed on over generations, and are only activated by specific external trigger mechanisms. Others reside in

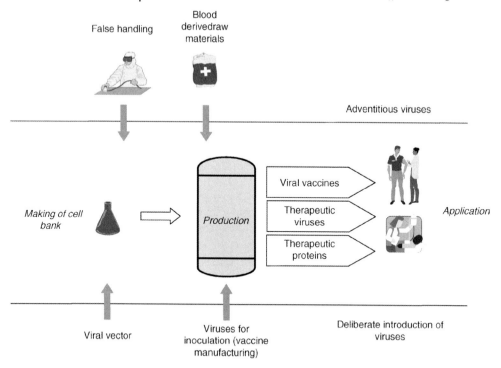

Figure 1.12 Intentional and unintentional addition of viruses (or phages) in the general process scheme. Upper = unintentional contamination; lower = intended functionality.

the cell, and let the cell produce and secrete viral successors continuously. In both cases the contamination can remain undetected for an extended period of time.

In biotechnology the ability of viruses to inject DNA into biological cells is used for genetic manipulation. Viruses function as 'vectors' to introduce genetic information into the genome of cells, which has been implanted beforehand into the virus (transfection). For that purpose the virus is deprived of its pathogenic and cell-destructive effects. Transfection can be used to generate of a host cell line for protein production, to inoculate a culture for vaccine production or for direct injection into humans for the purpose of gene therapy (Section 1.2.4).

1.3.1.5 Protein Biosynthesis

The vast majority of therapeutically active substances in biotechnology are proteins. How can these proteins, which naturally are generated in the human body, be produced outside man?

The answer is as simple as it is surprising: nature has laid down the building plans in the form of the genetic code. This code, which is present in any biological cell as a DNA molecule, is universal – a certain sequence of links in the molecular chain, be it in a bacteria, plant or human cell, leads to the synthesis of essentially the same protein molecule. If the human genetic code for a protein can be introduced into a biological cell, this cell produces the desired protein, which can later be separated and administered as a therapeutic agent.

The important mechanisms of protein biosynthesis are illustrated in Figure 1.13. DNA consists of four molecular elements, the so-called bases: adenine (A), guanine (G), thymine (T) and cytosine (C). The whole genome is composed of these four molecules. A combination of three bases, called a triplet or codon, encodes for one amino acid (i.e. a sequence of codons encodes a certain sequence of amino acids and thus a protein). There are also signal codons that, for example, mark the beginning and the end of a protein assembly.

Protein synthesis starts with the transcription of the DNA into RNA, which is unique since each base of the DNA has a corresponding complementary base in the RNA molecule. The RNA transfers to the

Transcription and translation

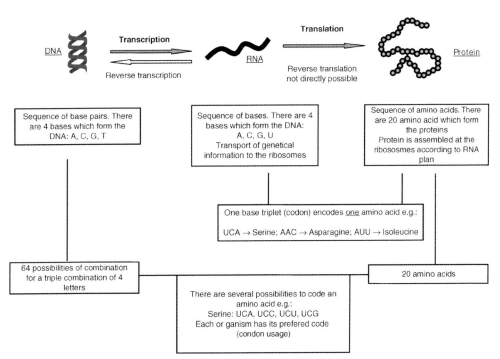

Figure 1.13 Basic principles of protein biosynthesis.

ribosomes, where it is decoded and delivers the information for the protein sequence (translation). After being synthesized at the ribosomes, the proteins then assume their role in and outside the cell.

Gene expression

Thus, the goal of genetic manipulation is to introduce the genetic code for the protein into the cell and let the ribosomes produce the amino acid chain. This realization of genetic information into functional protein molecules is called 'gene expression' and the intensity of the production is called 'expression strength'. The gene can either be integrated into the chromosome or exist extrachromosomally as a plasmid. Since there are 64 possible triplets, but only 20 amino acids, some triplets encode for one and the same amino acid. It appears that different organisms have different preferences for the genetic code. The expression strength is higher if the one code is selected instead of the other. This 'genetic dialect' is called 'codon usage'.

Post-translational modifications

Regardless of which type of cell (plant, bacteria, human or animal) is chosen, the sequence of amino acids in the protein can be generated anywhere; however, the functionality of many therapeutic proteins not only depends on the amino acid sequence, but also to a great extent on the so-called post-translational modifications (Section 1.3.2.1). Usually these modifications can only be realized in eukaryotic cells (yeasts, plants, animals, humans).

1.3.2
The Four Molecular Building Blocks of Biochemistry

Composition of a cell

The technological and therapeutic approaches of biotechnology require knowledge of the basic fundamentals of biochemistry. After all, the fascinating functionality of biological systems is founded on biochemical processes. A typical cell of the simple bacterium *E. coli* consists to about 70% water; the rest is called the dry weight: 55% of the dry weight is a subset of the 2500 known proteins, 20% is RNA and 3% DNA, and 17% polysaccharides and lipids. Approximately 5% is not macromolecules, but substances of low molecular weight like ions, and precursors of amino acids, sugars and nucleotides.

1.3.2.1 Proteins

Proteins function in the body as enzymes, transport proteins, receptor proteins for cellular communication, messenger (hormones), defense agents (antibodies), contractible fibers in muscles, and structural molecules in bones, cartilage and connective tissue.

Amino acids, peptides and proteins

This broad functionality is provided by combinations of only 20 amino acids that are lined up in the protein as a linear chain. Up to 2300 amino acid building blocks are tied together by chemical peptide bonds. These bonds are located between the amino end and the carboxyl

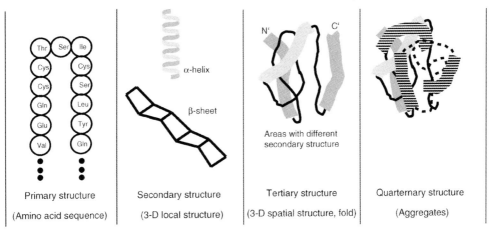

Figure 1.14 Protein structures.

end of two amino acids. The beginning of the chain is called the N-terminus and the end is called the C-terminus. Shorter amino acids molecules of up to 50 amino acids are also denoted as peptides or peptide chains.

The linear chain folds itself into a three-dimensional structure that is of essential significance for the biological functionality of the protein. The structures are illustrated in Figure 1.14. Folding is dominated by the formation of disulfide bridges between cysteine molecules, other post-translational modifications and ambient conditions (pH value, salt content, temperature). As the term indicates, 'post-translational modifications' are changes that are made to the protein after translation. They can have a decisive influence on the biological activity, stability, solubility, plasma half-life and immunogenicity of the protein. There are several types of post-translational modifications; the most important are disulfide bridges and the attachment of sugar molecules to the primary structure (glycosylation). Sugar molecules can make up to 45% of the molecular weight of the protein (Table 1.1). Proteins with similar primary and secondary structure, but different tertiary structure, usually show completely different functionalities. The grouping of equal molecules – denoted as aggregate formation – can also have a positive or negative effect on therapeutic biological activity.

Protein folding

Aggregation

Unfolding of the protein is also referred to as 'denaturation', meaning the destruction of the three-dimensional structure, and typically goes along with a loss of biological activity. As long as primary and secondary structures are intact, the denaturation is usually reversible.

Denaturation

Figure 1.15 shows a typical example for a non-glycosylated protein – human insulin. The protein is composed of two chains linked by

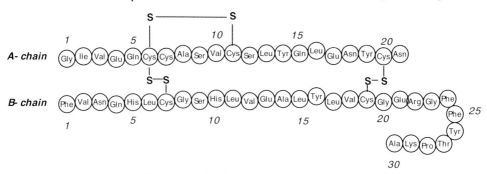

Figure 1.15 Molecular structure of human insulin. S–S = disulfide bridges between cysteine molecules.

disulfide bridges at the cysteine amino acid molecules. The A-chain has 21 amino acids and the B-chain has 30 amino acids. With a molecular weight of 5.7 kDa, it is one of the smaller proteins.

An example of the schematic structure of one of the biggest serum proteins – immunoglobulin G (antibody IgG) – is shown in Figure 1.16. The molecular weight lies between 146 and 165 kDa depending on the molecule type. The molecule has carbohydrate chains (glycosylation) that are essential for the stabilization of the structure.

The boxes in Figure 1.16 indicate folded protein structures. The antibody consists of a so-called crystalline fragment (Fc, bottom) and an antigen-binding fragment (Fab, top). The variable regions are located in the upper part of the Fab, which are adapted to the foreign protein to be attacked. Both fragments are connected by protein chains. In the vertical orientation the antibody is symmetrical and connected by disulfide bridges. The molecule consists of two so-called heavy and two light chains that are connected by disulfide bonds. If antibodies are manufactured from one defined biological cell line they are called monoclonal antibodies. In contrast, polyclonal antibodies are recovered from immunized animals.

In order to understand biological production processes and analytical methods after this short introduction into the structure and function of proteins, it is important to become familiar with some of the physico-chemical properties of proteins.

Amino acids determine the physical properties of the protein

Individual amino acids have special features that determine protein solubility in water or organic solvents and their interaction with solid surfaces. Both properties are important for production as well as medical applications. Neutrally charged, water-adverse (hydrophobic) areas in the molecule are formed by the amino acids glycine, alanine, valine, leucine and isoleucine. Polar, water-friendly (hydrophilic) structures result from the acidic amino acids asparagine acid, asparagine, glutamine acid and

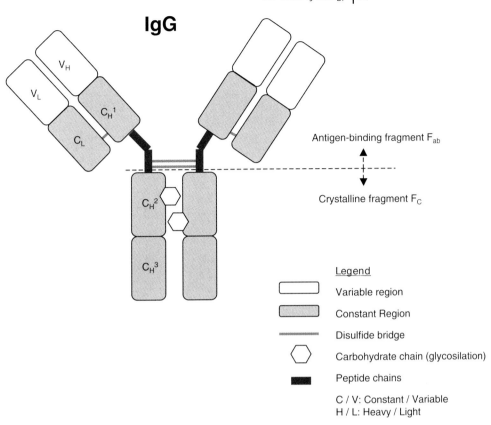

Figure 1.16 Schematic of antibody IgG.

glutamine, and the alkaline amino acids lysine and arginine. The other amino acids exhibit special functional groups or heterocyclic moieties.

Proteins are unstable with regard to temperature, mechanical shear and chemical stress (e.g. contact with oxygen), and are sometimes light sensitive. This has an effect on the applicability of production, sterilization and storage technologies. The types of instabilities can be classified as follows:

Proteins are sensitive to thermal and mechanical stress

Physical instabilities
- *Adsorption:* protein attaches to container material. This results in a decrease of dissolved protein concentration or film formation.
- *Denaturation:* destruction of three-dimensional, primarily tertiary structure.
- *Aggregation:* formation of protein aggregates that cannot only have a negative impact on drug efficacy, but can also be toxic.
- *Precipitation of protein from the solution:* happens by aggregate formation or supersaturation due to temperature shifts.

- *Association:* chemical bonding between proteins that can lead to aggregation and precipitation.

Chemical instabilities
- *Oxidation:* separation of hydrogen atoms and attachment of an oxidizing agent (methionine, histidine, tryptophan, cysteine).
- *Hydrolysis:* clipping of peptide bond.
- *Deamidation:* separation of the amide group (e.g. from asparagine or glutamine).
- *Disulfide-bridge clipping* (cysteine).
- *Beta-elimination:* clipping of other functional groups of the amino acids.

Amphoteric behavior

Proteins show amphoteric behavior – their total charge is influenced by the pH of their environment. This forms the basis for the selectivity of several protein separation processes. Figure 1.17 illustrates this behavior. A protein in acid bulk phase binds to a cation exchanger; the same protein in alkaline conditions binds to an anion exchanger. The pH value where positive and negative partial charges equalize each other is denoted as the 'isoelectric point' (pI). Under these conditions, the protein does not bind to polar matrices; hence, its solubility in water is minimal at the pI, limiting the freedom to change the pH value.

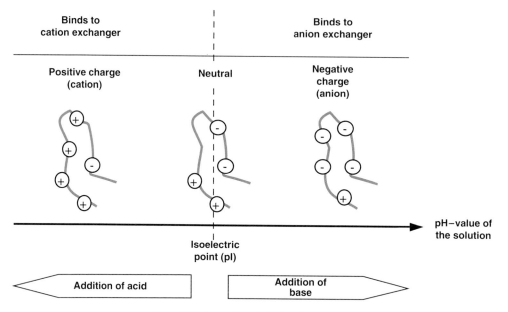

Figure 1.17 Amphoteric behavior of proteins. The ambient pH value determines the charge of the protein.

1.3.2.2 Nucleic Acids

The genetic information of the cell is stored in the ribonucleic acids DNA and RNA. The DNA constitutes the genome of the cell and the RNA is responsible for the translation of the information stored in the DNA into functional proteins.

The DNA is constructed as a chain of nucleotides, which on their part are composed of three elements, out of which the most important is the base. There are four bases – and consequently four nucleotides – which make up the DNA: A, G, T and C. The RNA has an additional base: uracil (U). DNA is structured as a helical double strand in which complementary bases are located opposite to each other. Its size is declared in numbers of complementary base pairs. RNA is a single nucleotide strand with no paired or helical arrangement.

The genetic information can be displayed as a sequence of bases. Each amino acid is encoded by a combination of three subsequent bases (base triplet); for example UCA stands for serine.

The entirety of the genetic information of an organism is called the 'genome'. Each cell of an organism contains the complete genome of *Genome* the organism. The genome of a bacterium is a simple ring-shaped DNA molecule; higher organisms have distributed their genome over several chromosomes each of which is a lengthy DNA molecule. For example, the human genome holds 23 chromosomes with a total number of approximately 3 billion base pairs. The genome of the protozoa *E. coli* has 4.6 million base pairs. At cell division the genome has to be replicated – this last approximately 20 min for *E. coli*, higher organisms need much more time.

1.3.2.3 Polysaccharides

Polysaccharides are molecular chains composed of sugars. Sugars (carbohydrates) are molecules that contain the elements carbon, hydrogen and oxygen in the ratio $1:2:1$ (e.g. glucose $C_6H_{12}O_6$).

Sugars play two important roles in biology: they function (i) as an energy source (glucose, fructose) and (ii) as building material for the cell (glucose, ribose). What role the sugar plays depends on the type of bonding between the links in the sugar chain (glycosidic linkage); one linkage type generates the building material cellulose, the other the energy source glycogen, which is ultimately decomposed by the cell. Another class of polysaccharides is the riboses, which play an important part in the formation and stabilization of the DNA double helix.

Polysaccharides can connect with proteins to form glycoproteins. *Glycoproteins* The vast majority of therapeutic proteins are glycoproteins (Table 1.1). The sugar structures attached to the proteins are manifold. They determine the functionality of the protein since the location and type of integration dominates the folding and flexibility of the protein. Both the type and function of this glycosylation is species specific – a protein

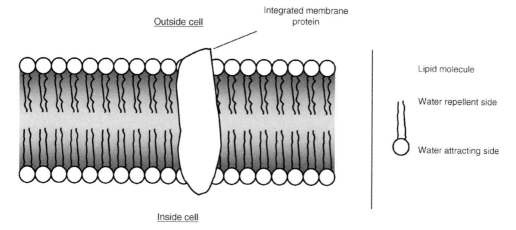

Figure 1.18 Schematic of a cytoplasmic membrane with lipids.

expressed in a yeast cell displays sugar structures different from those expressed in an animal cell. Since this glycosylation pattern does not coincide with that of humans, those proteins usually show different biological activity and immunogenicity from those expressed in human cells.

Lipo polysaccharides

Molecular combinations of sugars with lipids (glycolipids) form a large part of the cell wall of Gram-negative bacteria (lipopolysaccharides).

1.3.2.4 Lipids

Lipids have a water-repelling and a water-attracting side, which makes them ideal structural elements of cytoplasmic membranes. Figure 1.18 shows the structure of such a cytoplasmic membrane and the role of lipids. The water-soluble parts are directed to the cytoplasm and the surrounding aqueous phase. The water-repellant parts are conjunct inside the membrane. Therefore, water-soluble substances cannot simply penetrate the cell wall. Proteins are channeled in a controlled fashion by transport structures embedded in the cell wall. Lipids are formed by fatty acids connected by alcohol molecules.

Part Two
Technology

2
Manufacturing Process

In biotechnology, the manufacturing process determines the quality of the product. First, this chapter will give an overview of the general process scheme. After that the functions of the important manufacturing areas of cell banking, fermentation, purification, formulation, and finally labeling and packaging will be described and their technologies introduced.

2.1
Role of the Manufacturing Process in Biotechnology

Compared to chemically produced substances, biotechnological products are special in two main aspects: they are very large and they are manufactured using living organisms. This results in remarkable variability:

- Molecules of equal primary peptide sequence can have many different structures (folding into secondary and tertiary structures, isoforms, glycosylation sites).
- The same molecule interacts with different substances at specific sites by different interaction mechanisms.
- Proteins are usually unstable with respect to time as well as thermal or mechanical stress.
- Biological products are difficult to characterize and describe.

Molecular weights of pharmaceuticals (kDa)

Aspirin	0.18
Insulin	5.7
Hirudin	7
Erythropoietin	about 34
Immunoglobulin G	about 150
rFactor VIII	about 300

The huge difficulty in characterizing a protein drug is the main reason for the special significance of the manufacturing process. It is virtually impossible to generate the target protein purely with respect to its byproducts or impurities.

Figure 2.1 illustrates how the variability of the process results in product quality fluctuations. The process originates in the host cell, which as a living system is prone to genetic mutations. The synthesis of the target protein is only one step of the metabolic system of the

Variability due to host cell and culture conditions

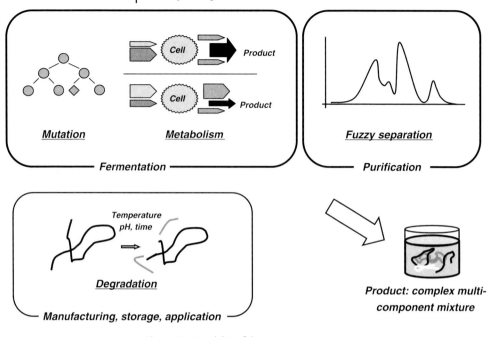

Figure 2.1 Variability of the process generates variable product quality.

organism; fermentation conditions can influence target protein production as well as expression of byproducts. An example for these conditions is the oxygen dissolved in the fermentation medium – in the case of non-homogeneous mixing, it can become unequally distributed and lead to unintended local fluctuations in metabolism.

In purification, the selectivity between the desired main products and undesired side-products is usually small due to the comparable sizes and overall similarity of the molecules. The complex interactions are responsible for the sensitivity of the purification steps with respect to variations in the process parameters. Moreover, the choice of applicable separation methods is limited due to the poor chemical stability, the range of high and similar molecular weights, and the uniformity of interaction mechanisms.

Variability due to poor selectivity

Biotechnological pharmaceuticals are complex mixtures

The product emerging from this complex process chain is more than the actual target protein. It consists of a mixture of substances in which indeed the target protein constitutes the essential ingredient, but whose adjacent components are, amongst others, determined by the process itself.

An additional issue with biologics is that due the complexity of the substances, it is generally impossible to provide an exact analysis of the generated mixture. If significant deviations from the original and described manufacturing process occur, byproducts will usually appear in the product whose identity has not been determined and, moreover, may not even be detected by the implemented analytical methods. Since the clinical trial has been conducted with the original process, it is very difficult to make a statement on the safety of these hitherto unseen substances. That means that the quality of the product cannot be established undoubtedly *ex post* by means of analytical methods. Therefore, in biotechnology the adage applies: 'The process determines the product'.

The progress in analytical characterization has led to a significant improvement of the *in vitro* quality statement; however, the high potential for variations due to the manufacturing process will remain an essential factor for quality assessment of biological products.

2.2 Process Schematic and Evaluation

Figure 2.2 shows the general flow scheme of pharma-biotechnological manufacturing. Essentially there are two main areas: the production of the drug substance [active pharmaceutical ingredient (API)] and the drug product (pharmaceutical production).

In API manufacturing, the drug intermediate is produced. The two main areas of fermentation (creation of target protein) and purification (purification of target protein) can be distinguished. API manufacturing results in the unformulated drug substance in which the protein is presented in a solution suitable for long-term storage.

API manufacturing

Subsequently, in pharmaceutical manufacturing, the API is turned into a pharmaceutical drug (i.e. into a safe and applicable dose and

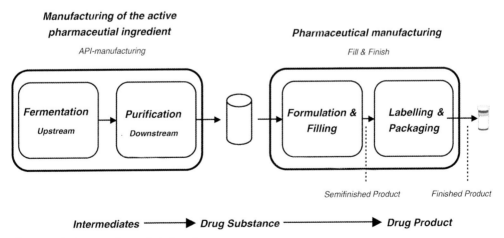

Figure 2.2 Basic terms of manufacturing of biopharmaceutical products.

pharmaceutical form). Here, aspects of clinical use (tolerance, dose), safety (purity, risk of mix-up), logistics (transport container, storage) and patient friendliness (pharmaceutical form, packaging) have to be considered.

In pharmaceutical manufacturing, the areas of formulation/filling and packaging can be distinguished. Formulation usually involves a reworking step like buffer exchange or lyophilization (freeze-drying). When filled into the pharmaceutical primary container (ampoule, vial, syringe), the final drug product is generated. As long as the primary container has not been labeled, the product is denoted as the 'semi-finished product'. The last step in drug manufacturing encompasses the labeling and packaging into the secondary packaging materials.

Pharmaceutical manufacturing

The following sections will provide a more detailed description of the different steps. The variables influencing the economic and technological performance of the manufacturing process steps will be discussed. The subsequent sections go one level further and provide the technical details of the individual unit operations.

2.2.1
Drug Substance Manufacturing

Unit operation: actual technical solution for the realization of a process step

Figure 2.3 presents a typical flow scheme of drug substance manufacture of a therapeutic protein. The process steps are framed black and the targets of the steps are framed gray. The dotted boxes show typical unit operations performed to achieve the targets. Essential raw materials are indicated by solid arrows.

Manufacturing starts with cell cultivation. An aliquot (a vial containing cell material) is taken from the cell bank and incubated on a small scale (shake flask). This is followed by a sequence of scale-up steps, which are typically different in volume by a factor 10, to generate the inoculation culture for the production fermenter. Cells are propagated and the target protein generated in the fermentation step. The nutrients for cell metabolism are supplied via the medium and aeration. After harvesting the cells from the fermenter, the supernatant (the aqueous water phase) is separated from the cell mass; this can be done by centrifugation or filtration.

If the product is expressed intracellularly in inclusion bodies, the cells have to be disrupted, the cell debris separated and the target proteins dissolved in a suitable aqueous solvent. Since after that they may be denatured (Section 1.3.2.1), a refolding step may become necessary that restores the three-dimensional structure and therefore the therapeutic functionality of the protein.

Extracellularly expressed product

After refolding, a solution exists that contains the correctly folded protein and impurities from the preceding process steps. This is comparable to the situation of extracellularly expressed proteins after

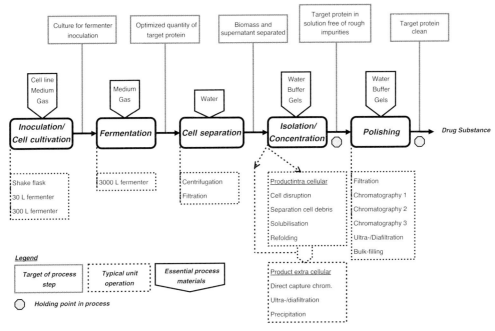

Figure 2.3 Schematic of the manufacturing process of the drug substance. Chrom = chromatography.

abscission of the biomass. In this latter case the product does not reside in the cells, but in the watery supernatant. Isolation and concentration can be carried out by a chromatography step (direct-capture chromatography) and ultrafiltration; additionally, the aqueous buffer can be exchanged by diafiltration in order to condition the solution for the subsequent purification. Before downstream processing starts, there is often a holding point in the process to decouple the processes of fermentation and purification, therefore the buffer is supplemented with protein-stabilizing additives. Apart from buffer solutions and water, chromatography requires gels as the stationary phase. After isolation/concentration the protein is obtained in solution, free of crude impurities. In its entirety the process up to here is denoted as an 'upstream' process.

Ultimate purification is done in polishing ('downstream'), in which a whole range of chromatographic processes can be applied. The target of this section is to remove impurities akin to the product, such as host cell proteins (HCPs), denatured forms of the protein or other byproducts. The end of the process chain is marked by ultra/diafiltration again to condition the product for further processing. Special consideration is given to the bulk filling since this operation requires enhanced hygiene

Purification

precautions. The target of purification is to obtain the dissolved protein as pure as possible and with optimal stability to make long-term storage at moderate conditions possible. The purification is again followed by a holding point in the process chain, decoupling API manufacturing from pharmaceutical manufacturing.

2.2.2
Drug Product Manufacturing

In pharmaceutical manufacturing, the API solution is transformed into a drug appropriate for therapeutic treatment, and that is easy to handle for the patient and doctor. Typically, biopharmaceuticals are delivered for administration in liquid or solid, lyophilized form as a powder. For parenteral injection the powder is reconstituted with water or a physiological solution. Figure 2.4 shows a typical schematic for the pharmaceutical manufacture of therapeutic proteins. The process steps are framed black, the targets of the steps are framed gray and the typical unit operations necessary to realize a process step are framed by the dotted line. Solid arrows indicate essential raw materials.

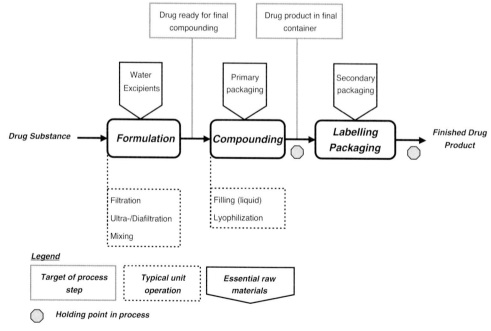

Figure 2.4 Schematic of the manufacturing process of the final product (pharmaceutical manufacturing).

In formulation, the storage buffer can be replaced by a buffer that is suitable for injection. If the product has to be freeze-dried (lyophilized), a cryoprotecting agent has to be added to the liquid to avoid protein degradation during the freezing process. Also, pH adjustment is very common. In the subsequent compounding the liquid is filled into the primary container, which is often a glass vial, but also other types of containers like pre-filled syringes, carpoules or cartridges are used. The product usually is lyophilized if efforts to improve stability of the liquid form have not been successful. *Lyophilization*

After compounding, the product is in its final container. Often at this step there is a holding point to avoid an overlap between the sterile filling process and the non-sterile packaging steps, and to enable storage of goods on a country-unspecific stage, which enhances logistical flexibility.

In the last step – labeling and packaging – the drug gets the country-specific make-up including safety-relevant information as well as the sales and transport packaging. *Labeling and packaging*

The process flow is supported by intensive testing and monitoring, which is described in Chapter 3.

2.2.3
Key Factors for Process Evaluation

In daily practice questions regarding the performance of a process are raised frequently. Main concerns are the costs and simple and robust reproducibility of the process. To answer these questions, in most cases only incomplete information is available; however, not all details are needed to give a fairly good answer to the most important questions. On the other hand, providing answers without knowing these key factors can entail surprises afterwards. The key factors that will be elaborated on in this chapter have to be known; when a production concept (Chapter 12) is put together, process development has to formulate clear goals addressing these factors:

- *Fermentation titer.* The titer is the achievable product concentration in grams per liter fermentation volume at the end of the fermentation. For a given process duration it depends on the cell growth rate, the achievable cell density and the productivity of the individual cell.

- *Overall yield.* Each of the downstream process steps is associated with a more or less significant product loss. The fraction of product that remains after the filling process related to the initial production in the fermenter is denoted as the overall yield.

- *Type and capacity of process steps.* The type and capacity of the process steps determine the type of the plant and the process time. The capacity of an individual step indicates how often such a step has to be performed for the manufacturing of a desired product quantity. It is the leading factor for bottleneck analysis.

- *Process time.* The process time is the main factor for the evaluation and optimization of plant usage.

- *Analytical effort.* The analytical effort for biologicals can be significant. It should be clear which and how many tests have to be performed for validation, in-process control (IPC), release, stability testing and environmental monitoring.

- *Process robustness.* A robust process delivers a consistent product quality within a wide range of process parameter variations. Thus, high robustness simplifies process validation and technical process control. Higher tolerance for variations decreases the risks of rejects. To evaluate the robustness of the process one should look at the specification ranges of the input parameters (e.g. stirrer speed, gas flow rate, temperature range, volume flow, etc.). These ranges should be broad and accomplishable with technical equipment. Ideally, these specifications and their ranges have been established in development studies ('critical parameter studies').

- *Raw materials.* The most important raw materials are media and buffer preparations, chromatography gels, pharmaceutical water, and packing material. Media and excipients of animal origin are sometimes difficult to procure due to the limited offer of certified TSE/bovine spongiform encephalopathy (BSE)-free material (Section 2.4.3.1); moreover, there is a remaining risk for contamination and high regulatory hurdles for approval of processes using these materials. Synthetic media can be expensive and may not support the process as well as their natural equivalents. The operation of high-performance liquid chromatography (HPLC) columns requires facilities suited to handle organic solvents. When using chromatography gels, attention should be paid to the service life and the ability to regenerate. Water quality for both process and cleaning should be adequate for the purpose. Finally, the transfer of processes is significantly simplified if standard packaging materials are used instead of custom-made vials or syringes.

- *Product stability.* Product stability means the ability of a product to retain its properties over a long period of time under defined environmental conditions. Usually, biopharmaceuticals tend to degrade with time and they are sensitive to extreme environmental conditions. Product stability has a significant impact on process design as well as transport and storage conditions. Poor stability enforces a quick turnover after fermentation; downstream processing at cold temperatures of 2–8 °C may be necessary. Virus or protease inactivation steps that purposely apply temperature or pH shifts can go along with high product losses. Intermediates may have to be stored and transported frozen. The less stable the product is, the less feasible it is to establish a safety stock; therefore, securing production may mean to add a back-up capacity. At the end of the supply chain, the distribution to the client is associated with risks, since the sensitive drug leaves the area controlled by the manufacturer. This can cause restrictions to the marketing profile and consequentially disadvantages compared to competitive products.

2.3
Cell Bank

The biological cell is the core of each biotechnological process. It is the 'microreactor' in which the active ingredient is produced. By means of its performance it determines the size of the fermentation process and by means of its byproduct spectrum it determines the type of purification process.

Being living organisms, cells are susceptible to natural mutation. They also possess their own metabolism that adapts to ambient conditions. This variability of the living system stands in direct contrast to the pharmaceutical requirement to deliver a drug with invariable quality.

The synthesis of a target protein by a cell is called 'protein expression'. The entirety of the biological host cell and the implanted modified genetic information is called the 'expression system'. Section 2.3.1 highlights the features of expression systems deployed in production.

The cells that are used to inoculate the fermenter are stored in a cell bank. Due to the potential variability, the manufacturing, storage and characterization of cells is of essence – this will be further illustrated in Section 2.3.2.

2.3.1
Expression Systems

Square one of biotechnological production is protein expression by the genetically modified host cell. This can be achieved with different expression systems, which can be classified into microbial, metazoan and transgenic systems.

2.3.1.1 Microbial Systems
Members of the microbial expression systems are bacteria (e.g. *Escherichia coli* and other bacillus species), yeasts (e.g. *Saccharomyces cerevisiae*, *Pichia pastoris* and *Hansenula polymorpha*) and molds (e.g. *Aspergillus niger*). The leading protagonists are *E. coli* and *S. cerevisiae*.

Microbial systems: bacteria, yeasts and molds

Bacteria and bacillus species are prokaryotes; molds and yeasts are eukaryotes (Section 1.3.1.1). This allows molds and yeasts to glycolyze (Section 1.3.2.1) the proteins, while the prokaryotes cannot. The glycosylation, however, is not comparable to the human sugar pattern – this is a clear disadvantage of microbial systems compared to metazoan systems.

Expression vector

The genetic code for the expression (expression vector) of the target protein can either be active outside of the host cell genome (episomal plasmid DNA) or integrated in the genome. The integrated form often is more stable (i.e. it does not degrade as much over the reproduction cycles as the episomal form). On the other hand, the episomal transformation generally yields a higher number of copies in the cell, which increases the synthesis performance. *E. coli* and *S. cerevisiae* have an episomal expression vector; bacillus species have both integrated and episomal expression vectors.

Intra- and extracellular expression

All microbial systems are characterized by high robustness and unpretentious, inexpensive media. They can be fermented up to scales

---

Box 2.1

Intracellular and extracellular expression

If a cell expresses intracellularly, the protein is not secreted to the surrounding water, but aggregated in the cell in so-called inclusion bodies. The inclusion bodies can be located in the cytoplasm and for *E. coli* also in the periplasmic space the between inner and outer cell wall. Intracellular expression has three distinct disadvantages:

i. The cell has to be disrupted to release the proteins. Therefore, the solution contains many impurities like HCPs or DNA. Gram-negative bacteria like *E. coli* additionally release cell wall fragments, so-called lipopolysaccharides (endotoxins), which cause fever when administered parenterally to humans.

ii. The proteins in the inclusion bodies are aggregated and difficult to dissolve. Their release usually leads to denaturation, after which the molecules have to be refolded into their native structure. This refolding can be associated with significant product losses.

iii. In many cases the transport mechanism out of the cell is a part of the folding process of the proteins (e.g. formation of disulfide bridges in the endoplasmic reticulum). Incorrect folding usually leads to a loss of biological activity (and hence therapeutic efficacy) and cannot always be restored completely.

With extracellular expression, the protein is guided out of the cell into the surrounding water phase, which significantly simplifies purification.

of 100 000 l and reach titers of up to 1000 mg/l fermentation broth. E. coli usually expresses intracellularly into the cytoplasm or the periplasmic space, while the other systems can express either intracellularly or extracellularly.

The advantages of microbial systems lie in their modesty, their fast growth and that they are easy to manipulate. In contrast stands the distinct disadvantage that they cannot assemble post-translational modifications like glycosylation. A further drawback of E. coli is that it expresses into inclusion bodies that are difficult to dissolve. Yeasts coexpress proteases that attack the target protein and express proteins that can cause immune reactions in humans. Products from E. coli include Humulin®, Protropin®, Betaferon® and Proleukin®; products from S. cerevisiae include Novolin®, Refludan®, Leukine® and Revasc®.

Microbial systems are modest

Yeasts produce proteases and immunogenic proteins

2.3.1.2 Metazoan Systems

Cells from insects and mammals are denoted as metazoan (both are eukaryotes). Since mammalian cells can approach the human glycosylation pattern, they have prevailed for the expression of glycoproteins. The most commonly used systems are Chinese hamster ovary (CHO) and baby hamster kidney (BHK) as well as mouse myeloma cell lines (e.g. NS0) for the production of monoclonal antibodies (Section 1.3.2.1). In their natural habitat, mammalian cells are, in contrast to bacteria, embedded in the animal organism and obtain nutrients via blood vessels. Consequently, cell cultures require a much more complex medium than microbial systems. These media contain hormones and growth factors, and often components of animal origin. A further very important difference is their long doubling time. While it takes E. coli 20 min for cleavage, the mammalian cell replicates on average in 20–30 h. Almost all the prevalent cell lines can be cultivated (i.e. they can be almost endlessly proliferated). Moreover, they can be cultivated to relatively high cell density in suspension culture in bioreactors and do not depend on attachment to a solid matrix (non-adherent growth). Mammalian systems express extracellularly and can reach titers of several 1000 mg/l. Typically, the maximum titers reported from production cell lines used for commercial processes currently do not reach over 2000 mg/l; fermentation volumes can be up to 25 000 l. The expression vector of metazoan systems can be integrated or episomal. The disadvantages of mammalian systems are the slow growth rates and the demanding culture conditions; in contrast, they are advantageous when post-translational modifications are essential for protein functionality.

Glycosylation is an advantage of mammalian expression systems

Cell doubling times

Examples of products from CHO cell lines are Epogen® and Recombinate®, and the antibodies Herceptin® and Avastin®. Examples for BHK cell lines are Kogenate® and NovoSeven®.

2.3.1.3 Transgenic Systems

Transgenic systems are genetically modified plants and animals. Examples of plant systems are protein-producing corn, tobacco plants (leafs, segregation through roots), potatoes (accumulation in tuber) or moss cells (segregation through roots). Plants are of benefit as they are of the eukaryotic cell type and therefore principally able to provide the human glycosylation pattern. Examples of animal systems are transgenic sheep or goats, which segregate the proteins into the milk, or chickens, which release the proteins via their eggs. A significant drawback with using animals is the risk of viral transfer to humans.

Transgenic systems are not yet as far developed as microbial or mammalian systems. Their high productivity leads them to be regarded as being competitors to mammalian systems. There are, however, questions of genetic stability and GMP compliance of animal husbandry or plant breeding respectively.

In recent years, mammalian cell systems have dramatically developed and platform technologies with high expression rates have been established, so that it can be questioned whether transgenic systems will prevail in the long run.

2.3.2
Manufacturing and Storage of the Cell Bank

The manufacturing of a cell bank stands at the onset of product development. A change of the expression system in the midst of the clinical trial is associated with significant effort for process and analytical development, and may trigger a restart of the clinical program (Section 4.3). This is due to the fact that the cell substantially influences product composition. An unintentional variation may result in the loss of the marketing authorization for the affected protein drug. Therefore, manufacturing, storage and analytical characterization of the cell bank are of immense importance.

Figure 2.5 illustrates the typical method of the manufacture and use of a cell bank. The expression system suitable for large-scale production is identified very early on in the development program. In laboratory fermentation, this cell is proliferated, aliquoted into a number of cryovials and frozen in the liquid or vapor phase of liquid nitrogen typically between -196 and $-70\,°C$. This cryoconservation stops all cell activities; it not only protects the cell and the genetic information, it also puts the biological clock of the cell on hold, which is important for non-continuous cell lines that have a limited number of cell divisions.

Cryoconservation

Master cell bank

The frozen cryovials of this fermentation are denoted as the 'master cell bank' (MCB). The MCB usually remains unchanged for the entire lifetime of the product, since a change in the expression system

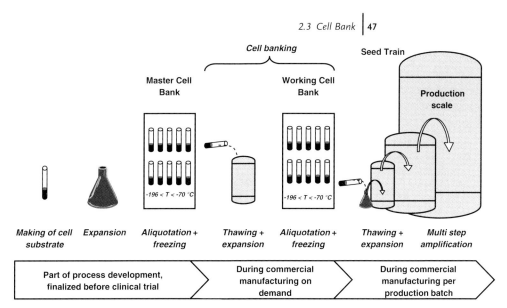

Figure 2.5 Typical path for the generation and use of the cell bank.

would result in significant investments for demonstrating product comparability.

The MCB should be fermented as quickly as possible to guarantee optimal homogeneity of the culture. Extended fermentation times up to higher generation numbers can result in genetic instability (mutation) or contamination of the cells, thus the number of aliquots generated for the MCB typically is too small to inoculate all the production batches throughout the lifetime of the product.

Inoculation is made from the so-called working cell banks (WCB), *Working cell bank* which are generated from time to time out of aliquots of the MCB, typically on the scale of 5–10 l. How often this process has to be run depends, amongst others, on the capacity of the WCB fermentation, the cell density at inoculation and the genetic stability of the expression system up to the end of the final production fermentation.

To start production, one or more cryovials from the WCB are thawed and transferred through a cascade of lab fermenters or shake flasks and one or two further amplification steps into the production fermenter. Typical volume ratios are at a factor of 10 (3–30–300–3000 l).

The time of subcultivation from thawing of the MCB through WCB and amplification adds to the cell lifetime as well as production fermentation.

The cell bank is the most important starting material of the biotechnological process. Therefore, frequent repeated testing of the content, identity and stability of the cells is important.

2.4
Fermentation

In fermentation, the cells that produce the desired protein are cultivated in bioreactors (fermenter). After fermentation the product is attained from the harvested fermenter content by isolation and purification of the target protein.

This section begins with the basic principles of fermentation, and continues to the essential production technologies and features of sterile technology as well as the nutrient media that the cells need to grow. The section closes with an overview over the entire fermentation process.

2.4.1
Basic Principles

The product is generated by cells in fermentation. The more cells are available in the fermentation volume, the higher the product yield. Therefore, two main goals have to be accomplished:

Cell density and cell productivity

- Creation of an environment favorable for cell growth with the aim to reach a high cell density.
- Creation of optimized expression strength with the aim to achieve high productivity per cell.

Apart from of these principal goals, pharmacy and technology dictate additional requirements:

- The spectrum of byproducts expressed by the cell should be as narrow as possible. This is achieved by the creation of highly homogeneous fermentation conditions.
- The spectrum of process-related concomitant substances should be defined, narrow and easily separable. This is achieved by the use of a chemically defined nutrient medium and the reduction of processing aids.

Product yield depends on impurities

The product yield achievable at the end of the overall process to a large extent depends on these impurities since they have to be depleted in elaborate purification steps, each of which is associated with product losses.

2.4.1.1 Cell Growth and Product Expression

Cell growth depends on many parameters like nutrient supply, pH value, temperature, cell density, concentration and toxicity of cell products, and oxygen availability. Cell activity (i.e. growth and metabolism) adapts to environmental conditions. Cells proliferate by division;

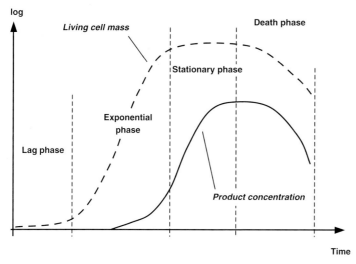

Figure 2.6 Development of the cell mass and the target protein concentration over time.

after having reached its lifespan the cell dies leaving dead biomass in the reactor. Product generation is a part of the metabolism.

Figure 2.6 shows a typical course of the genesis of a cell culture and the product expressed from the cells over time. After inoculating the fermenter at time point zero, the cells need some time to adapt to the new environmental conditions; no growth can be detected here. This lag phase is followed by an exponential growth phase in which the biomass increases by means of cell division. The exponential phase merges into the stationary phase, which is characterized by a dynamic equilibrium between new and dying cells; that is why the living cell mass does not any more change after this point. The limitation of growth can be caused by overconsumption of a specific nutrient (substrate) or formation of a toxic byproduct. In the closed system (batch culture), the course ends with the death phase in which the cell number decreases and, due to the limitation of substrates, no new cells are generated.

The solid line shows a typical course of the product concentration. *Proteolytic degradation* Only once a reasonable cell number is achieved can a measurable quantity of product be detected in the fermentation volume; the curve rises in parallel with the cell mass and reaches a peak once a high cell number with the desired expression rate is present. The subsequent decline of product concentration is usually a consequence of the poor stability of the protein. The protein can lose its three-dimensional structure or be attacked by enzymes breaking the protein chains apart (proteolytic degradation).

The indicated course is only one of several options. Product expression can also start after the lag phase by itself or be induced purposefully by the addition of expression-activating agents.

2.4.1.2 Comparison of Batch and Continuous Processes

The hitherto introduced fermentation course corresponds to the well-established lab experiment in shake flasks. In production, the batch mode means relatively high effort for change-over activities, including equipment cleaning, between the batches. These disadvantages can be circumvented using the continuous mode. Figure 2.7 shows a comparison of both modes. In continuous culture, the fermenter is supplied constantly with fresh medium, which means that the same volume has to be extracted at the outlet. This results in a constant nutrient supply and dilution of growth-inhibiting byproducts. The target protein can be recovered continuously and transferred into a stabilizing medium.

An important prerequisite for continuous culture is the stability of the expression system over the duration of fermentation. A typical application for continuous culture is on hand when a cell line has long doubling time and the product could be damaged if it remains too long in the fermentation broth. The long doubling time results in a long-lasting inoculation culture, making the batch mode less attractive. On the other hand, the long doubling time also means that the number of living

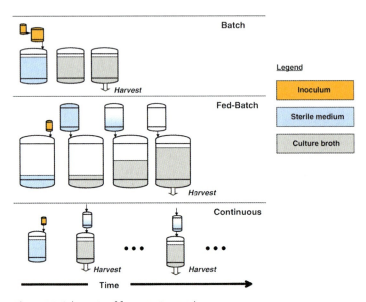

Figure 2.7 Schematic of fermentation modes: batch, fed-batch and continuous mode.

Box 2.2

Calculation example for the difference between continuous and batch culture

A volume of 80 000 l of fermentation supernatant has to be harvested per year to obtain a certain quantity of a product from mammalian cell culture. The cell doubling time is 22 h and the plant can be run 220 days per year for 24 h.

Two technologies are available:

i. Batch culture using an inoculation culture of 1/10 of the production volume.
ii. Continuous culture in which the cell retention system allows an exchange of 10% of the fermenter volume per hour.

For batch culture, it can be calculated that approximately a 4-fold cell doubling is necessary to achieve the inoculation cell density in the whole fermenter. Thus, the time to reach the stationary phase is about 100 h. If 1 day is calculated for each, pre- and post-processing, the entire process lasts 7 days, which results in a possible number of 31 fermentations per year (220 operating days). The volume of the fermenter has to be approximately 2580 l.

For continuous culture, an equivalent consideration results in an hourly extractable volume of 15 l/h [80 000 l/(220 days × 24 h/day]. Multiplication by 10 gives the theoretical fermenter volume of 150 l. In practice, pre- and post-processing times have to be added, moreover in the 220 days, more than one run (here two to three runs) will be made. These times can be considered with a 20% increase of the theoretical volume which ultimately results in the required fermenter volume of 180 l for the continuous operation.

The size difference is remarkable. The batch reactor (2580 l) is 14 times bigger than the continuous reactor (180 l).

---

cells in the fermenter decreases if the cells proliferate slower than being washed out. In order to keep the fermentation in a steady state, dead cells have to be washed out; separating those from the living ones is achieved by cell retention systems, based on differences in cell density or weight.

A comparison between the batch and continuous modes reveals that the continuous mode has significant advantages over the batch mode. This also becomes apparent when analyzing the required fermenter sizes – in general, the continuous fermenter for the same product in the continuous mode can be much smaller than the batch reactor.

Table 2.1 shows a comparison between the batch and continuous cultures. Despite the described advantages of the continuous mode, its disadvantages (i.e. the issue of flexibility) are so significant that up to

Table 2.1 Comparison between batch and continuous culture.

	Batch	Continuous
Disadvantages	– discontinuous product expression – downtime for change over and cleaning – large tanks	– risk of insterility over run time – genetic stability of cell has to be secured – alteration of cells should be precluded by in process controls – high volume flow
Advantages	– facility is more flexible due to shorter process time – easy to compare with lab fermentation	– consistent product quality – fast and continuous product harvesting – smaller fermentation volumes required

now it has not prevailed over the batch mode. The most significant commercial biotechnological product manufactured in the continuous mode is Kogenate® (recombinant Factor VIII).

Fed-batch mode

The advantages of batch and continuous cultures are combined in the fed-batch fermentation mode. Often the fed-batch mode is simply denoted as 'batch mode'. In the fed-batch mode the fermenter is supplied with medium; however, no outlet stream is drawn off. This mode allows for a good control of the process, since nutrients and (if necessary) expression-inducing agents can be added. Fed-batch processes are also run with withdrawal flow and cell retention, approaching the continuous mode fairly closely.

Typical process times for closed-batch cultures are between 20 and 100 h, fed-batch processes are typically between 10 and 30 days, while continuous cultures can reach up to 100 days.

2.4.1.3 Sterility and Sterile Technology

Sterility denotes the absence of undesired living organisms. It is important to operate the complete process sequence that contains living cells sterile – from cell bank generation up to harvesting of the fermenter. However, also after that, in purification, bioburden has to be controlled as much as possible, yet complete sterility is not of essence here. For example, in fermentation, an undesired and fast-growing organism can overgrow the actual host organism and spoil the run. Looking one step further downstream to the filling process of parenteral drugs, sterility is again required. In this latter case the protection of the patient against pathogenic germs is the driving force for sterile processing.

Since our environment is full of microorganisms and viruses, sterile conditions have to be generated. Sterility is achieved by sterilizing process equipment and by operating in facilities specially designed for clean processing. The more 'open' a process is to the environment, the more stringent are the requirements as to the surrounding atmosphere. For example, in cell bank manufacturing, liquids are filled in open glass vials; this process has to be carried out in a work bench with highly purified air to avoid access of foreign germs. When closed vessels like fermenters are operated, it is sufficient to guarantee sterility inside of the vessel, while a certain microbial load may be accepted for the ambient. It should be noted that all product-contacting surfaces and media have to be steamed or filtered, including the tank itself, and the supplied materials like water, nutrient medium and gases. Three main sterilization methods are applied in biotechnological processes:

Sterilization methods

- *Heat sterilization.* This is applied to heat-insensitive equipment and media that do not contain thermal instable substances. Heat sterilization can be performed with dry air or steam. Dry air requires higher temperatures than steam to achieve the same death kinetics; while sterilization with steam is completed after 15 min at 121 °C, dry air sterilization takes 32 min at 170 °C.

- *Sterile filtration.* This is applied to heat-sensitive liquids like nutrient media containing hormones.

- *Irradiation sterilization.* Some liquids like blood serum cannot be filtered or heated. They are exposed to high γ-irradiation in order to deplete bioburden (γ-irradiation is also used for transplants or medical devices, e.g. syringes).

Which method is used depends on the application case. Small parts without temperature-sensitive gaskets (e.g. glass vials or laboratory equipment) can be sterilized with hot air. A typical application is the sterilization tunnel in aseptic vial-filling lines, where the vials are exposed to hot dry air prior to being filled with the product solution. One major advantage of dry air is that it can be brought to high temperatures without having to pressure rate the sterilization equipment, enabling continuous sterilization. High temperatures (250 °C for 30 min) are also necessary for depyrogenation, which means that bacterial endotoxins are inactivated.

Depyrogenation

Where large technical installations made of stainless steel and temperature-sensitive gaskets need to be sterilized, steam is the best choice. Not only does it provide favorable death kinetics, but by transferring its condensation heat to remote spots it also guarantees that all parts of the installation are heated up to the desired temperature relatively fast. Its downside certainly is the remaining condensate water. Process liquids like media or buffer solutions can be heat-sterilized if they do not contain temperature-sensitive substances, otherwise they have to be filtered or irradiated.

The reduction of microbial count during heat-sterilization is qualitatively shown in Figure 2.8. Total elimination cannot always be

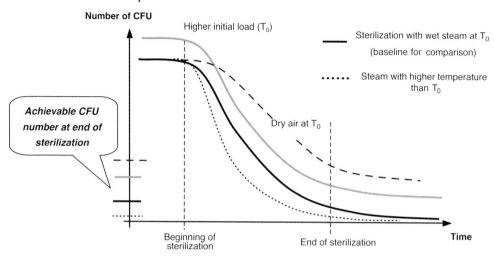

Figure 2.8 Sterilization kinetics. The achievable germ reduction depends on different factors like temperature, time and initial load. CFU = colony-forming unit; T_0 = defined temperature.

achieved in technical processes. The success of sterilization depends on the four parameters time, temperature, initial microbial burden, and sterilization atmosphere. In order to reduce equipment downtime, sterilization should be a short as possible. The remaining three parameters are in part determined by the construction of the tanks, that means how well they can be flushed prior to being steamed or whether they have heat bridges built in resulting in cold spots.

In equipment design, sterile technology deals with the special requirements of sterile processing. Aspects of this special technology are that vessels and piping are designed without dead spaces and heat bridges, and that special gasket materials are identified resisting the sterilization conditions.

Sterile technology

2.4.1.4 Comparison of Fermentation with Mammalian Cells and Microorganisms

Section 2.3.1 gave an introduction into expression systems; the main differences between microbial and mammalian cell cultures were identified:

- Mammalian cells need more and more pretentious nutrients than microorganisms.
- Mammalian cells are more susceptible to variations of their environment as to their metabolism and growth rate.

- Mammalian cells are more sensitive to shear forces applied through stirrer or bubble movement.
- Some mammalian cells grow adherently – they bind to a certain structure and cease to grow after reaching a monocellular layer by the mechanism of contact inhibition. Only genetically manipulated cells as well as naturally non-adherent cells (e.g. blood cells) can grow in suspension culture. The most established mammalian cells, CHO cells, can be adapted to non-adherent growth conditions; all microorganisms can be cultivated in suspension culture.

Suspension culture: detached biological cells float in culture medium

These differences are reflected in the technology and the scale of the used equipment.

2.4.2
Technologies and Equipment

2.4.2.1 Fermentation in Suspension Culture

Suspension cultures are usually run in stirred reactors (Figure 2.9). This type of stirred reactor is by far the most widely used one, for both

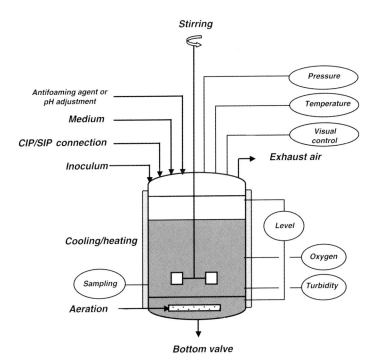

Figure 2.9 Typical fermenter for suspension culture with microorganisms or animal cells.

microbial and mammalian systems. Stirred tanks are available in a volume range from several liters lab-scale up to 75 000 l for industrial production.

The fermentation process starts with steam sterilization of the vessel [sterilization (sanitization) in place (SIP)]. After having cooled down, the sterile medium and the inoculation broth are transferred from upstream tanks (Section 2.4.4). Oxygen is supplied throughout the fermentation by an aeration installation; the waste air is purged over the head of the tank. Media for the fed-batch mode is fed from the media-hold tank. Often an additional port is needed for anti-foaming agents or pH adjustment with sodium hydroxide or phosphoric acid. The fermenter is harvested through the bottom valve.

Purpose of the stirrer

The stirrer serves to homogenize fermentation conditions, to disperse gas bubbles and to improve convective heat transfer by moving the fluid. Moreover, the fluid movement counters aggregation of cells and reduces fouling of the tank walls.

The reactor shown is equipped with a heating jacket that can be used for cooling after sterilization and during exothermal microbial fermentation. For tank sterilization it is charged with steam. During inoculation of all types of cultures and the main fermentation of cell cultures, reactors are moderately heated to create optimal growth conditions. Outside heating is preferred as the tank is easier to clean the less internal fixtures it has.

Tank cleaning

Cleaning agents like diluted sodium hydroxide or water are brought in via the cleaning in place (CIP) port (Section 8.5) and distributed in the vessel through spray balls.

Figure 2.9 shows the essential properties to be measured online. Apart from temperature and pressure, the filling level, dissolved oxygen and turbidity as a measure for cell density are usually of interest. A sight glass enables visual control. A sample port is indispensable for tracking the fermentation course, which is done for process implementation and validation (Section 3.3.1).

The reactors are designed as pressure vessels. High mechanical stress during intended use is caused by the vacuum generated by the condensing sterilization steam. The high number of ports that can be realized from Figure 2.9 means that cleaning and sanitization procedures need to be well defined.

The basic configuration of fermenters for microbial and mammalian systems is similar; however, the remaining differences are so significant that a dual use seems highly unlikely. Table 2.2 gives an overview of the main differences between the two fermenter types.

2.4.2.2 Adherent Cell Cultures

In adherent cell cultures, cells grow in a monocellular layer (monolayer) on a suitable surface and are surrounded by nutrient medium.

Table 2.2 Differences between suspension fermenters for cell culture and microbial fermentation. Modified from (Stadler, 1998).

	Microbial fermentation	Cell culture
System characteristics	– fast growing, 0.5–4 h doubling time – high oxygen demand – duration 18–80 h – cells are relatively robust as to mechanical and thermal stress	– slowly growing, 20–30 h doubling time – low oxygen demand – duration 30–40 days – cells are sensitive to mechanical and thermal stress
Vessel dimensions	– high headspace to accommodate foam formation caused by intensive aeration	– small headspace sufficient
Heating/Cooling	– heating jacket at vessel perimeter with high cooling capacity	– additional heating of vessel top and bottom side necessary to account for temperature sensitivity – moderate heating capacity
Number of ports	– 12–15	– 18–20; redundant ports for risk mitigation due to long runtime
Surface finish	– less stringent requirements, since fast growth and shorter duration minimize the effect of small contaminations	– high requirements, since clean vessel walls are mandatory for long and stable fermentation
Stirrer	– high energy desired to (i) disperse bubbles (high oxygen demand) and (ii) enable heat transfer, that means a special stirrer geometry, thick drive shaft, strong drive, vortex breaker in vessel	– low energy desired to avoid cell disruption – magnetic stirrer possible
Aeration	– high oxygen demand due to fast growth – strong oxygen insertion, big bubbles can be disperged by stirrer	– low oxygen demand due to slow growth – shear forces caused by rise and collapse of big bubbles have to be avoided. aeration equipment generates small bubbles (sintered metal plate)
Waste air system	– designed for 1–2 vessel volumes per minute – droplet separator/condenser needed due to strong fluid movement	– designed for 0.1 vessel volumes per minute – no droplet separator
Instrumentation	– single mounting generally sufficient due to smaller risk – instruments designed for higher flow rates – adjustment of pH value often by dosing of acid/base	– redundant mounting to reduce risk caused by long fermentation time – no foam control – adjustment of pH value often by dosing of carbon dioxide
Automation	– highly variable control system for adaptation to different conditions	
Cleaning	– many spray balls due to complicated stirrer geometry and internal fixtures (flow breaker)	– single spray ball often sufficient

This technology is important in tissue engineering; for the production of recombinant proteins, adherent cell cultures play a minor role.

The main issue with adherently growing cells is that their proliferation is limited due to contact inhibition, which means that they need a large surface to grow to reasonable numbers. Figure 2.10 illustrates the principles of cultivation of adherent cell cultures for protein production.

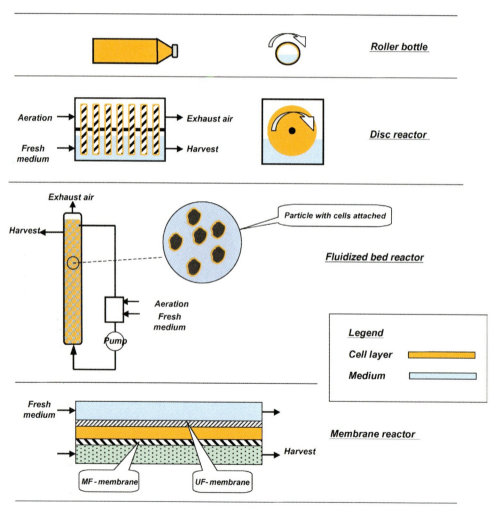

Figure 2.10 Principles of reactors for adherently growing cell cultures. UF = ultrafiltration; MF = microfiltration.

In the roller bottle, the cells grow surface-attached to the bottle wall. *Roller bottle*
By rotating the bottle, the cells come into contact with the liquid medium and the gas phase alternately. For harvesting, the liquid is poured out and thereafter the bottle can be further used by refilling with new medium. The roller bottle is a direct transfer from the lab and huge racks holding hundreds of roller bottles have been constructed for commercial production.

A step towards industrial production is made by using disc reactors, where cells grow on a disc rotating in a horizontally positioned cylindrical vessel, filled half with liquid. Similar to roller bottles, the cells are exposed alternately to the medium and the gas phase.

A further possibility for adherent culture is to immobilize the cells on small particles that can be kept in flotation by a recirculating liquid stream (fluidized bed reactor). The cells are aerated by disperging gas in the recirculating flow.

The membrane reactor is another alternative. The membrane pro- *Membrane reactor*
vides a surface for adherent growth and can also serve to retain cells in the non-adherent case. Figure 2.10 shows a three-compartment membrane module. Cells grow between two membranes. One of them, the ultrafiltration membrane (Section 2.5.2.3), separates the cells from the medium; the other, the microfiltration membrane, separates the cells from a buffer solution that absorbs the product. Thus, two separation steps are coupled with the reaction. On one hand, undesired impurities from the medium can be kept away by the membrane; on the other hand, the ultrafiltration membrane does not allow the product to pass. Issues with this system are limitations for nutrient and product transport through the membranes, and the cleaning.

2.4.2.3 Transgenic Systems

Technologies for product generation in transgenic systems (Section 2.3.1) are very different from each other and only in some cases can they be compared to the technologies described above. From the technological point of view the entire plant or animal is the bioreactor. The main difference to monoclonal cell cultures is that transgenic systems are living structures of many different cells that are embedded in their natural habitat. The systems have in common that protein production is followed by protein purification.

Keeping the cells in their natural environment is an advantage since the organism takes care of nutrient supply by itself; the expensive supply of complex media is obsolete.

Different technologies are available depending on the expression system.

Moss cells grow just like cellular systems in stirred bioreactors that have to be light penetratable to enable photosynthesis (photo-bioreactor). This option cannot be used when working with tobacco plants since their roots have to be in the nutrient media, while their leaves have to be in an atmosphere with air and light. When secretion happens via the roots, GMP-grade greenhouses with filtered air and defined nutrient media can be considered. Rhizosecretion (secretion in roots) is related to membrane reactors. Non-secreting systems like the potato or corn do protect the product better; however, issues analogous to inclusion bodies in microbial systems arise when trying to recover the protein.

GMP plant breeding

Outdoor cultivation of plants raises serious pharmaceutical and environmental concerns. First, the product is exposed to uncontrollable natural influences and the manufacture of seeds could be affected by genetic mutation by pollen dispersal. Second, the carry-over of such genetically modified material into the natural food chain, again by pollen dispersal, should be avoided.

GMP livestock husbandry

Products from animal systems present substantial contamination risks. Apart from potential virus transfer from animals to humans, if milk is chosen as the product carrier, there is *per se* a microbial contamination by the gut flora. Additionally there are risks associated with the hygiene of the animals (excrement, fur, etc.).

2.4.3
Raw Materials and Processing Aids

In fermentation the cells are surrounded by the nutrient medium – an aqueous solution that contains the nutrients important for cellular activity. Its composition is essential for cell growth and metabolism. Apart from optimal productivity, other factors have to be taken into consideration like the introduction of new substances that need to be removed again in purification.

Water quality and generation (Section 8.4.1.1)

Water is by far the most used raw material in fermentation. The following sections provide a short overview over the essential materials used in upstream processing.

2.4.3.1 Nutrient Media

Chmiel (1991) has formulated four technological requirements of cell culture media that also apply to microbial media and are given here in an abbreviated version:

- *Cell integrity criterion*: the medium should support the vital functions and the cellular reactivity of the target cell.
- *Analytical criterion*: the medium components should not interfere with the analytical methods used in the process.
- *Interaction criterion*: the medium components should not interact with the segregated target protein or other segregated proteins.
- *Preparative criterion*: the medium should be composed in such a way that purification is not unnecessarily complicated.

The requirements of the media are very different for cell cultures and microbial fermentations. Microorganisms only need a carbon source, essential salts, amino acids, vitamins and suitable buffers to stabilize the pH value. These media can be obtained from simple sources like yeast extract. Cell culture media need additional growth factors and hormones since the cell culture medium has to resemble the *in vivo* conditions of the cell as far as possible.

At present the high exigencies to cell culture nutrients can often only be satisfied with complex media containing components of animal or human origin. In many cases bovine (calf) or human blood serum has been used; that is why they are called 'serum-containing media'. They are associated with the risk of being contaminated with virus or prions (TSE issue); therefore, these media can only be used under high safety standards (certificate of origin, analytical testing) and new products can hardly be licensed if serum-containing media are used. The limited access to appropriately qualified material also becomes an important economic factor. *TSE issue*

The search for serum-free (i.e. fully synthetic) media to replace complex media is often difficult, but has made some progress in recent years. For example, the fully synthetic medium BM86 contains 94 different chemical substances.

The high-molecular-weight components in cell culture media are thermally instable. Therefore, these media cannot be heat-sterilized inside the reactor as is commonly done with microbial fermentation media. Cell culture media usually are mixed in a media preparation tank and sterile filtered into a sterile media-hold tank or directly into the bioreactor

2.4.3.2 Water, Gases and Other Processing Aids

Apart from the medium, other agents are used in fermentation, such as water, gases and excipients for pH adjustment, induction of protein expression or foam depletion. Here, there are also differences between microbial and mammalian cell culture.

Due to the high water consumption of biotechnological processes, the water quality used is economically relevant. Microbial fermentation can be run with dechlorinated drinking water; cell culture water should go through additional purification steps (desalting, deionization). Purified water [or water for injection (WFI) in pharmacopoeial quality (Section 8.4.1.1)] is neither mandatory nor, in most cases, necessary. Usually more pathogenic substances are introduced by the medium or the cellular metabolism than by the water. Since purified water exists at many production sites already, it is often used for fermentation in order to achieve a high degree of process control, although it may not be necessary in all cases. *Water quality*

Oxygen supply

In microbial fermentations, oxygen is supplied through filtered ambient air. Cell cultures are often supplied with synthetic mixtures of oxygen and nitrogen. Carbon dioxide is added for pH adjustment.

The pH value in microbial systems is controlled with diluted acids (phosphoric acid) or bases (NaOH). The problem of foam formation also exists in microbial cultures due to the high aeration rate and this is controlled with foam-depleting agents. Systems with induced expression require the addition of an inducer.

2.4.4
Overview of Fermentation

The entire fermentation process is composed of the following steps:

i. Preparation.
   - Pre-cleaning.
   - Sterilization.
   - Medium dispensing and preparation.
   - Pre-fermentation.
   - If required: filter tests of downstream harvest processing.
   - If required: pressure test in the case of assembly of parts or cleanliness control.
ii. Actual fermentation process.
iii. Equipment cleaning.

In batch processes, the steps of preparation and cleaning take more time than in continuous processes. Usually the preparation steps can be made in parallel, so that in batch processes an estimated 25% has to be added to the net processing time to calculate plant usage.

Figure 2.11 provides an overview over the typical schematic and the periphery of a fermentation installation in the fed-batch mode. Note the extensive installations for media preparation and storage, pre-fermentation, and cleaning that are essential to operate the main fermenter. For the sake of clarity cleaning installations are not shown – they can make up the biggest part of the tank periphery.

The arrows indicate in/outgoing flows. Fermenter operation is supported by media supply, CIP tanks out of which the main vessel is cleaned and the pre-fermenters for preparing the inoculation culture. The processing aids are media powder, water, air or clean gases, and chemicals for preparation of the cleaning agents and several other purposes. Clean steam is used for heat-sanitization of vessels, piping and prepared heat-stable media. Cold and hot heat-transfer loops support temperature control of the vessels.

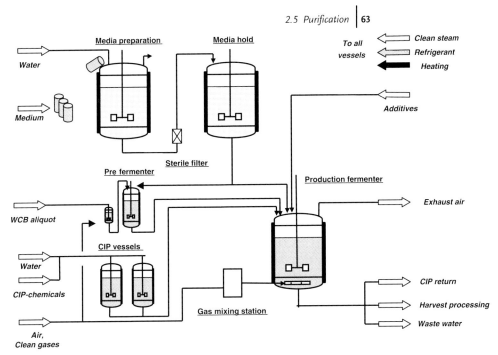

Figure 2.11 Typical installation for a batch or fed-batch process for mammalian cell culture.

The outgoing streams indicated here are the recirculating cleaning agent, the outlet to the harvest treatment and the waste water, which plays a role for cleaning, dumping of contaminated fermentation batches or the discharging of steam condensate after sterilization.

Notably the pre-fermenters require the same periphery as the main fermenter, but on a smaller scale. Therefore, media preparation and pre-fermentation can only be made in parallel if separate media-prep installations support the pre-fermentation. This is an important prerequisite for optimized usage of the plant.

2.5
Purification

Purification follows the harvesting of the fermenter. The goal of purification is to isolate and purify the target protein to achieve a pharmaceutical-grade active ingredient from the biological fermentation broth. This section is dedicated to the pathway that leads from the 'dirty' harvest to the 'pure' product. It starts with a look at typical impurities and basic separation principles in Section 2.5.1. Technologies for cell separation and primary product isolation are described in Section 2.5.2, followed by technologies for final purification in Section 2.5.3. The last two sections deal with processing aids (Section 2.5.4) and give an overview over the process (Section 2.5.5).

2.5.1
Basic Principles

2.5.1.1 Basic Pattern of Purification

In purification, three essential sections can be considered:

i. *Cell separation*: separation of cells from the harvest.
ii. *Concentration/isolation*: reduction of the volume to be purified and separation of easily separable impurities.
iii. *Final purification*: purification of the protein from substances difficult to separate.

Figure 2.12 schematically shows the path of the product solution from harvest to purified protein. For segregating expression systems (yeasts and mammalian cells, see Section 2.3.1) the primary step after separation of the cells is to eliminate easily removable fermentation residues. After that the volume of the protein-containing solution is diminished by removing the impure water, resulting in increased protein concentration. In the subsequent final purification the more persistent substances are removed, namely molecules that are similar to the target protein in size and solubility like HCPs or DNA and degradation products of the protein itself.

Non-segregated proteins exist in inclusion bodies at the end of fermentation. The cells have to be disrupted and the cell debris has to be separated. After being released from the cell the protein usually is aggregated and folded incorrectly. The aggregates are dissolved in a solubilization step with the help of a suitable agent. Subsequently, the protein is refolded into its biologically active form. Now the final purification – similar to the segregating type – can follow.

The letters in Figure 2.12, indicating the process steps, are placeholders for several unit operations (e.g. the polishing can indeed encompass between six and 10 individual steps including chromatography, ultra/diafiltration, precipitation, virus inactivation/filtration and filtration for bioburden reduction). All this will be further described in the subsequent sections.

Despite the fact that selectivity for the target protein is the main goal of downstream development, separation of impurities without loosing product is not possible. As a rule of thumb one can say: 'The more

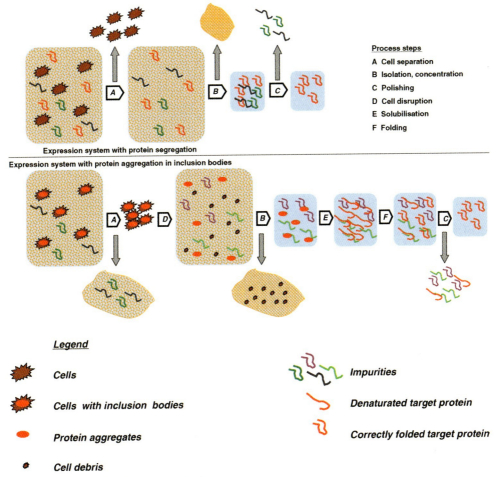

Figure 2.12 The path from harvest to purified target protein. Upper = host cell expresses in the surrounding medium; lower = host cell expresses in inclusion bodies.

impurity and target protein differ from each other, the lower are the product losses at the separation step'.

For example, cells can be easily separated from the aqueous solution by mechanical processes (filtration, centrifugation), since these processes only barely affect the protein dissolved in water. Admittedly product losses also occur here since the product is dissolved in the water portion dumped together with the discharged cells. In this example the product losses can be minimized by reducing the residual water in the discharged cell remainder.

Process yield

Product losses caused by a separation step are expressed by the process yield. It represents the share of product which can be found at the end of the process step relative to the quantity which was fed to the step. Since product losses always occur, the yield is always smaller than 100%.

$$\text{Yield of separation step in\%} = \frac{\text{Product quantity at outlet}}{\text{Product quantity at inlet}} \times 100. \quad (2.1)$$

The step yields accumulate in a sequence of separation steps and have to be multiplied to get the overall yield. Product losses of a separation sequence can be significant. Figure 2.13 illustrates the overall yield of a purification process for different step yields and varying numbers of separation steps.

Obviously the number of separation steps should be kept as low as possible and the step yield as high as possible. Indeed, often the step yield stands in competition to product purity. For example, in chromatographic separations (Section 2.5.3.1), product and impurity are separated, but in most cases both overlap, which means that a high yield can only be achieved by accepting impurities that are present in the overlapping part. Typical industrial overall yields are between 30 and 70%.

The volume of fermentation is dominated by the titer; the volume of purification is dominated by product quantity

As previously discussed, the volume of the process flow is reduced to minimize the size of downstream equipment. An important finding in this context is that the equipment volume in final purifi-

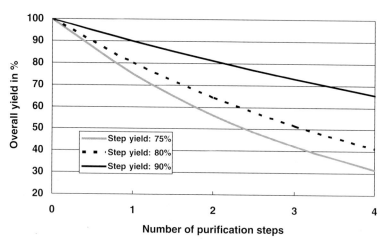

Figure 2.13 Interaction between overall yield, step yield and number of separation steps.

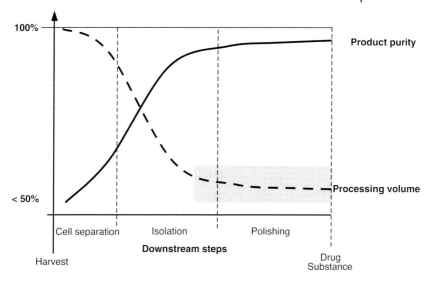

Figure 2.14 Product purity and volume from harvest through active agent. The shaded area indicates that in polishing different volumes can occur.

cation only depends on the product concentration and the loading capacity of the chromatography gels, not on the fermentation volume. The latter is solely determined by the achievable titer. Both volume streams are connected by the concentration step in which the titer-dominated volume is reduced to the capacity-dominated volume of final purification.

The decreasing volume and the increasing concentration are illustrated qualitatively in Figure 2.14. Other than in fermentation, the steps in polishing have nearly equal volume. Furthermore, it becomes apparent that product purity does not change as much in polishing as it does upstream. In polishing, the separation processes are challenged by the similarity of target protein and impurities, which can lead to high yield losses. In this area it should be carefully considered whether a separation step is necessary or not. The introduction of an additional step – just to be on the safe side – can potentially result in minute purity improvement, but high yield losses, longer process times and extended development timelines. Of course the product should be recovered as pure as possible; however, more important in the sense of patient safety is that the product is consistent from the early safety studies down to marketed goods and, furthermore, throughout the whole lifecycle. Therefore a certain byproduct spectrum can and must be accepted as long as it has been shown that the use of this product is safe.

2.5.1.2 Types of Impurities

When trying to understand protein purification it is important to become familiar with the main types of impurities. Apart from the actual biomass (cells and cell debris) fermentation solutions contain various low- and high-molecular-weight substances. They have to be removed insofar as they impact the safety and efficacy of the drug. It is important to realize that neither the final product nor the active ingredient are pure binary solutions of target protein and water. Indeed, functional additives play a very important role in the stabilization and pharmacological functionality of the product.

Many of the substances that accompany the product in the production process would be life-threatening if they reached the human blood system, therefore the best possible removal of these pathogenic substances is an important goal of purification.

The impurities can be classified in three types:

i. Process-related impurities.
ii. Product-related impurities.
iii. Contaminations.

While impurities cannot be avoided, contaminations are avoidable.

Process-related impurities
Cellular DNA

Figure 2.15 shows the main sources of process-related impurities. Being a living reactor – besides producing the target protein – the cell generates other low- and high-molecular metabolites (e.g. HCPs) including proteases (enzymes attacking the target protein). Cellular DNA finds its way into the product solution by natural cell lysis at the end of the cell lifetime as well as by enforced lysis by cell disruption. A

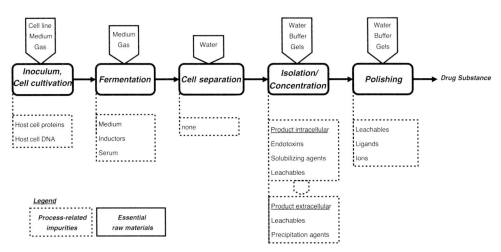

Figure 2.15 Sources of process-related impurities.

further essential source of impurities is the fermentation medium which may contain serum, inducers, anti-foaming agents, hormones and other growth-promoting ingredients.

Even purification itself vitiates the solution. Several inorganic salts and organic solvents are added to the buffers to operate the chromatographic steps. The construction materials are not always completely inert to the product solution and molecules can be dissolved from chromatography gels (leachables). Ligands of affinity chromatography can dissolve protein A; ion-exchange steps leave a high ion concentration in the product flow. *Purification adds impurities*

Expression systems that require cell disruption, flood the solution with cell debris. In particular, the high DNA and endotoxin freights are challenging. Endotoxins are lipopolysaccharides that stem from the lysed cell wall of Gram-negative bacteria (*E. coli*) and cause fever when administered parenterally ('pyrogenic effect'). *Endotoxins*

Figure 2.15 shows the processing aids like water and air that can also be sources of impurities. These substances should normally be present in adequate purity (Section 8.4) and not exhibit sources of additional process-related impurities; however, they have to be closely monitored to avoid unintentional contamination.

Proteolytic degradation products of the target protein are protagonists of the class of product-related impurities. This class comes along with another subclassification: if the degradation does not change the biological activity, the substances are denoted as 'product-related substances'; however, if the activity changes, they are called 'product-related impurities'. Lepirudin can serve as an example of this. Arvinte (2003) showed that Lepirudin changes its molecular parameters when exposed to thermal stress; at the same time, it does not change its biological activity as verified by an *in vitro* test. One could conclude that the molecular variations are unimportant; however, such an activity test does not prove the safety of the application since product-related impurities can potentially have unexpected side-effects in the human body. Likewise, proteolytic degradation products behave in the same way; the change of only one terminal amino acid can have significant effects on the pharmacological properties, while the proteins still are pretty similar with respect to their size and binding affinity – and therefore their separation behavior. An example of this is insulin (Section 4.2). *Product-related impurities*

Other members of the group of product-related impurities are modified target proteins that have a glycosylation pattern different from the target. Protein aggregates (i.e. groups of several proteins) also count as impurities.

Process- and product-related impurities are at the focus of process development. The primary goal is avoidance; if this is not possible, separation from the product by purification steps is required. A *Contaminations are avoidable* prerequisite for this strategy is that the accompanying substances are all known and recognized, which is not always the case, because if complex media are used, not all components are known. By means of accurate and diversified methods (i.e. the use of different methods for the same target) they can be differentiated from the actual product. Essentially the same applies to product-related impurities. After all, the product that is administered to the patient will always be a mixture of the target protein and process- and product-related impurities. For example, a purity specification of 98% means that 2% of these impurities are allowed. Their harmlessness has to be demonstrated in clinical trials.

While process- and product-related impurities are unavoidable, unintentional impurities (contaminations) can in principle be avoided by using suitable measures. Contaminations can be of biological and particulate nature. The group of biologicals is made up by microorganisms and viruses. Microbial contamination during fermentation usually can be detected fairly easily by early product analysis or when unexpected process courses show up. Both lead to rejection of the fermentation batch.

Microbial contamination

Microbial contamination by air-borne microorganisms in the purification process should be avoided; however, a low and controllable level of contamination can be accepted here. The downstream process usually is not declared 'sterile', but 'bioburden controlled'. This is justified by the fact that the product solution is led through 0.2-μm filters at the end of purification and before filling, which ensures that no germ originating from the product solution ends up in the parenteral product.

Microbial contamination in the area of pharmaceutical manufacturing bears the risk that a germ is carried over into the blood of a patient. Sources of contamination are air, water, operators and insufficiently cleaned equipment. The discovery of a microbial contamination can result in a loss of the whole product batch if it cannot be shown that only the tested sample and not the other containers were affected.

Viral contamination

Viral contaminations are much harder to detect than microbial contaminations. They are especially harmful for mammalian cell cultures since they can be introduced by contaminated cell substrates from the cell banks into the process and proliferate inwardly. Even virus testing of cell banks does not give ultimate security, since there are viruses which hide in the cells and are only expressed after a certain generation number (retroviruses). Viruses can also be introduced by contaminated media ingredients of human or animal origin.

2.5.1.3 Principles of Separation Technologies

Solid suspended matter

Biosuspensions accrue at the end of fermentation processes. Suspensions are characterized by undissolved impurities that float in the

> **Box 2.3**
>
> **Overview: impurities and contaminations**
>   Process-related impurities
> i. Cell substrate: HCPs, host cell DNA, virus fragments
> ii. Fermentation: medium components, serum, inducers, anti-foaming agents, antibiotics
> iii. Downstream: leachables, ions from ion exchange, endotoxins from cell disruption, ligands like protein A, inorganic salts, biochemical and chemical reagents, cell disruption, solvents
>
>   Product-related impurities
> i. Degradation products
> ii. Modified products (e.g. glycosylation forms)
> iii. Aggregates
>
>   Contaminations (unintended impurities)
> i. Biological impurities (bacteria, virus, mycoplasma)
> ii. Particulate impurities

aqueous environment making the liquid cloudy. After having separated this solid suspended matter, the solution can still be colored, but no longer be turbid. The solid particles are separated by filtration (microfiltration) or by chemically or mechanically accelerated settling (e.g. flocculation or centrifugation).

In the remaining clear, aqueous ambient phase there are numerous molecules dissolved chemically in the water. The most general principle of thermal separation technology applies to these substances: the more similar the molecules are, the more difficult it is to separate them from each other. There is a choice of unit operations for separating these molecules:

Dissolved molecular impurities

- *Ultra/diafiltration*: large and small molecules are separated by filtration.

- *Chromatography*: depending on the type there are different principles of separation based on size differences [gel-permeation chromatography (GPC)], affinity to unspecific surfaces [hydrophobic interaction chromatography (HIC), HPLC], affinity to specific molecules (affinity chromatography) and polar moieties (ion exchange chromatography).

- *Extraction*: the separation is based on different solubility in two coexisting liquid phases as can be seen in milk (a mixture of aqueous and organic fatty phase). The different polarities of the molecules cause a different solubility in the two phases.

- *Crystallization*: the separation happens by precipitating a solid phase.

The similarity of proteins is defined in terms of molecular size, the polarity and the affinity to auxiliary materials which are used for the separation:

Similarity of molecules

- *Molecule size.* The molecule size is given as the weight in Daltons (grams per mole). Sometimes the molecular unit for length is used: Ångström ($1 \text{ Å} = 10^{-7}$ mm). One can also find the notation 'aa' for proteins, which stands for the number of amino acids and the notation 'bp' for the number of base pairs in nucleic acids. This latter size description can be misleading for proteins as the molecular weight and size can be very different from the peptide structure due to glycosylation.

- *Polarity.* Polarity is determined by the distribution of charges within the functional groups of the amino acids or the nucleotides (for DNA). Usually in a long and folded molecule there are polar and non-polar areas. Polarity dominates water solubility.

- *Affinity to auxiliary materials.* Affinity is based on molecular interactions of the dissolved molecules with the surface of solid particles. For solid surfaces these are not only determined by force fields, but also by spatial accessibility – the molecules may not have enough room to position themselves optimally ('steric hindrance'). There is no clear measure for affinity. Thermodynamics of phase equilibria, which describes the theoretical basis for chromatography technologies, in fact provides models for binding affinity of individual substances; however, the influences of the concomitant substances are so prominent that only experiments provide a reliable real-world picture of the separation behavior.

2.5.2
Technologies for Cell Separation and Product Isolation

The first steps of purification encompass cell separation and product isolation. The following sections will introduce the principal technologies and auxiliary materials for this separation task.

2.5.2.1 **Cell Separation**

Cell separation immediately follows harvesting of the fermenter. This separation of solid particles from the aqueous liquid is achieved by filters, centrifuges or decanters. Filtration processes can be operated in the classical way as flow-through or normal-flow filtrations (dead-end filtration) or alternatively in the tangential-flow mode. The latter will be described under the keyword 'ultrafiltration' in the following chapters.

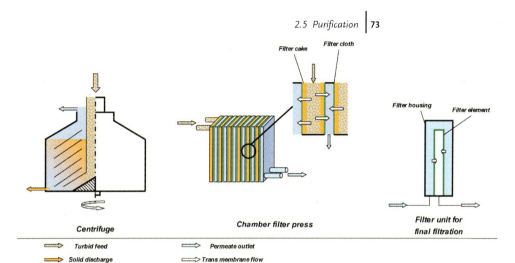

Figure 2.16 Methods for cell separation (centrifuge, chamber filter press) and subsequent depletion of solid suspended matter.

In normal-flow filtrations the filter residue forms a filter cake that has to be discharged mechanically after filtration. Usually, due to the resulting insufficient cleaning, simple reflushing is not recommended. Figure 2.16 shows a so-called chamber-filter press in which the disposable filter cloth is clamped between hollow cassettes. The harvest and the filtrate flow in the cassettes. Modules of this type are used to adapt the filter surface to the process conditions. The apparatus has to be disassembled for cleaning. *Normal-flow filtration*

Centrifuges and decanters are much easier to handle than the filter units, and have therefore prevailed as the first cell separation step. Product losses during cell separation are caused by the residual water bound to the discharged cells in which the protein is dissolved. Some proteins also like to bind to the solid surfaces of cells and are therefore unintentionally discharged. To avoid this, the remaining slurry can be flushed with water. Typical yields of this step are between 95 and 98%. *Centrifuge/decanter*

The filtrate from the centrifuge usually still contains residual solid particles that have to be removed in a particle or microfiltration unit. Both filtration processes differ in the pore size of the filters. Due to the low particle freight, the risk of blocking is small; therefore, these filters are designed as simple normal-flow filters.

2.5.2.2 Cell Disruption, Solubilization and Refolding

In non-secreting systems the protein is recovered from the cells. The sequence of process steps (cell disruption, cell debris removal, solubi-

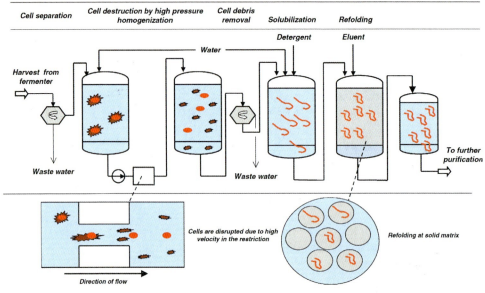

Figure 2.17 Primary treatment of proteins expressed intracellularly.

Cell disruption

lization of aggregates, refolding) has already been described in Section 2.5.1.1.

Cells can be disrupted mechanically, using mills or homogenizers, and physicochemically using enzymes, detergents or bases. One disadvantage of mechanical processes is that the friction generates heat, which can degenerate the protein. Chemical operations work with additives, which can be harmful to the product and have to be removed in purification. High-pressure homogenization (outlined in Figure 2.17) has prevailed in most applications.

Just like cell separation, the separation of the cell debris uses centrifuges. The protein aggregates are solubilized with the help of surface-active substances, while the refolding step can be performed at a solid matrix or by providing suitable ambient conditions. Altogether, the cascade of cell disruption, solubilization and refolding significantly contributes to the process time.

The yield (share of recovered and refolded proteins relative to the total of intracellularly expressed protein) for the whole sequence typically is below 30%. The high losses can be explained by incomplete disruption, deficient solubilization and incomplete refolding. The high performance of intracellularly expressing systems is often gambled away by the high effort and the significant product losses in primary recovery.

2.5.2.3 Concentration and Stabilization

Now that the protein exists in its biologically active form in a solution free from solid impurities, two essential tasks remain for primary recovery:

- Prevention of product degradation by proteases (stabilization).
- Reduction of the volume to be processed in final purification to make the equipment smaller and more cost-effective.

Stabilization of the protein

The protein can be stabilized by inactivating or separating the disintegrating substances. Inactivation can be supported by adding a chemical protease inhibitor, but as always: 'What goes in has to come out'; thus, this entails further yield losses at subsequent purification steps aiming at reducing the concentration of this specific agent. Thermal inactivation of the proteases by short heating can be successful; however, the protein is also thermally unstable and the process is therefore associated with product losses. Moreover, protein degradation products are still present in the solution. It is more efficient to deplete the proteases quickly. Since they are also proteins they are relatively similar to the product molecule. A real chance for separating the two only exists if there are significant differences either in surface activity or water solubility. In the first case, adsorption or chromatography processes can be considered; in the second, a two-phase extraction (Section 2.5.3.2) could help. These operations at the same time concentrate the product.

Adsorption processes

Adsorption processes make use of the fact that, if the conditions are favorable, molecules prefer to attach to surfaces rather than stay in the liquid surrounding phase. Thereby the distribution of the molecule between the solid surface (adsorbent) and the liquid depends on the molecular interactions and is specific for each substance. Adsorption is performed by leading the solution through a bed of adsorption particles; ideally at the end of the process the molecule type to be bound sits on the surface, while the residual molecules are carried away with the flow. The adsorbent has to be chosen specifically for a separation task.

Adsorption can either bind the target molecule or a disturbing impurity like proteases or colorings (chromophoric substances) that are present in the clarified harvest. In the first case, the product has to be dissolved from the solid – a process called 'desorption'. In desorption, a second liquid is led through the particle bed, and conditioned in a way that the molecules detach from the surface and dissolve in the liquid. If impurities were bound, desorption is made to regenerate the solid particles and prepare them for the next run. These processes are illustrated in Figure 2.18.

Batch adsorption

Figure 2.19 shows two different designs for adsorption which can be found far upstream in biotechnological processes. In batch adsorption the particles and the harvest liquid are mixed in a stirred tank. After

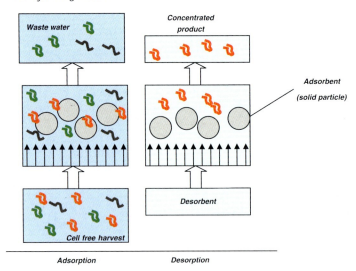

Figure 2.18 Adsorption and desorption mechanism.

Expanded-bed adsorption

Direct capture step

Tangential-flow filtration

mixing the liquid phase is drained and the bound molecules are desorbed using a second liquid.

The lower part of Figure 2.19 shows expanded-bed adsorption. Here, the particles initially are packed into a vessel. A liquid flows through this bed from below lifting the particles (fluidization). The essential difference compared to the packed, non-fluidized gel layer which is described in the section on chromatography (Section 2.5.3.1) is that the expansion of the bed results in interstitial space between the particles. Therefore, solid particles that may still be in the liquid can pass the bed and the apparatus does not plug. This can save one clear-filtration step. If the expanded-bed mode is coupled with an affinity gel, one step can suffice to separate suspended particles as well as dissolved impurities.

Adsorptive or chromatographic processes that bind the target protein and are actioned before concentration close to harvesting are denoted as direct-capture steps. As outlined, expanded-bed adsorption (EBA) can be used to spare a clear-filtration step; however, EBA requires suitable adsorbents and fluidization is a challenge for process control. A different direct-capture strategy is a filtration followed by ion-exchange chromatography (IEC) (Section 2.5.3)

One of the most important methods for concentration and clarification of cell-free harvest is tangential-flow filtration (TFF, also known as cross-flow filtration). Other than in normal-flow filtration, in TFF processes the product to be filtered passes the filtration membrane. Figure 2.20 shows the difference in the flow pattern. Both processes can be designed as surface filtration, in which the particles are retained at

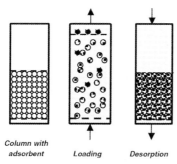

Figure 2.19 Adsorption in batch and expanded-bed mode.

the surface, and as depth filtration, in which the particles are retained inside the membrane.

There are two types of filtration, depending on the pore size of the membrane:

i. In particle filtration or microfiltration the product is pressed through the membrane together with the liquid solution. Larger particles are retained.

Microfiltration and ultrafiltration/ nanofiltration

ii. In ultra- and nanofiltration, low-molecular-weight impurities are pressed through the membrane together with the aqueous liquid, while the product remains on the primary side.

The cross-wise flow prevents formation of a layer of impurities which would impede the trans-membrane flow (so called 'membrane fouling'). Despite the flow pattern, a thin layer of impurities forms at the membrane surface. This 'filter-cake' is desired, since it enhances filtration efficiency.

Membrane fouling

Figure 2.21 illustrates the typical operating modes of TFF processes. The flow through the membrane causes a decreasing liquid volume on the primary side. This can be desired; however, it can also lead to undesired effects like precipitation of the protein due to supersaturation

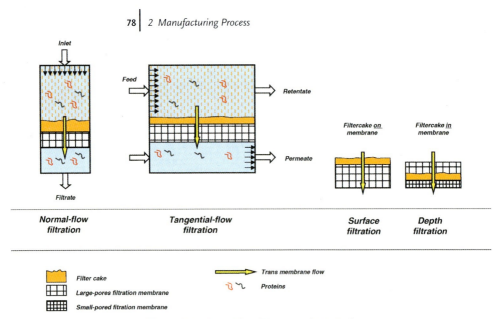

Figure 2.20 Normal-flow filtration and TFF. Both can be operated as surface or depth filtration.

of the solution. The latter can be circumvented by continuously adding fresh liquid to the primary side. This process is called diafiltration and is used to exchange the liquid at the end of purification. The outward design of the equipment resembles the chamber filter presses (Figure 2.16), but often the modules are designed as round cartridge filters.

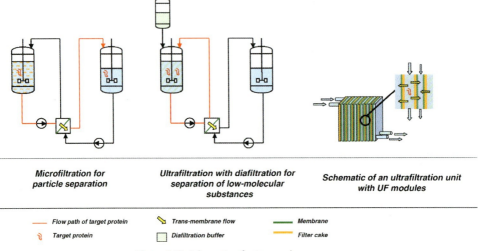

Figure 2.21 Schematic of micro- and ultrafiltration (UF) in the TFF mode and the design of a membrane module.

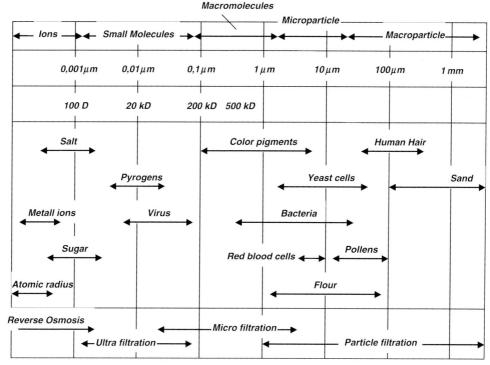

Figure 2.22 Membrane processes and their potential applications for the separation of impurities (Wheelwright, 1991).

Figure 2.22 (Wheelwright, 1991) describes the molecule sizes of typical impurities and the filtration processes which can be used for their treatment.

The most important parameters of the membranes are the trans-membrane flow and the molecular weight cut-off (MWCO) (Section 2.4.5.2). The trans-membrane flow is the flow through the membrane. It depends on the pressure differential between the membrane sides (trans-membrane pressure) and is the key parameter for the duration of the filtration step. The MWCO is a measure for the maximum molecule size that can pass the membrane; it strongly depends on the filter cake and can only be determined experimentally for the actual application.

Trans-membrane pressure

2.5.3
Technologies for Final Purification

The preceding sections have eluded to the first steps of purification of a suspension-type liquid coming from the fermenter. Product concentration and isolation have the task to remove the process-related substances from fermentation. At the end of these steps usually a solution is presented that is free of cells and solid particles, and has a clear (not necessarily colorless) appearance. This solution contains the product as well as low- and high-molecular-weight impurities. The task of final purification is to deplete these impurities.

Final purification primarily deploys chromatographic processes, but crystallization, precipitation and membrane processes can also be used.

Polishing generates the ultimate purity in which the active ingredient is used for patient treatment. Safety aspects are very important in this area. Therefore, another crucial task of final purification is to secure the biological purity before the product is transferred to formulation and sterile filling. One section is therefore dedicated to the safeguards from biological contaminations like microorganisms or viruses.

2.5.3.1 Chromatographic Processes

Chromatography processes are the spine of final purification. Preparative chromatography ensures the selective separation of high-molecular-weight substances without chemical or thermal stress. These advantages are bought by accepting relatively small capacity and high costs. The chromatography for the technical recovery of proteins is denoted as preparative chromatography (in contrast to analytical chromatography).

This section starts with elaborating on the general principles of chromatography, then the differences in the chromatography types are highlighted and, finally, a typical process with the required equipment is shown. Section 2.5.4.1 describes the materials used in chromatography.

Separation processes summarized under the term 'chromatography' are based on two principally different physicochemical effects. One of them is an adsorption–desorption mechanism in which the molecule is either bound or not bound to a solid surface (binding chromatography); the other effect is based on different migration velocities of molecules when transported in flowing liquid due to their affinity to solid surfaces. The latter is denoted as flow-through chromatography.

Binding and flow-through mode

The binding mode works in ion-exchange columns and in strongly binding affinity columns. Flow-through chromatography works for separation processes based on differences in hydrophobicity or molecule size.

Figure 2.23 illustrates the binding mode. The starting point is the equilibrated column. In the equilibrated state the column packed with gel is conditioned for the following loading step that can be achieved by flushing the column with a solution resembling the product solution in

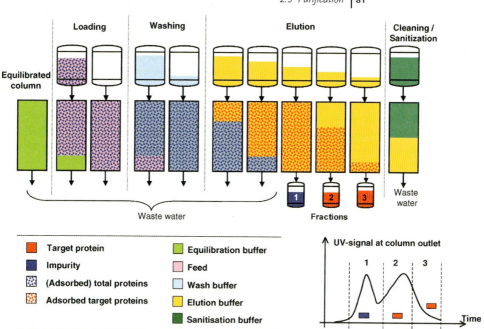

Figure 2.23 Schematic of the chromatography operation in the binding mode (IEX, affinity chromatography).

pH value and salt content. After equilibration the column is loaded (i.e. the product solution is fed from a tank). The complete column will be loaded with protein. After loading, the column is flushed with a washing liquid displacing the residues of the feed solution. Washing is followed by the elution step in which the target protein is solved from the gel as selectively as possible and thereby separated from the retained impurities. In the binding mode, the discrimination of the components is based on their different solubility in the elution buffer. The buffer properties (e.g. pH value) can be changed stepwise so that initially the impurity (blue in Figure 2.23) is eluted and after that the target protein (red in Figure 2.23). At the column outlet the eluate is captured in several fractions. In Figure 2.23 the product is in the fractions 2 and 3. At the end of the process the column is washed (sanitized).

The chromatogram at the right bottom edge of Figure 2.23 shows a possible course of the protein concentration at the column outlet [measured by ultraviolet (UV) detection] as well as the separation into different fractions. These methods are concentrating methods, which means that large solution quantities can be fed into the column and relatively small, highly concentrated amounts of solution can be recovered.

Flow-through chromatography does not have this concentrating advantage. Here, the liquid also passes a bed of solid particles. Two proteins which float dissolved in the liquid are separated by the fact that they migrate at different speeds through the bed. This is due to the different affinity of the molecule for the gel material. The molecule having a higher affinity for the gel is decelerated relatively to its counterpart with less affinity. The slower molecule leaves the column later than the faster molecule. The target protein is obtained by separating the different fractions at the outlet.

Affinity is based on different molecular interactions between the proteins and the gel. Types of flow-through chromatography differ in the type of gel.

The general process is equal for all types of flow-through chromatography and is illustrated in Figure 2.24. In the loading step the affinity for the gel causes the protein mixture to attach to the upper part of the column while the liquid passes the column. The difference to the binding mode is that the proteins are not bound in the whole column, but only in a small separation zone. In the subsequent elution the

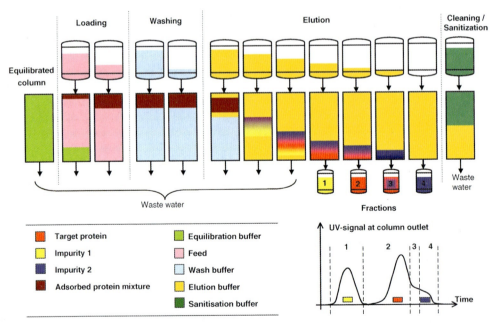

Figure 2.24 Schematic of chromatography operation in the flow-through mode.

proteins are dissolved from the solid phase and start to migrate to the column outlet.

Thereby, molecules with higher affinity are decelerated more than molecules with less affinity for the gel. The speed of the liquid is in any case higher than that of the proteins. The chromatography is designed in a way that at the outlet the target protein is separated as sharply as possible from the others. The outgoing flow is collected in several fractions; the target protein usually is distributed over multiple adjacent fractions.

The chromatogram in the right bottom edge of Figure 2.24 shows a possible course of the protein concentration at the column outlet (measured by UV detection) and the separation in fractions. Fraction 1 contains the impurity 'yellow'. The separation from the target protein 'red' is very good (baseline separation); the target protein is in fraction 2. Unfortunately there is an overlap with the impurity 'blue', that means that 'red' and 'blue' eluate together. Fraction 3 is therefore collected separately from the 'blue' fraction 4. One could increase the yield for 'red' by mixing fraction 3 and 2; however, this would mean contaminating 'red' with a certain share of 'blue'. This conflict of interest between yield and purity is typical of preparative chromatography. A complete baseline separation is hard to find here, since besides the selectivity, cost aspects are also taken into account. They have a decisive impact on gel selection, processing times and the technical feasibility of scaling-up.

For the outside observer it does not make any difference whether the separation works on the basis of the binding or the flow-through mode. Indeed, both operations are similar in design, equipment and process operation. If the differences in affinity are only weak, affinity chromatography can be run in the flow-through mode. Therefore, often no distinction is made between the mechanisms and both together are denoted as chromatography.

What are the causes for the different affinities of the proteins? The proteins are chains of amino acids that are folded into spatial structures. The 'surface' of such a structure displays chemical moieties that can be ionized, fit sterically to complementary structures or attract other non-polar surfaces by means of their hydrophobic character. The processes along a purification sequence mostly use several of these features, since the impurities may be different from the target protein in one or other property. The different affinity mechanisms have lead to the development of different chromatography principles:

Differences in gels

i. *IEC: also know by the abbreviation IEX (ion exchanger):* separation due to different electrostatic forces. The cation IEX (CIEX) and anion IEX (AIEX) exchangers belong to this group.

ii. *Affinity chromatography*: separation due to different structures.
iii. *HIC*: separation due to hydrophobic interactions. Reversed-phase (RPC) and reversed-phase HPLC (RP-HPLC) belong to this group.
iv. *Size-exclusion chromatography (SEC)*: separation due to different molecule sizes at an inert and porous phase. Also denoted as GPC or gel-filtration chromatography (GFC).

All four types will be further described in the following sections.

Ion-exchange chromatography

IEC works on basis of the different electrostatically charged molecular moieties in a protein. They are capable of binding to oppositely charged molecules that are immobilized on the solid gel particles in IEC. Figure 2.25 illustrates the physical background of IEC. A column charged with ions is loaded with the proteinaceous feed stream. The ionized proteins bind to the particles displacing their counter ions 'A'. In the next step the more weakly attached impurities are separated by a buffer of low concentration in molecules 'B$^-$'. This first fraction of the elution step can be discharged. With rising concentration in B the stronger bound target proteins can be detached and captured at the column outlet as the product fraction. CIEX (negatively charged immobilized ions) and AIEX (positively charged immobilized ions) depend on the polarity of the bound proteins.

Affinity chromatography

In affinity chromatography, binding to the molecules is achieved by a very specific complementary structure between ligand and protein. An example for this process is the separation of recombinant Factor VIII by a column which uses a gel of Factor VIII antibodies. The antibodies bind the target molecule and let the others pass. Obviously, this type of gel is

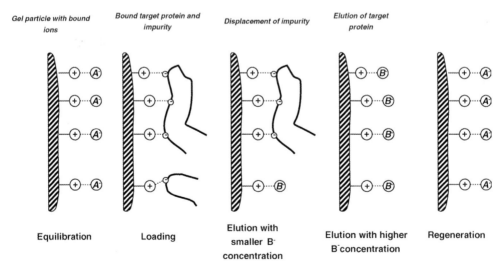

Figure 2.25 Mode of operation of an IEC (here anion exchanger).

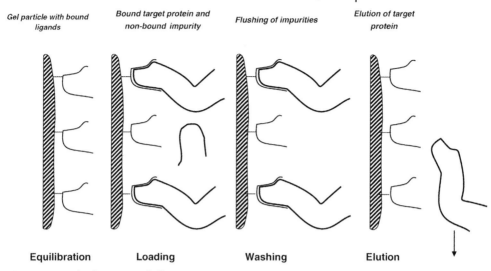

Figure 2.26 Mode of operation of affinity chromatography.

expensive. Another example is the commonly used protein A chromatography for purification of antibodies. The physical background of affinity chromatography is illustrated in Figure 2.26. The analogy to IEC is obvious – in both cases the loading of the column is followed by flushing impurities and elution of the target protein.

The pivotal disadvantage of affinity chromatography is the expensive gel. It requires a biotechnological manufacturing process by itself, which has to be performed according to the same quality standards as the protein drug process. Moreover, biological ligands bear the risk of contaminating the solution.

While IEC uses the polar parts of the protein and affinity chromatography uses the spatial structure, HIC targets the non-polar, 'water-repellent' areas of the molecule. These areas are characterized by displaying linear or cyclic hydrocarbon chains that occur at several amino acids (Section 1.3) and interact by means of unspecific molecular forces. The more moieties of this kind the protein has, the better it dissolves in organic solvents that also display these linear or cyclic hydrocarbons. The same applies to affinity to solid surfaces which is dominated by non-polar hydrocarbon molecules; it is this mechanism that separation by hydrophobic interactions is based on. The binding is weaker than in IEC and less specific than in affinity processes. HIC and their adjacent processes function according to the 'flow-through' chromatography principle.

Hydrophobic interaction chromatography

Reversed phase: the name 'reversed phase' originates from the distinction from the 'normal phase chromatography' known from chemistry and, as opposed to reversed phase, is based on polar interactions

Processes such as RPC and RP-HPLC follow the same principle; their difference to HIC is gradual. In RPC the hydrophobicity is higher than

Size-exclusion chromatography

in HIC. While in HIC an aqueous buffer can be used for elution, RPC needs organic solvents to separate the protein from the hydrophobic matrix. Large proteins sometimes undergo denaturation when in organic solvents. Therefore, RPC is mainly suited for smaller proteins. The difference between RPC and RP-HPLC lies in the packing density of the gel.

The high pressure in HPLC is needed to press the liquid through the dense gel packing. The density of the gel correlates with the gel particle size. The smaller the particles are, the higher the surface exposed to the liquid; therefore, a column with smaller gel can provide higher capacity or selectivity. On the other hand, the density limits the flow, resulting in a special pressure-rated construction of the column. Therefore, the HPLC process comes with long process times, high investments and narrow technical limits. Normal low-pressure chromatography columns are built with diameters up to 200 cm, while the limit for HPLC columns is currently reached at 80 cm.

In SEC, the gel should ideally not interact via molecular forces with the molecules to be separated. The separation principle is based on the pores in the particles. When passing the particles, small molecules diffuse easier and deeper into the pores than big ones. Consequentially the small molecules lose speed relative to the large ones and arrive at the column outlet later. Figure 2.27 shows the operating principle of SEC and HIC.

The following sequence of steps is run to operate chromatography processes:

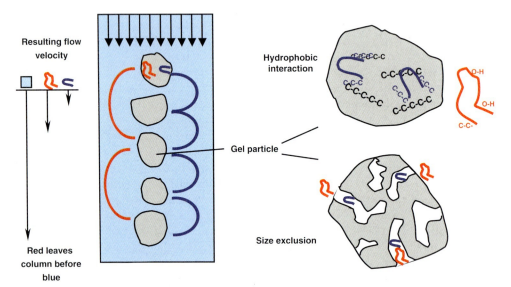

Figure 2.27 Mode of operation of SEC and HIC.

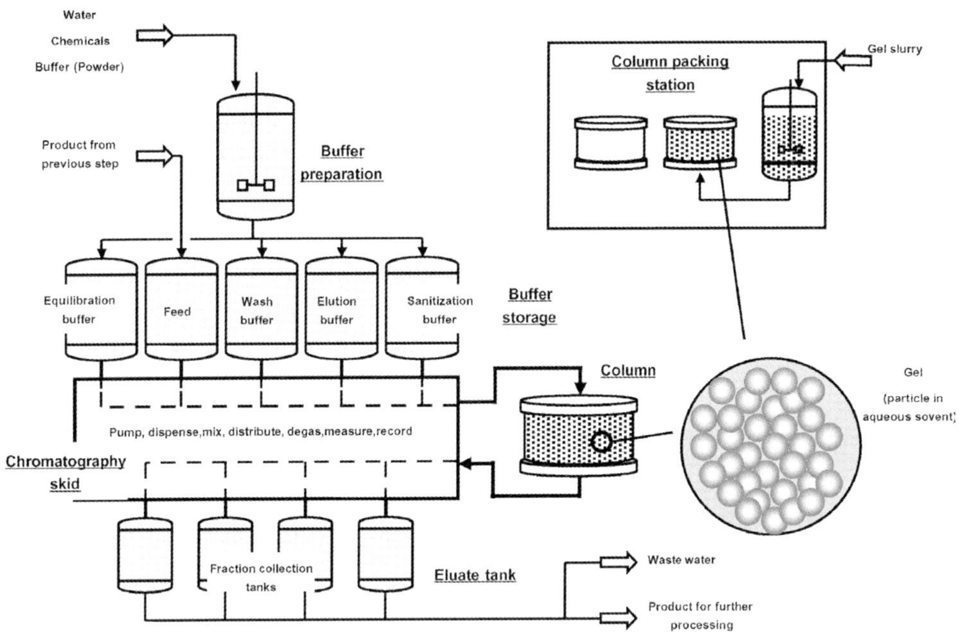

Figure 2.28 Design of a chromatography installation. Enlargement shows the gel.

- Equilibration.
- Loading.
- (Washing).
- Elution.
- Cleaning/sanitization.

Figure 2.28 illustrates the equipment installation to perform this operation. For the sake of clarity the installations for equipment cleaning are not shown.

The chromatography column is hooked up to an installation for pumping, distributing, mixing (for gradient elution), dosing, measuring and degassing. These installations can be purchased as a package and are denoted as 'chromatography skids'. Buffer solutions and the product-containing feed flow into the skid. The buffers are prepared in a buffer preparation area using water, buffer powder and chemical auxiliaries. The product coming from the column is collected in different fraction tanks and dispatched for further treatment after being tested.

Correct packing of the gel is of crucial importance. The gel consists of small particles that are bloated in an aqueous solution. If the empty column is filled ('packed') with the gel, irregularities in packing density or homogeneity can occur, impacting separation characteristics negatively. Therefore, after repacking, and sometimes prior to operating the

Chromatography skid

column, packing homogeneity is tested. If the column is correctly packed, the signal peak of the test agent leaves the column as a symmetrical Gauss curve; in the case of deficient packing this curve has an asymmetric shape, which is why this test is also known as a 'symmetry test'. The gel material is renewed and repacked with a packing station after a certain number of runs for which the material is qualified.

Symmetry test (margin note)

2.5.3.2 Precipitation and Extraction

The generation of a solid (crystalline) fall-out of a component from a solution is called precipitation. In contrast, in extraction, a second liquid phase, providing different solubility for the target molecule than the coexisting phase, is generated from a homogenous solution. In extraction as well as precipitation the basic separation principle is that the solubility of the protein or the impurity is reduced in the aqueous product solution. It depends on the physical properties of the component in question as to the measures that need to be taken to achieve this reduction.

A protein is precipitated by temperature shift, change of salt content or addition of an organic solvent. Solid and liquid can be separated by filtration.

A second phase in extraction can be obtained by adding dextran or poly(ethylene glycol) (PEG) (aqueous/aqueous phase) or butanol (aqueous/organic phase). The target protein and the impurities should distribute as unevenly as possible in the two phases, so that one of the phases can be recovered as the product-rich phase and the other be discarded or forwarded to another separation step. The liquid phases can be separated by settling or centrifugation.

Figure 2.29 schematically shows the processes of precipitation and extraction.

Disadvantages of precipitation are:

- The solid, often sticky precipitate is difficult to handle in the filtration unit and sometimes has to be taken out manually in a process open to the atmosphere.
- Scale-up is limited due to difficulties in the homogenization process in the large scale; local concentration peaks cause local precipitation.

Disadvantages of extraction are:

- Insufficient selectivity. The protein does not migrate completely to one of the phases but will distribute in a certain percentage to both. In a one-step process this results in high product losses.
- The addition of a process auxiliary increases the number of impurities. Moreover, this substance should, from a medicinal standpoint, be as harmless as possible – a fact that dramatically limits the choice of agents.

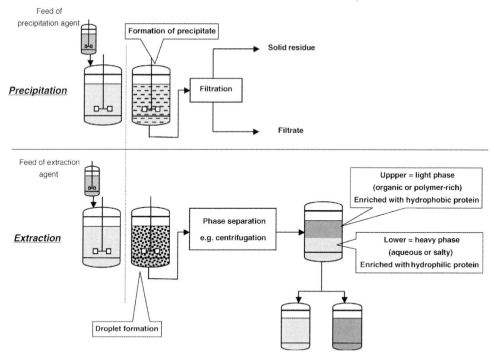

Figure 2.29 Principles of precipitation and extraction.

These disadvantages are only partly compensated by the advantages, namely that the precipitation yields high purity and that extraction can be easily automated.

2.5.3.3 Sterile Filtration and Virus Removal

Viruses and bacteria are biological contaminations. They can find their way into the process stream either unintended or, in the case of viruses, as process-related impurities (Section 1.3.1.4).

Viruses can be removed from the process by inactivation or purification. If products are made in animal cell culture or materials from animal or human origin are used, the ability of the process to remove the viruses has to be proven. For products of microbial origin, the risk of viral contamination is only given by external sources; this does not justify a comprehensive virus removal effort. Viral load is reduced by physical separation in chromatography or filtration steps and additionally by chemical or thermal inactivation. Virus filtration is popular; however, the sizes of virus particles and target proteins sometimes overlap (Figure 2.22).

Virus filtration

Virus inactivation

In chemical inactivation the product solution undergoes a high and low pH value for a short time (pH swing) or a solvent treatment. Thermal inactivation means heating the solution to the highest possible temperature at which the protein is not harmed too much and the virus is sufficiently attacked.

Virus validation

Viruses can origin from the MCB by: (i) the original cells from infected animals, (ii) the use of a viral vector for the modification of the cell line, (iii) the use of contaminated biological reagents and (iii) contamination during cell handling. They can also appear unintentionally during the process; by (i) use of contaminated biological reagents (serum, protein in affinity chromatography), and (ii) contamination during cell and media handling. When, which type and in what quantities they appear is usually unknown. The efficiency of virus removal is therefore proven with model viruses in a virus validation study ('spiking study'). The reduction factors are analyzed for each separation step.

0.2-μm filtration

Bacteria (germs) can access the product solution during processing. Purification processes are not completely germ-free (sterile), but bioburden controlled, which means they have very low germ counts. Microbial control is achieved by diligent handling and sterilization of buffers and auxiliaries. A sterile-filtration (0.2-μm filter, sometimes 0.1-μm filter) at the end of the process leaves the solution essentially germfree. Indeed the buffers for chromatography are filtered with the same pore size; however, sterility is not claimed here, but a reduction and control of bioburden, in order to have more leeway for the evaluation of potential later microbial observations.

2.5.4
Raw Materials and Processing Aids

2.5.4.1 Gels for Chromatography

The particles constituting the solid phase in chromatography are denoted as the chromatography gel or stationary phase. The gel is essential for the separation task as it determines the separation behavior as well as the hydrodynamic conditions. This section highlights the process parameters that are defined by the gel as well as the structure and main types of gels.

The main process parameters are:

- *Resolution*. The resolution explains to what extent two substances can be separated from each other. It determines the achievable purity and yield of the separation step.

- *Throughput*. The throughput is a measure for the capacity and the time in which this capacity can be used. The higher the capacity (loading capacity) or the lower the processing time, respectively, the higher the throughput. The process time is determined by the flow

Flow rate

rate (i.e. the liquid volume flowing through the column per time unit). It should be noted that apart from loading and elution, the process encompasses other steps (Section 2.5.3.1) that can add to the overall time and can be optimized.

- *Pressure drop.* The pressure drop or pressure gradient across the column packing results from the flow resistance of the gel. The higher the resistance, the higher the pressure has to be at the column inlet in order to push the solution through the bed at reasonable velocity. The pressure drop is an essential factor for column construction and therefore investment costs.

Pressure drop

Chromatography gels consist of porous particles with a size of 10–300 µm. One particle is made up of a structure (matrix) and attached ligands or ions specific to the application. The structures can be built from silica gel, agarose dextran components, cellulose or polymers.

The fields of application can be identified by the name of the gels. IEX gels carry an abbreviation for the type of ions they exhibit and whether they are strong or weak. Affinity gels often carry the name of their ligand; hydrophobic interactions are recognized by the name of the hydrophobic group. Only gels for SEC cannot be immediately identified (Tables 2.3 and 2.4).

The smaller the particles are, the higher their flow resistance and therefore pressure drop across the gel packing; on the other hand, the separation performance is better, since they display a higher surface for molecular interactions. Often hydrophobic columns are designed as HPLC columns, since the relatively weak hydrophobic interactions depend on a high surface area to result in a good separation.

It is of utmost importance that the gel is equally distributed in the column. Any irregularity of the column packing has a direct influence on column performance (Section 2.5.3.1).

Table 2.3 Examples for matrix materials and their brand names.

Brand name	Material	Application
Lichroprep®	silica gel	HPLC
Sepharose®	agarose dextran	IEX, HIC, affinity
Sephadex®	dextran-based	SEC, IEX, affinity
Cellufine®	cellulose	IEX, SEC, HIC, affinity
Source®, Amberlite®	polystyrene – co- or monopolymer	HIC, IEX
Fractogel®, Toyopearl®	methacrylate – copolymer	IEX, Affinity, SEC

Table 2.4 Labeling of chromatography gels.

Type of chromatography	Label	Remark
IEX	C, CM	weak cation exchanger
	S, SP	strong cation exchanger
	A, DEAE, QMA,	weak anion exchanger
	Q, QA, QAE, TMAE	strong anion exchanger
HIC	phenyl, butyl, C4, C8, C18	
Affinity	protein A, heparin, chelate, arginine	

Lifespan

Another important parameter for the use of gels is their lifespan – how many production and cleaning cycles the material can undergo before the performance drops unacceptably. In the pharmaceutical arena the ranges of this performance have to be determined before routine production by evaluating the operating limits for purity and yield of the specific separation step.

Leaching

Gels can pass their components into the product solution ('leaching'). The column material is exposed to pretty high chemical stress due to changing pH values and sanitization, so that parts of the material can dissolve in the liquid. When using affinity columns whose ligands are derived from biotechnological processes with animal cells (antibodies, protein A), even viruses can end up in the product solution.

Scale-up of chromatography

The capacity of a chromatography column can be upgraded by increasing its diameter. Increasing bed height would result in longer process times and higher pressure drops, and has no effect on column capacity. Due to their pressure rating, the scale-up potential of HPLC columns is limited; thus, HPLC steps can easily become bottlenecks of an industrial purification sequence.

Chromatography gel is decisive for quality

When designing the chromatography process, choosing the right gel is essential. Apart from the separation, performance aspects like throughput, useful lifespan and availability have to be taken into consideration. Changing the chromatography gel in an existing process is a variation that needs regulatory approval prior to implementation.

2.5.4.2 Membranes for TFF

In ultra- and microfiltration the molecules are transported through the pores together with the liquid. Pore size and distribution are decisive factors for separation.

Molecular weight cut-off

The pore sizes of microfiltration membranes range from 0.08 to 10 µm. For ultrafiltration membranes this pore diameter is difficult to measure; in lieu of that the penetration is given as the MWCO (with the physical unit of Daltons). This value describes which molecules of a certain mass are retained by the membrane up to 90 or 95%,

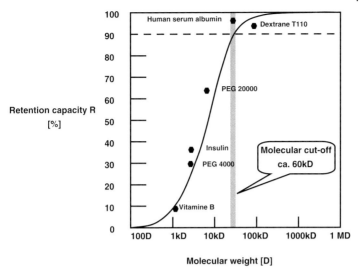

Figure 2.30 Separation characteristic of an ultrafiltration membrane (Rautenbach, 1997).

respectively, disregarding the effect of the filter cake. Figure 2.30 shows the typical characteristics of a separation membrane (Rautenbach, 1997). The retention capacity R is the percentage of the target protein remaining in the retentate.

In real application cases the cut-off is affected by the filter cake, the molecular structure and the specific interactions between the molecules; after all, whether a membrane fits the purpose always has to be tested with the real product solution.

Material selection should take into account the chemical or thermal stress during membrane cleaning. Typical construction materials are polymers or inorganic materials like aluminum oxide (Al_2O_3).

In symmetrical membranes, the pore size is equal over the membrane cross-section; in asymmetrical membranes, the pore size changes along this line. Membranes are delivered in prefabricated modules that can be combined to give the desired membrane surface.

Symmetrical and asymmetrical membranes

2.5.5
Overview of Purification

Although there are indeed typical paths for protein purification, there are a huge variety of downstream processes.

This section only gives two examples for typical processes. The figures are intended to give an impression of equipment installations used for industrial downstream processing; they are reduced to core equipment since cleaning installations are not shown.

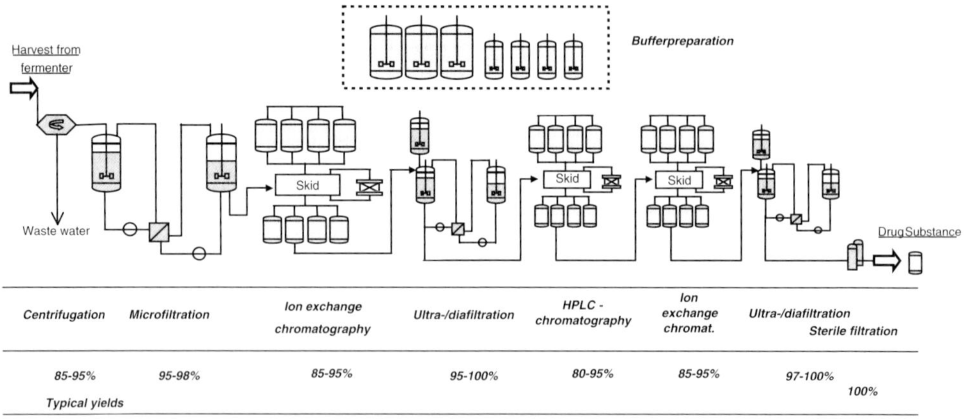

Figure 2.31 Typical separation process for an extracellularly expressed protein from microbial culture.

Centrifugation	Microfiltration	Ion exchange chromatography	Ultra-/diafiltration	HPLC - chromatography	Ion exchange chromat.	Ultra-/diafiltration	Sterile filtration
85-95%	95-98%	85-95%	95-100%	80-95%	85-95%	97-100%	100%

Typical yields

Figure 2.31 illustrates a typical purification process for an extracellularly expressed protein from microbial culture. After having separated the cells by centrifugation, solid particles are separated from the solution by microfiltration in the tangential flow mode. After filtration a direct-capture IEX column for concentration and protease depletion is shown. The reduced volume is further narrowed down by ultrafiltration and conditioned for the next chromatography step by buffer exchange (diafiltration). The HPLC chromatography purifies the solution from essential HCPs and degradation products. The last chromatography step is IEC. The final ultra/diafiltration again concentrates the solution and renders the buffer suitable for freezing and storage, before the solution goes through a 0.2-μm filtration.

Typical overall yields for this process result from the multiplication of the indicated step yields and range from 45 to 80%. In the case of intracellular expression, instead of the centrifuge, the installation from Figure 2.17 would be necessary.

Figure 2.32 shows a somewhat different example for the purification of a protein expressed from mammalian cells. Centrifugation is followed by normal-flow filtration. The filtrate is led into a vessel and exposed to a chemical protease inactivation. The subsequent ultra/diafiltration concentrates the solution. Three chromatography steps are needed for further purification (affinity chromatography, HPLC and SEC) followed by a chemical virus inactivation. The final ultra/diafiltration again serves to deplete salts from the acid/base reaction and to exchange the buffer for storage. The end of the sequence is given by nanofiltration and 0.2-μm filtration.

Typical overall yields result from the multiplication of the step yields and range from 30 to 70%.

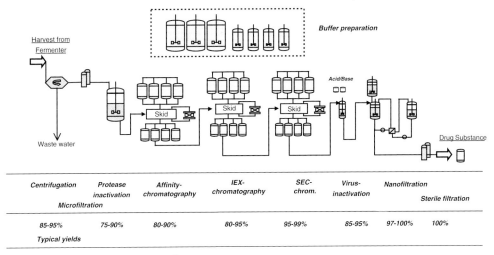

Figure 2.32 Typical separation process for an extracellularly expressed protein from cell culture.

2.6 Formulation and Filling

2.6.1 Basic Principles

In formulation, the API is transformed into a drug. Formulation encompasses the chemical composition of the drug as well as the pharmaceutical form and has to make sure that the drug:

- Is presented in a form suitable for therapeutic treatment (application/bioavailability).
- Has sufficient stability with respect to ambient conditions (temperature, shaking, light, etc.) since on route to the patient it leaves the controlled area of the manufacturer.
- Is competitive on the market. The acceptance by patients or doctors can depend on application features like form of administration or storage conditions and shelf-life.

In order to comply with these requirements formulation development deals with the following questions:

- Which route of administration can the drug take?
- Which pharmaceutical form (liquid, powder, aerosol) can support this route?
- Which chemical additives (excipients) are necessary to comply with both the stability as well as the medical requirements?

Administration route and form, excipients, and primary container

- Which primary container (container with product contact, e.g. glass ampoule, vial, pre-filled syringe) is adequate?

Chapter 4 highlights the pharmacological principles that are indispensable for the understanding of these questions. Here, in a first step, we will further elaborate on the essential technologies for protein formulation.

Of all the hitherto discussed process steps, formulation is closest to the therapeutic application of the patient. Although protein formulation is a science on its own, at the same time it is the least 'biospecific' step since the origin of the API – be it from chemical synthesis or biological fermentation – plays a minor role in formulation. For this classical pharmaceutical field many guidelines and experiences exist from the area of small chemical entities. The biospecific challenges for formulation result from the limited stability of the proteins, which has an effect on both the process and the form of the drug.

Figure 2.33 shows a typical scheme for formulation. Typically, the drug substance is presented in frozen form. After thawing and sterile filtration, the formulation buffer is prepared. In easy cases, this is done by simple pH adjustment of the present storage buffer; in less advantageous cases, a complete buffer exchange using diafiltration has to be performed. After that the product is filled into the primary pharmaceutical container, which can be vials, ampoules, cartridges or pre-filled syringes. If the product is formulated as a liquid, the chemicophysical process ends with this step. The container with the liquid drug is closed, labeled and packed into the secondary packaging (folding box, patient information leaflet) and is ready for shipping. However, due to stability issues most of the proteins are freeze-dried

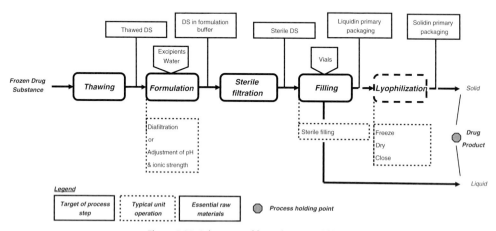

Figure 2.33 Schematic of formulation and filling. DS = drug substance.

(lyophilized) (Section 2.6.2) and supplied as solid powder or so-called lyophilization cake. This solid has to be dissolved (reconstituted) prior to injection.

Due to their temperature sensitivity proteins cannot be terminally sterilized in the final container; all steps after sterile filtration are performed under the strict rules of aseptic manufacturing. Note that in the preceding steps a bioburden-controlled environment was required, but not a completely sterile one.

Terminal sterilization

2.6.2
Freeze-Drying

In freeze-drying (lyophilization) the water is extracted from the protein solution. Targets of lyophilization are:

- Increase of storage stability.
- 'Good' and homogeneous visual image of the lyophilisate ('lyo-cake')
- Good reconstitutability (i.e. dissolution time and behavior after mixing with water for injection).
- Unchanged product quality (purity, identity, activity).

The process of lyophilization has three steps:

i. Freezing of the solution.
ii. Main drying by evaporation of water in the ice state (sublimation).
iii. Secondary drying of the then solid matrix.

Due to the temperature sensitivity of the protein the process cannot use high temperatures. Therefore, open vials are held under vacuum,

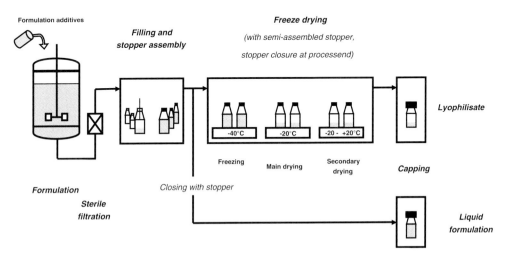

Figure 2.34 Position of freeze-drying in the formulation process.

Lyophilization cake

which leads to relatively fast sublimation of the ice at low temperatures. The process can last between several hours and days.

The result of lyophilization is a solid so-called lyophilization cake. The majority of this cake consists of structure-forming agents like amino acids (glycine, alanine), sugars (saccharose, trehalose) or polymers [dextran, hydroxy propyl methyl cellulose (HPMC)] in which the active ingredient is embedded. These substances also serve as cryoprotectants during the freezing process by replacing the water molecules and thus inhibiting structural changes of the proteins.

Figure 2.34 illustrates the position of lyophilization in the process sequence. After adding the formulation excipients, the solution is sterile-filtered and filled in vials. These are closed by stoppers incompletely, so that vapor can leave the container. A complete closure with caps only happens after passing the three-step lyophilization process; thereafter, they are ready for labeling and secondary packaging.

The lyophilized product typically is tested for identity, activity, purity, content, residual moisture, filling height, appearance, sterility and endotoxins.

2.7
Labeling and Packaging

Labeling and packaging are quality-relevant processes in the pharmaceutical industry. The packaging material contains safety-relevant information for patients and doctors, including the expiration date

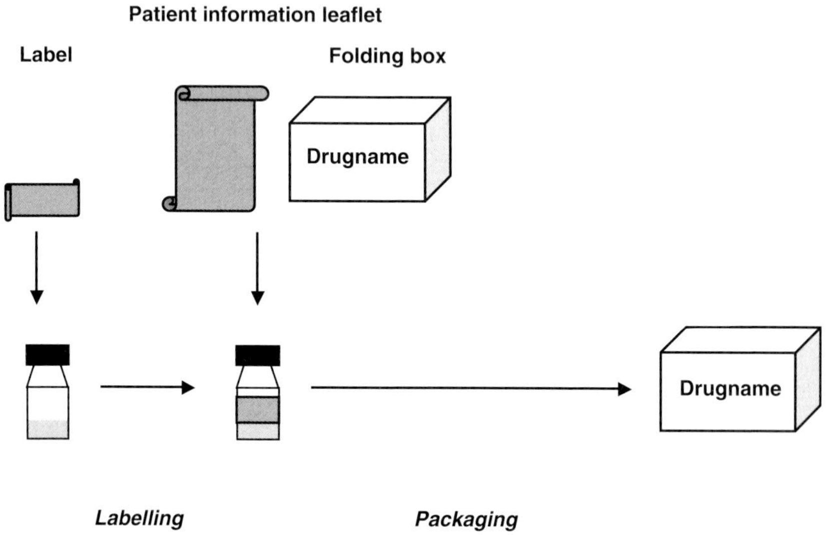

Figure 2.35 Labeling and secondary packaging.

and instructions for use. It also guarantees traceability when quality defects occur. Therefore, packaging is carried out under stringent quality requirements. In addition to the safety aspects, the package provides important protection for the sensitive proteins.

While the unlabeled vials are denoted as the 'semi-finished product', labeled and packed vials ready for delivery to market are known as the 'finished product'. Figure 2.35 shows a schematic of the labeling and packaging process.

3
Analytics

While the preceding chapters elaborated on process control as a significant part of product quality, this chapter is dedicated to analytics. Analytical methods guarantee consistency and quality, and hence the safety of the pharmaceutical product.

Quality control is based on four pillars (Figure 3.1):

- Testing product quality.
- Monitoring process parameters.
- Control of clean rooms and process media (environmental monitoring).
- Testing raw materials.

First, Section 3.1 provides an overview over the characteristics of analytics in the biotechnological field. The essential terms of product testing are presented in Section 3.2, where the link between pharmaceutical requirements and analytical methods is also shown. Some methods are contemplated here in more detail. The areas of IPC, environmental monitoring and raw material testing are covered in Sections 3.3–3.5. Finally Section 3.6 is dedicated to questions of product comparability playing a role after process changes. The important part of method validation will be explained in the context of validation in Section 6.1.5.

The International Conference on Harmonization (ICH) guideline Q6B (Section 7.1.3.5) has been an essential source for this chapter.

3.1
Role of Analytics in Biotechnology

Chapter 2 showed that biotechnological pharmaceuticals are characterized by their heterogeneity in terms of composition and molecular structure. This heterogeneity creates special challenges for the pharmaceutical application of bioanalytics, which are better understood if one becomes familiar with the term 'specifications'.

Manufacturing of Pharmaceutical Proteins: From Technology to Economy. Stefan Behme
Copyright © 2009 WILEY-VCH Verlag GmbH & Co. KGaA, Weinheim
ISBN: 978-3-527-32444-6

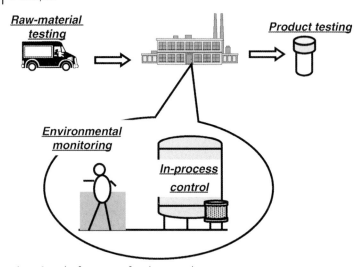

Figure 3.1 The four areas of quality control.

Specifications

The most important link between the analytical lab and the manufacturing of the medicament is the product specification. It contains the requirements for which analytical methods should be deployed and the expected results (acceptance criteria) that need to be met in order to release the product to the market. The specification guarantees quality consistency and, therefore, product safety.

The inherent heterogeneity of protein drugs results in a number of principles for the design of specifications:

- Specifications are coupled to the manufacturing process. The final specifications should be fixed together with the finally established and consistent production process.

- Specifications should take into account the instability of the product. The product changes while being stored; however, product characteristics should not change to an extent that clinical safety and efficacy are altered.

- Specifications have to be based on pre-clinical and clinical studies. The tolerated deviations from the target value of a parameter should be justified by pre-clinical or clinical experience.

Specifications are bound to analytical methods

- Specifications have to be seen together with the analytical methods. Quality-relevant parameters like biological activity, content and identity of product as well as product- and process-related impurities can show different results with different methods. Since analytical methods are developed in parallel with the process and the clinical trials, it can be challenging to prove how data from early development and from the commercial phase fit together.

Basically these requirements also apply to smaller molecules that are easier to characterize; however, in biotechnology – due to the complexity of the molecules – the analytical accessibility is eminently challenging. This aspect finds special consideration in the question whether there can be biological generics or not. In the small-molecule world, successful analytical testing is sufficient to claim comparability; usually no clinical studies are required to prove that safety and efficacy profiles of two similar products are equal. In the biotech world, this question is still heavily debated, since the analytical options as well as the ability to provide an identical copy of the original product are very limited.

3.2
Product Analytics

Bioanalytics is a highly complex science on its own. Having made enormous progress in the last decades, it has opened breathtaking insights into the properties and functionality of biological molecules. The newcomer approaching this important field sees them self confronted with confusing nomenclature. Typical terms like 'identity', 'polymerase chain reaction (PCR)', 'chromatography', 'characterization', 'DNA analytics', 'purity', 'immunological methods', 'blot', 'sequencing' and others are difficult to comprehend since they can be used for different questions depending on the application. Chromatography can be used for nucleic acids as well as for protein analytics. The same applies to blot technologies. In all cases, testing can be focused on impurities or the target protein itself.

The field can be approached in three different ways, each of which offers a framework for understanding pharmaceutical bioanalytics:

- Following the 'method path', analytical technologies are explained according to their physicochemical nature (chromatography, PCR, blot methods, immunological methods). This path addresses the mainly scientifically interested reader.

- The 'substance path' looks at the substances that are targeted by the methods (proteins, nucleic acids, carbohydrates, lipids). This path offers an overview of the methods for a particular class of molecules.

- The 'pharmaceutical path' focuses on the pharmaceutical requirements with respect to product quality (identity, purity, activity, stability). Here, the methods are categorized under their pharmaceutical target. This approach introduces the pharmaceutical framework and essential terms of quality control.

This book primarily takes the 'pharmaceutical path'; however, since analytical methods cannot always be unambiguously dedicated to one or the other pharmaceutical quality factor, the description of methods is partly anchored in the method path.

There are three main points in biopharmaceutical manufacturing where product analytics play a major role: the cell bank, the API and the finished product.

Cell bank analytics

The cell bank has to be characterized and the stability of the cell substrate has to be proven. It is beyond the scope of this work to elaborate on the methods for the pheno- and genotypic characterization of cell banks, but it should be mentioned that the process of cell banking for both the WCBs as well as the MCB, and the repeated re-validation of the cell banks, requires additional biological methods for demonstrating cell identity and purity.

API analytics

Often the API step marks another holding point in the process. This holding point can be characterized by a formal intermediate release for that material to continue processing, requiring comprehensive testing resembling that applied for release testing of the final product. The purity of the substance at this stage usually allows using the same test methods as for the final product.

Finished product analytics

The finished product step undergoes the most intensive scrutiny. Obviously the analytical program can be reduced for process-related impurities if their absence has been demonstrated in the API step. This does not apply to product-related impurities since they can occur even after the intermediate step.

There are six pharmaceutical features that are used to characterize a drug:

Identity, content, purity, activity, appearance and stability

- *Identity*: is the protein that I have the one that I want?
- *Content*: how much target protein is there?
- *Purity*: what else is there and in what quantity?
- *Activity*: how active is the product with regard to a defined pharmaceutical property in the biological system?
- *Appearance*: what does the product look like?
- *Stability*: how do these five preceding parameters change with time or varying physical conditions?

The following sections deal with these features.

3.2.1
Identity

Identity testing is mostly based on physicochemical properties, *inter alia*: (i) determination of composition of the product solution, (ii) the physical properties and (iii) the structural data of the protein.

Structural data include the primary structure and information regarding post-translational modifications (e.g. glycosylation patterns). Usually also structural data of higher order including protein folding have to be considered. Three-dimensional structural data can be obtained using spectroscopic methods, which cannot be applied economically in routine production. Therefore, the labor-intensive methods would typically be used to qualify an easier method for the identification of the higher-order structures. This can, for example, be an activity assay if the deficient structure clearly affects biological activity; analytical affinity chromatography may also be applicable.

When considering the inherent heterogeneity of the product the question arises: 'What needs to be characterized to show product identity?'. Ideally, the mixture consists of a dominant lead component (usually above 95%) and main side components (e.g. between 5 and 0.5%) as well as other side components. The main side components have to be identified and essentially characterized like the main component. The remaining components should at least be specified in their total. Application safety of the specified side components has to be demonstrated together with the main active component in clinical trials.

The biological product consists of the target protein and the side components

Identity is often tested by checking protein structures (Table 3.1) as well as general physical parameters of the product solution (Table 3.2).

Table 3.1 Methods for structural analysis of proteins.

Purpose	Description	Analytical method; remarks
Amino acid sequence	– order of amino acids	– from combination of methods for amino acid frequency, amino acid terminal sequence and peptide mapping. – the found amino acid sequence in the protein can be compared to the sequence expected from the DNA sequence
Amino acid frequency	– quantity of amino acids	– amino acid analysis
Terminal sequences	– order of amino acids at the N- and C-termini	– protein sequence analysis
Peptide pattern	– identification of peptides forming the protein	– peptide mapping
Sulfhydryl groups, disulfide bridges	– number and position of free sulfhydryl groups and/or disulfide bridges	– peptide mapping; only necessary if the gene sequence hints at the existence of cysteine groups
Carbohydrate structures for glycoproteins	– type and quantity of sugar molecules; structure of sugar chains; glycosylation sites at the protein	– glycoanalytics
Secondary and tertiary structures	– share of helical or sheet structures; denaturation	– circular dichroism, IR spectroscopy
Tertiary and quaternary structures	– spatial folding, share of aggregates	– DSC, crystal structure analysis, nuclear magnetic resonance

Table 3.2 Methods for physical data of protein solutions.

Purpose	Description	Analytical method; remarks
Molecular weight	– weight of 1 mol of the target protein (in g/mol or Da)	– SEC, SDS–PAGE, MS
Extinction coefficient	– weakening of a (light) beam by absorption and diffraction	– UV spectrometry
Electrophoresis pattern	– typical 'pattern' (separation pattern) of the product solution	– PAGE, SDS–PAGE, IEF, Western blot, CE
Liquid chromatography pattern	– typical chromatograms (retention time, peak number, peak height)	– SEC, RP-HPLC, IEX, affinity chromatography
Spectroscopy profile	– molecule-specific UV and IR absorption spectra – mass spectrum	– UV/Vis and IR spectroscopy – MS

3.2.2
Content

The methods introduced in the previous section allow qualitative determination of identity parameters. In contrast, content measurements aim at quantifying the substance in question. The content can be specified relative to the total (percentage of total) of dissolved molecules or absolutely (in mass or moles).

Absolute quantification usually encompasses impurities, since the methods do not allow differentiating between product and related byproducts. The relative quantification then gives further information about what share of the total protein content is the target protein.

One method for analyzing absolute protein content is the method of Kjeldahl (Section 3.2.8.4); relative measurements can be carried out by analyzing the peak areas of a chromatogram (Section 3.2.8.8).

3.2.3
Purity

As outlined in Section 2.5.1.1 there are different types of side products that accompany the target protein in the product solution.

Requirements regarding purity result from the need to provide a safe and efficacious product. The acceptable impurity of the product is specified during the clinical trial. Amongst others, the target protein is defined by the fact that it fulfils expectations as to the biological mechanism of action, and that it has demonstrated safe and efficacious administration in the clinical trial. It can be present in a more or less broad mixture of different subtypes (e.g. in different glycosylations or aggregations). In such a mixture of active substances it is exactly this

Table 3.3 Methods for identification of impurities.

Purpose	Description	Analytical method; remarks
Proteins	– degradation products, proteases – HCPs	– chromatography, SDS–PAGE – HCP ELISA, SDS–PAGE
DNA/RNA	– genetic information from host organism released at cell lyses	– PCR, RT assay
Endotoxins/pyrogens	– lipopolysaccharides – other pyrogens	– LAL test – *in vivo* pyrogen test
Low-molecular-weight substances	– salts/ions, e.g. from medium, buffers, chromatography gels – Acid/base, e.g. from media, buffers, inactivation steps – chromophores from fermentation – residual solvents	– conductivity, osmolality – pH measurement – visual control – solvent testing from classical chemistry, e.g. TOC method

mixture that constitutes the product, since the clinical data have been obtained for it. A deviation in the direction of one or other protein subform is considered to be a variation of the product.

Some substances evoke fever if administered parenterally. The most important pyrogens are endotoxins

Consequently, purity means to demonstrate that the solution is free from undesired impurities and that the profile of therapeutically active substances is equal to the one tested in the clinical trial. This profile is measured by means of the methods described in Section 3.2.1.

The most important substance groups of impurities are proteins, DNA/RNA, endotoxins, other pyrogens, lipids and low-molecular-weight substances (Table 3.3). For these substances it has to be proven that they are sufficiently depleted from the product in purification; moreover, testing of these substances is a prerequisite for market release of each product batch.

Microorganisms, viruses or other particles are the most important contaminations (Table 3.4). Methods to detect them are described in the various pharmacopoeias.

The remaining amount of water (residual moisture) in lyophilized vials is also an important parameter, since degradation is in most cases enhanced by the presence of water.

Pyrogens: *substances that evoke fever if administered parenterally. Most important pyrogen: endotoxins*

Endotoxin: *lipopolysaccharides from the cell wall of Gram-negative bacteria (e.g. E. coli.)*

3.2.4
Activity

Biological activity denotes the ability of a product to cause a certain biological effect. It is measured with an activity assay. While the content

Table 3.4 Methods for identification of contaminants.

Purpose	Description	Analytical method; remarks
Microorganisms	– bacteria, molds; from open product handling or from auxiliaries like water and air	– diverse methods (bioburden, sterility, identification of organisms)
Visible particles	– particles of various origin in water or air, e.g. by abrasion	– visual control
Virus	– in mammalian cell cultures from cell bank or from the unintended introduction by contaminated reagents	– PCR, RT assay – *in vivo* with animal models: antibody production, infectivity – *in vitro* with mammalian cell culture (plaque test) – TEM
Bacteriophages	– in microbial cultures at the end of fermentation; phage infection of host cells	– like virus
Mycoplasma	– bacteria without cell wall	– tests based on reaction with specific mycoplasma enzymes or PCR

measurement gives the quantity of protein, activity is a measure of the share of biologically proteins active based on correct folding into the secondary, tertiary or quaternary structure. It is therefore provided in active units per milligram protein, whereas the activity of units is related to a reference substance. Figure 3.2 illustrates the relationship between the units of the reference substance and the quantity of test substance.

Reference standard

The reference material (standard) can be an internationally established testing substance; this is usually the case if the drug targets a

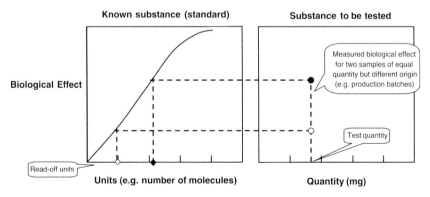

Figure 3.2 Concept of measure for activity. The effect achieved with a defined quantity of test substance is compared to an effect achieved with the reference standard. Depending on the state of the substance to be tested (e.g. denatured protein), the same quantity can be more or less active.

prominent indication with a known mechanism of action. However, often such a standard is not known and has to be created specifically for the product. In this case the reference material should be produced during product development. It is distinguished from non-reference material by tightly controlled small-scale processing and very comprehensive characterization. Testing of the ongoing routine production batches against this standard guarantees batch-to-batch consistency. In order to have an additional assurance of efficacy in the patient, the standard should be clearly linked to the material that has been used in the clinical trials. Since the reference standard is consumed for testing it has to be resupplied and the new material has to be compared to the old one prior to use.

The biological effect can be defined in different ways. Activity assays can be performed in animals (*in vivo*) or *in vitro* in cell culture assays or even cell-free biochemical assays. Naturally *in vivo* tests are more elaborate and come along with higher variance than *in vitro* tests, so that *in vitro* methods are preferred. Examples of activity assays can be very different for different therapeutic indications:

- Prevention of viral infection of animals or cells (in cell culture assays).
- Dissolution of blood clots.
- Binding of antibodies to cells displaying certain antigens.
- Inhibition of protein synthesis in cellular or cell-free systems.

Since activity assays should demonstrate desired biological activity, an essential part of their conclusiveness is that they relate to a known biochemical mechanism of action. Ideally, the activity assay mimics this mechanism. The assay is highly product specific and usually one of the more complex methods of product characterization. Reproducibility of the test depends on the deployed analytical equipment, the reagents and the routine in testing. It can last up to several days and vary in range up to 40%.

Activity assay/mechanism of action

3.2.5
Appearance

After filling, the vial is checked for outer appearance; tests include color, clarity of liquids or constitution of the lyophilization cake. Visual control is performed to detect visual particles or other defects like integrity of the primary container and the closure system. Clarity and color are assessed by comparing them against reference standards in defined light conditions.

Appearance testing can include conditions before and after reconstitution of lyophilized products.

3.2.6
Stability

In stability investigations, the sensitivity of the product to primary container materials, temperature, humidity, light, mechanical stress (shear forces in product solution by shaking) and storage time is tested. Stability is essential for product shelf-life during production as well as in the market place. Long shelf-life in modest conditions can be an essential advantage for marketing.

Instability of the product can show up in denaturation (unfolding), degradation, aggregation (with clouding), discoloration or formation of a precipitate (Section 1.3.2.1).

Stability profile
A single analytical method for stability testing is not sufficient; in fact, adequate selection of stability-indicating methods is required to demonstrate that product properties are retained. This stability profile should contain activity, identity, purity, content and appearance measurements.

Stability data for market approval
At the time of market approval, stability data should be available that support the claimed storage time and conditions as well as administration conditions. It can, for example, result from these data that the product has to be consumed in a certain time after opening the vial or that it has to be stored refrigerated. Another application condition is the solution behavior in the vial prior to administration and the absorption of water of freeze-dried powder formulations.

Stability studies before market application contribute to the time for approval ('time to market') since they have to be performed with the product from the final commercial launch process. If a shelf-life of over 6 months is claimed, at least 6 months of real-time data at the time point of license application have to be presented. At the same time it must be shown that such stored material is comparable to the that used in clinical studies. A rigorous enforcement of these requirements would result in a significant delay of market entry. The submission can be supported by data collected earlier from non-GMP batches (research data) or by data collected under extreme conditions (accelerated data) showing that the degradation pathways of the drug are well understood. Using these supporting data, extrapolations of real-time data to achieve a higher shelf-life is generally accepted, provided that the applicant continues to collect real-time data and presents them once they are available.

Follow-up studies
After market authorization, ongoing or follow-up studies are performed. The intervals for testing are dictated by regulatory guidelines [ICH Guideline Q5C, Section 7.1.3.5].

Stability of intermediate products
Stability data also have to be presented for intermediate products in order to justify storage conditions and periods between the process steps. The product has to be in specification at the beginning and the

end of the storage period. Since storage is almost always associated with slight product degradation, the specification for the beginning of the storage period has to be tighter that at the end. This tight internal specification allows releasing the product at the end of the storage period within the limits of the approved specification.

Stress stability studies are undertaken to simulate extreme storage or transport conditions. They should definitely be made to (i) investigate degradation processes, (ii) to identify stability-indicating methods and (3) be able to assess unexpected deviations from the approved storage and transport conditions (e.g. temporary elevated temperature). *Stress stability study*

Stability is essential for the application properties of the product. Amongst others it determines the time to market and comes along with high analytical effort. Product sampling must be well organized in order to be prepared for follow-up studies. A deviating stability test can trigger elaborate testing and jeopardize the approved shelf-life. In unfavorable cases it can lead to recall of the product.

3.2.7
Quality Criteria of Analytical Methods

Many analytical methods follow the same basic principle: first, the components of the test substance are isolated from each other and, in a subsequent step, they are made detectable.

Qualitative methods allow identification of a substance (e.g. gel electrophoresis); however, the amount of a substance can only be measured with quantitative methods that at the same time provide the qualitative information (e.g. chromatography). The procedure to follow when performing the method is called the 'protocol'. *Qualitative and quantitative methods*

The performance of an analytical method is expressed by means of parameters that also have to be verified during method validation (ICH Guideline Q2, Section 7.1.3.5):

- *Accuracy*: the closeness of the measured value to a known reference value. *Accuracy*

- *Precision*: the dispersion of resulting data when making several equal measurements. One can differentiate measurements which are made: (i) back-to-back with the same equipment at the same place (repeatability), (ii) in the same lab on different days and/or with different equipment and/or different operators (intermediate precision, and (iii) in different labs following the same protocol (reproducibility). *Precision*

- *Specificity*: the ability of the method to detect exactly the substance in question and differentiate it from other substances. *Specificity*

Range	• *Range*: the range between the lowest and highest detectable sample quantity.
Linearity	• *Linearity*: the attribute of a measured value to vary linearly with the quantity to be measured. This is the prerequisite for the exact designation of values which lie between the calibration points.
LOD	• *Limit of detection (LOD)*: the lowest limit of the quantity to be measured at which the method delivers a signal. The LOD is smaller than the limit of quantification (LOQ).
LOQ	• *LOQ*: the lowest limit of the quantity to be measured for which the method allows quantification.
Robustness	A further important feature of a method is its robustness, which means how the results depend on changes in the surroundings in general. Robustness is in part already described by precision, but it includes further effects like consistency of sample preparation and how narrow the parameters of the method have to be set in order to obtain a reproducible result.
Reference material	Some analytical methods need a reference material for calibration. For that purpose a reference substance is generated in a designated and well-defined lab process. This standard represents a particularly defined and comprehensively characterized product solution against which the solutions from routine production can be compared.
Sample quantity	Last, but not least, the required sample quantity can become an important issue especially in early stages of a development process when only low quantities are available.

3.2.8
Analytical Methods

The preceding sections were focused on biophysical and pharmaceutical parameters of bioanalytics; this sections introduces schematic outlines of selected analytical methods.

It should be noted that the selection of methods used for product testing should be geared to the intended use. During development, a very comprehensive characterization using different methods (orthogonal methods) for the same property is justified. When it comes to routine production, a subset of representative methods should be selected in order to balance effort and benefit. Structural data (Table 3.1) of proteins, for example, usually would not be analyzed for standard release purposes – they would probably be covered by a suitable activity assay. The selection of methods for stability investigations may be between these two poles.

Table 3.1 shows analytical methods as well as the main parameters that comprise the structural analysis of a protein. Table 3.2 details the main physical data of protein solutions. Table 3.3 provides an overview over methods for detection of prominent impurities and Table 3.4 provides an overview over methods for detection of prominent for contaminants.

The left columns of the tables indicate the property to be measured (i.e. the purpose of the method), the middle column provides a short background of the property and the right column lists methods applicable to measure the property. The list of methods is certainly not comprehensive, but should provide guidance for most cases.

3.2.8.1 Amino Acid Analysis

Purpose To measure the quantity and type of present amino acids.

Working Principle In a first step, the protein is cleaved by hydrolysis and the individual amino acid – after being derivatizized (connected to a specific molecule) – separated by chromatography (Section 3.2.8.8). The signal at the outlet of the chromatography column is detected qualitatively by comparative measurement with known amino acids (identification of amino acids by retention time) and quantitatively by identification of the peak area in the chromatogram. Disadvantages in the accuracy of this method arise from the degradation of certain amino acids (serine, threonine, cysteine, tryptophan) during hydrolysis. See Figure 3.3.

3.2.8.2 Protein Sequencing

Purpose To determine the primary sequence of the amino acids in a protein.

Working Principle The most widely used technique is Edman degradation, named after its Swedish inventor. In the Edman process, the

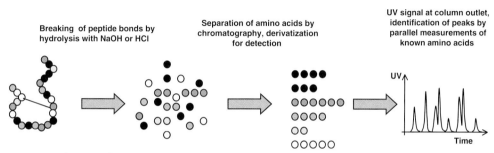

Figure 3.3 Schematic of amino acid analysis. AA = amino acid.

protein is decomposed stepwise beginning from the N-terminal side (Section 1.3.2.1). This works best if the protein has been cleaved into peptide fragments smaller than 7 kDa before, which is achieved by reagents specifically attacking certain bonds between amino acids. Once the peptide fragments are isolated from each other, stepwise degradation starts. The isolated amino acids are identified by chromatography. See Figure 3.4.

3.2.8.3 Peptide Mapping

Purpose To identify a specific pattern (map, finger print) of the peptide fragments of a protein. The map can be a chromatogram, a sodium dodecylsulfate–polyacrylamide gel electrophoresis (SDS–PAGE) pattern (Section 3.7.2.5) or a mass spectrogram.

Working Principle As outlined in Figure 3.5, the protein is split up into fragments and afterwards analyzed by SDS–PAGE, chromatography or mass spectrometry (MS), resulting in characteristic patterns.

In SDS–PAGE the pattern is mainly specific to molecular weight, whereas chromatographic patterns also depend on the molecular interactions between the peptide fragments and the chromatography material.

3.2.8.4 Protein Content

Purpose To measure the total content of protein in the solution – it should not be confused with the measurement of target protein content (the latter is mostly detected by chromatography; Section 3.2.8.8).

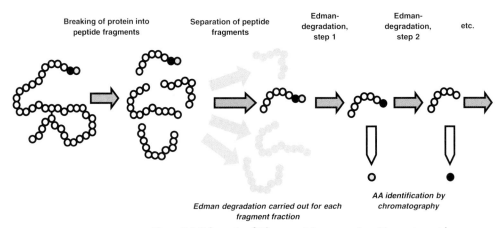

Figure 3.4 Schematic of Edman protein sequencing. AA = amino acid.

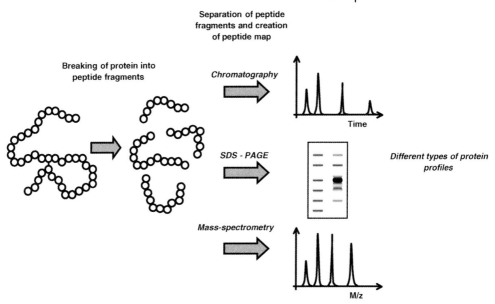

Figure 3.5 Schematic of peptide mapping. m/z = mass-to-charge ratio.

Working Principle The content of protein can be measured with the methods of Kjeldahl and Dumas. In both methods the amount of nitrogen bound in the amino groups of the amino acids is detected. Thereby, the nitrogen of other substances like nucleic acids is also measured and cannot be differentiated from the nitrogen bound in the proteins; thus, the methods require diligent sample preparation.

3.2.8.5 Electrophoresis

Purpose To measure the identity and purity of a given solution by analyzing the component according to their molecular weight or charge.

Working Principle The basic principle of electrophoresis is that molecules migrate through a gel bed driven by the force of an electrical field. After completing their movement the molecules are made visible and result in 'bands' in the gel bed that can be compared with known reference molecules, allowing conclusions regarding their molecular weight. The gel can consist of a highly cross-linked polyacrylamide polymer or agarose. Electrophoresis can be applied to proteins as well as to nucleic acids. Depending on charge or molecular weight, different molecules migrate different distances in the gel bed.

Electrophoretic methods include SDS–PAGE, isoelectric focusing (IEF) and capillary electrophoresis (CE).

SDS–PAGE In SDS–PAGE the proteins are treated with SDS and subsequently separated by electrophoresis. Due to the treatment with SDS, the separation is driven by differences in molecular weight only.

SDS causes the denaturation of the protein structure by cleaving the hydrogen bonds. Additionally, other reagents like β-mercaptoethanol or dithiothreitol can be used to clip the disulfide bridges that are responsible for links between protein chains (e.g. in antibodies).

SDS–PAGE separates according to molecular weight

The main effect of SDS is that it binds to the peptide chains and effectuates a strong negative charge superseding the natural charge of the proteins. Consequently, separation of the molecules is independent of their charge and molecular weight is the only driver for different migration velocities in the gel. The size (molecular weight) of the molecules is determined by comparison with known reference molecules.

After having migrated, the proteins have to be made visible. This is achieved by methods of gel staining: in a first step the proteins are fixed in the gel using acetic acid; subsequently they are colored by adding Coomassie blue or silver nitrate solution (silver staining). See Figure 3.6.

Isoelectric focusing separates according to isoelectric point

IEF In IEF the molecules in the gel are separated exclusively due to their isoelectric point.

Figure 3.7 illustrates the basic principle of IEF. A stable pH gradient is generated in a gel bed by suitable immobile gels carrying acid or alkaline groups. The protein mixture is then added and an electromagnetic field applied. Driven by the electrical field, the proteins migrate through the pH gradient until they reach the isoelectric point, which is characteristic for each individual protein. At this point the net charge of the protein is zero, therefore the electromagnetic forces are balanced

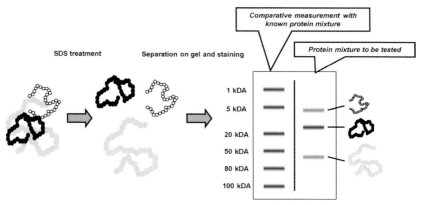

Figure 3.6 Principle of SDS–PAGE.
Left = mixture of different aggregated proteins linked by hydrogen bonds; right = typical picture with protein band and comparison with standard.

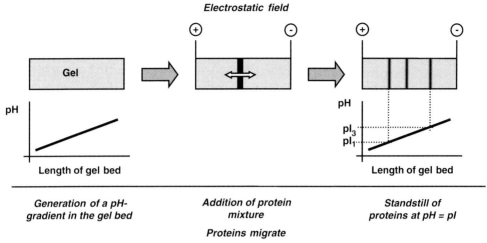

Figure 3.7 Isoelectric focusing.

and the protein stops moving. The bands are visualized by staining as described for the SDS–PAGE method.

CE In CE the molecules are separated in capillaries (diameter about 50–75 μm, length about 20–100 cm) instead of a flat gel bed. The capillaries can be empty or filled with HPLC phases or gel. At the outlet of the capillaries the proteins are detected with methods known from chromatography, such as UV detection (Figure 3.8).

3.2.8.6 Western Blot

Purpose To measure the identity and purity of a given solution by analyzing the component according to their molecular weight or charge.

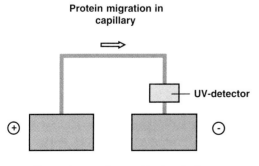

Figure 3.8 Schematic of capillary gel electrophoresis.

Western blot: *proteins*
Southern blot: *DNA*
Northern blot: *RNA*

Working Principle The term 'blotting' indicates the transfer of molecules from a separation gel onto a nitrocellulose matrix. The molecules are fixed immotile on this matrix and can be visualized by applying chemical reagents. Western blotting is a method for isolation and detection of proteins, while Southern and Northern blotting are used for DNA and RNA, respectively.

The following steps are run for the Western blotting (Figure 3.9):

(i) Separate the protein mixture by gel electrophoresis.
(ii) Transfer the proteins on a nitrocellulose matrix ('blotting').
(iii) Douse with enzyme-marked antibodies that bind to the protein being investigated.
(iv) Flush to remove the non-bound antibodies in order to avoid that in the next step all antibodies are detected and not only the ones bound to the protein.
(v) Detect the bound antibodies by an enzymatic reaction (e.g. by staining) that is exclusively caused by the enzyme bound to the antibody.

The antibodies specifically bind to the protein in question, thus the Western blot can be used to look for known proteins highly specifically – if antibodies for these proteins exist. Usually these antibodies are obtained by injecting the antigen (in this case the protein to be detected) into mice and isolating the antibodies from their blood.

Since the Western blot uses antibodies and electrophoretic separation techniques it is also called an immunoelectroblot. Closely related to

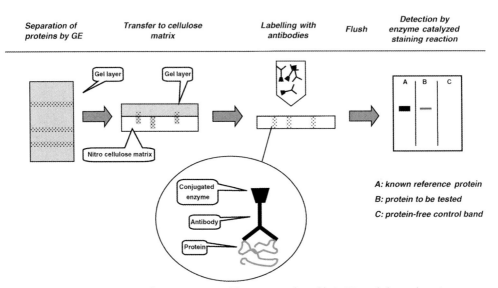

Figure 3.9 Western blot (immunoelectroblot). GE = gel electrophoresis.

that is the radioimmunoassay in which instead of using an enzyme, a *Radioimmunoassay*
radioactive marker is attached to the antibody that is detectable by measuring the emitted radiation.

The great advantage of the enzymatic reaction is that it amplifies the detection signal. The enzyme can generate many more color molecules than enzyme molecules exist, facilitating the detection of very small amounts of protein.

3.2.8.7 HCP enzyme-linked immunosorbent assay (ELISA)

Purpose ELISA is a highly specific method to detect proteins and nucleic acids in a solution; it can also be applied to other molecules. Here, HCP ELISA will serve as an illustrative example. The HCP ELISA tests whether proteins of the host cell are present in the product solution.

Working Principle The assay's name is derived from the fact that the proteins to be detected are marked by specific antibodies, which are coupled to an enzyme that on its part catalyzes a detection reaction. Thus, the method is based on immunological principles like the Western blot. The binding of the antibodies to the antigen to be detected is based on an antigen–antibody reaction. In the HCP ELISA the HCP represents the antigen. The generation of the antigen and the corresponding antibodies is very labor intensive.

Figure 3.10 illustrates the steps of the HCP ELISA:

(i) The manufacturing process is run with the original host cell line in which product expression has been suppressed by genetic modification ('mock run'). Therefore the cell only expresses the 'undesired' proteins that will later be detected as HCPs.

(ii) A sample is taken on a process step in which the host cell proteins still are present.

(iii) The HCPs can be isolated and purified by chromatography

(iv) By injection of the isolated proteins into mice, antibodies are generated which are harvested and purified.

(v) An enzyme is coupled to the antibodies. After this step the conjugated antibodies are present and the detection of the HCPs from the actual process can start.

(vi) In order to detect the proteins, the product solution from the real, non-modified process is given on a solid matrix (e.g. microtiter plates) to which the proteins attach. The same type of proteins that had been previously used to inoculate the mouse now represent the antigen in the antibody reaction.

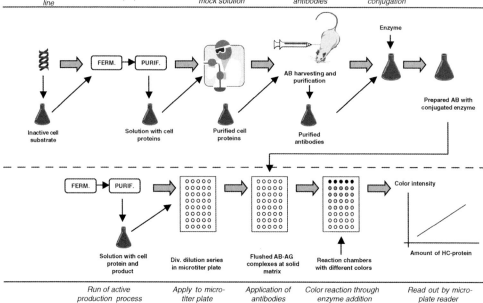

Figure 3.10 Schematic of a HCP ELISA. AB = antibody; HC = host cell.

(vii) After this step, the antibodies are put on the probe and bind to the HCPs. The non-bound antibodies are removed by flushing.

(viii) An agent is added that displays a chromophoric (color) reaction together with the bound enzyme. If the coloration of the solution is strong, there is plenty of HCP; if it is only weak, there is less.

During development and validation of the assay, these steps have to be repeated several times; the development lasts between 12 and 18 months. Sometimes the proteins do not bind to the solid matrix; in these cases an additional antibody is needed which definitely binds to the solid matrix and 'captures' the HCP out of the product solution. In this so-called 'sandwich ELISA', the protein is framed by two antibodies binding to two different protein domains.

Sandwich ELISA

3.2.8.8 Analytical Chromatography

Preparative chromatography: generation of large quantities

Analytical chromatography: detection of a certain substance

Purpose To measure the content and concentration and identity of the target protein and byproducts. Chromatography is one of the most important methods of product characterization and impurity detection. In analytics, SEC, HPLC and affinity chromatography are applied.

Working Principle The basics of preparative chromatography have been described in Section 2.5.3.1. Analytical chromatography follows

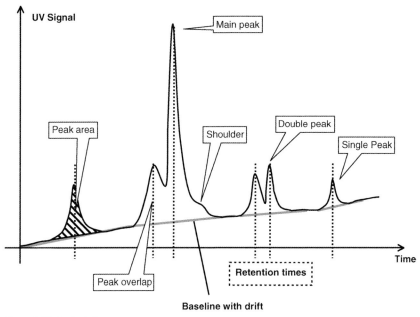

Figure 3.11 Typical chromatogram.

the same principles as preparative chromatography; therefore, they will not be considered here in detail again. Figure 3.11 illustrates essential terms around the chromatogram that is obtained as a resulting plot from the methods. The curve, which is recorded by an UV detector at the column outlet, shows different peaks indicating absorption of UV light by passing molecules.

Since each component migrates with a different velocity, the identity of such a compound can be determined by looking at the retention time of its signal peak. In the best case the retention times are so different from each other that their peaks do not overlap; this optimal case is called baseline separation. In the case that the components elute close to each other, their peaks overlap, which is unfavorable since the purity data of the main component is derived from its share of the overall area under the curve. For this purpose, the areas that are enclosed by the UV curve and the baseline are appraised and summed up ('integrator'). The concentration of the main component is then derived as the ratio between the main component's peak and the total area. If the peaks overlap with the main component peak, the area of the main component can only be determined imprecisely. In this case, assumptions have to be made regarding the peak symmetry and the slope of the UV curve in the overlapping area. Shoulders in the peak also point to byproducts eluting together with the main component.

Obviously the interpretation of a chromatogram is not clear-cut. Especially when smaller peak areas, which occur for byproducts, need to be evaluated, susceptibility to misinterpretation is high. Unexpected values can occur when changes are made to the column material, solvent, detector, type of equipment, temperature or integration software.

The relatively simple and fast conduct of experiments as well as accurate and reproducible results have made analytical chromatography the workhorse of release analytics; however, changes to both the equipment and the analytical protocol have to be treated with utmost scrutiny.

3.2.8.9 Infrared (IR) Spectroscopy

Purpose To measure protein structures.

Working Principle The physical principle is that IR electromagnetic radiation (wavelength 800–1 000 000 nm) activates oscillation of molecular bonds. This activation 'consumes' energy of a certain wavelength. From the measurement of the energy absorption of that wavelength it can be concluded what types of molecular bonds exist in a sample. This in turn allows us to draw conclusions about the structure. For example, the bonds between the CO and the NH groups are relevant for the α-helix or the β-sheet structures of a protein. The hydrogen bonds allow differentiating between these structures, since the stronger links in the sheet result in a shift of the absorbed spectrum relative to the helix. The IR principle also gives insight into the folding structures.

3.2.8.10 UV/Vis Spectroscopy

Purpose To measure the identity or presence of a molecule.

Working Principle UV/Vis spectroscopy uses the feature of molecules to diminish the intensity of UV and visible light by absorption. This is called extinction and is specific for each molecule. Extinction is measured by spectrophotometers that are positioned behind a glass tube filled with the sample substance. In order to compensate for disturbing effects, the extinction of the actual sample is measured relatively to the extinction of a reference sample of known composition (e.g. pure solvent without proteins). The characteristic spectrum is obtained by scanning a range of wavelengths.

In process analytics, UV spectroscopy with a constant wavelength of 280 nm is used at the outlets of chromatography column; it indicates the presence of proteins, controls the fractionation and helps to monitor process consistency.

3.2.8.11 Mass Spectrometry

Purpose To determine identity of unknown molecules or identify impurities in the product by measuring the molecular weights.

Working Principle MS characterizes ionized molecules based on their mass/charge ratio. Figure 3.12 shows the principle design of a mass spectrometer. In the ion source the sample to be investigated is ionized and mobilized (i.e. dissolved from its solid formation). After that, the ionized substance is accelerated in the so-called analyzer and captured in the detector. The different types of MS methods differ in the type of the three components: ion source, analyzer and detector.

For macromolecules like proteins, the most relevant technologies used for the ion source are:

- *Chemical ionization (CI)*: generation of ions by charge substitution with ionized reagents.

- *Fast-atom bombardment (FAB)*: ions are generated by 'bombardment' with noble gas atoms.

- *Matrix-assisted laser-desorption ionization (MALDI)*: ions are separated from the solid matrix and ionized by a laser beam.

- *Plasma-desorption ionization*: protein fragments are dissolved from a layer of proteins by bombardment with heavy ions (e.g. californium). The dissolution (desorption) of the ionized fragments is caused by the energy that is released by abrupt deceleration of the ions. This energy generates a plasma-like state for a short time at the site of impact which allows the fragments to escape from their formation.

- Other methods are the field desorption method and electrospray ionization (ESI).

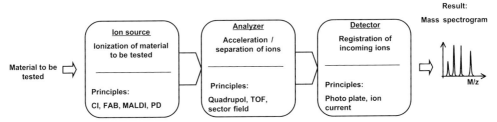

Figure 3.12 The essential elements of a mass spectrometer. m/z = mass-to-charge ratio. For other abbreviations, see text.

FAB, MALDI and plasma desorption ionization are denoted as desorptive methods since they ionize by desorbing from a solid formation. They are particularly well suited for large, non-volatile molecules like proteins and nucleic acids.

Analyzer The most relevant analyzer technologies are:

- *Quadrupole MS*: the ionized particles are at the same time accelerated and put into oscillations. Due to the mass dependence of the oscillation behavior, for a given activating oscillation, only particles of a certain molecular mass find their way to the detector. The ion current in the detector is plotted versus the changing stimulating oscillation. The resulting spectrum is characteristic for the mixture.

- *Time of flight (TOF)*: the TOF of the ions between the ion source and the detector is determined; conclusions are drawn from this time regarding the mass/charge ratio.

- In sector-field devices, the ions cross a magnetic field that is perpendicular to their flight direction. The ions are diffracted differently due to their different masses. The different sites of impact on the detector are a measure of the mass/charge ratio.

Detector In the detector, the ions are detected on a photo plate or by electrical registration of the ion current.

HPLC-MS MS can be combined with other methods; the combination with chromatography (HPLC-MS) allows simultaneous separation and characterization of the mixture.

3.2.8.12 Glycoanalytics

Purpose To provide information about the carbohydrate structures around molecules. Sugar molecules play a fundamental role in the biological activity of glycoproteins (Section 1.2.3.2). The tasks of glycoanalytics include identification of the type and size of the bound sugar molecules as well as determination of the protein heterogeneity as to glycoforms.

Working Principle Glycoanalytics comprises numerous spectroscopic, electrophoretic and chromatographic methods.

3.2.8.13 PCR

Purpose To enhance measurement of DNA and RNA. PCR can amplify the quantity and measuring signals of DNA strands. Often only copying DNA makes it possible that the lower LOD is reached that can, for example, be important if residual or viral DNA in protein solutions should be detected. The method can also be used to insert engineered mutations into gene strands.

Working Principle The functional principle of PCR is simple:

(i) DNA is heated up to 94 °C, which leads to disintegration of the double strands.

(ii) Primers are added. Primers are synthetic DNA fragments complementary to certain selected sequences in the DNA. By cooling down to 50 °C the primers bind to the disintegrated DNA strand (hybridization) and prevent that the DNA reassociates. *Primer*

(iii) Addition of nucleotide building blocks (desoxynucleic triphosphate: dATP, dCTP, dGTP, dTTP) and DNA polymerase – an enzyme that supports DNA formation from the bricks. Subsequent heating to about 75 °C yields optimal polymerase activity. Originating from the attached primers, the strands of the DNA are reconstructed from the nucleotide building blocks in the solution.

(iv) At the end of the cycle two identical DNA strands are present. The cycle can be multiply repeated; the DNA is doubled each time.

The use of the method is restricted to applications where the right primers are available. The primers are nucleotide chains of about 15 base pairs and have to be identified from a sequence analysis of a certain number of 'right' DNA strands. If the DNA to be analyzed is not known, certain restriction enzymes are used to disintegrate the DNA and a suitable primer is coupled to the ends of the DNA.

The detection of RNA is possible by transferring the RNA into DNA using reverse transcriptase (Section 1.3.1.5); after that the DNA can be analyzed and conclusions drawn for the corresponding RNA.

3.2.8.14 DNA/RNA Sequencing

Purpose To determine the order of base pairs in nucleic acids.

Working Principle The basic principle is identical for both DNA and RNA (Figure 3.13):

(i) The sample is distributed on four flasks. Each of the four vessels is filled with a base-specific reagent, which clips the nucleic acid at exactly one base. Therefore, each vessel contains fragments of different lengths with equal starting point.

(ii) In a second step the contents of the four flasks are separated by electrophoresis. The resulting band pattern is called sequence ladder – the base sequence can be directly read from this pattern.

Often there is not enough material for sequencing, thus PCR or reverse transcription (RT)-PCR (Section 3.2.8.13) is used to amplify the strands. RNA can be analyzed by conversion into DNA.

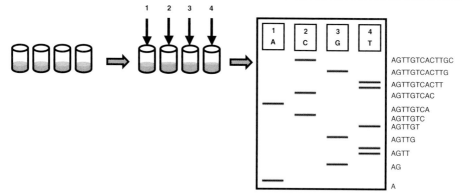

Figure 3.13 Principle of nucleic acid sequencing.
A, C, G, T = bases from the DNA molecule.

The difference between sequencing methods lies in the fragmentation. RNA can be disintegrated chemically, enzymatically or (after conversion) – as DNA. DNA can be fragmented either chemically using the Maxam–Gilbert method or with the chain-termination method (Sanger–Coulson method). The Sanger–Coulson method is the basis for the automated sequencing that was applied for decoding the human genome.

3.2.8.15 Endotoxins and Pyrogen Testing

Purpose Pyrogens are substances that, when parenterally administered in minute amounts, provoke fever in humans. Most prominent representatives of pyrogens are endotoxins – fragments of the cell wall of Gram-negative bacteria.

Working Principle
The test for pyrogens relies on *in vivo* testing in rabbits; therefore, its use has been widely discontinued. The important subgroup of endotoxins is covered by the *in vitro* limulus amebocyte lysate (LAL) test. To perform this test, the sample is mixed with a blood component of the horseshoe crab (*Limulus polyphemus*); if endotoxins are present, a gel is formed or a color reaction is visible.

LAL Test

3.2.8.16 Bioburden Test

Purpose Microbial testing aims at detecting the presence of bacteria or molds. Tests can be done for the complete absence of microorganisms (sterility) or to identify the remaining microbial burden (bioburden).

Working Principle The methods for detecting colony-forming units are described in the pharmacopoeias; the product solution is led through a membrane and subsequently transferred to an agar plate for incubation. The presence of microorganisms is apparent from their proliferation on the agar plate; one germ forms a circular colony. Once a microbial contamination is detected, the germ is identified by Gram staining and other pheno/genotypic methods.

The speed of growth of microorganisms cannot be accelerated, therefore bioburden or sterility tests take a relatively long time to complete (up to 2 weeks). For normal- to fast-growing organisms, however, quick tests have been developed, providing a reliable result after several days. *Quick sterility test*

3.2.8.17 Virus Testing

Purpose To detect viruses in the protein solution.

Working Principle Virus tests are manifold. Immunologic, microscopic and microbiological methods are applied as well as PCR.

An example of an *in vivo* method is the test for antibody formation. In order to find viruses, the solution to be tested is injected into a suitable animal model (mouse, rat); in the case of viral contamination, corresponding antibodies can be found in the blood of the animals.

Infectivity can be measured with animal models or *in vitro* with suitable cell cultures. For the experiment in cell cultures the solution to be tested is added to a monolayer cell culture. If the solution contains viruses, adjacent cells will die leaving defects in the tissue (plaque test). The result is obtained by counting the plaques (plaque forming units). *Plaque test*

PCR is an elegant method for virus detection. The approach is to amplify and thereby detect viral genes by using suitable primers. Virus search and identification also uses electron microscopy [transmission electron microscopy (TEM)].

3.2.8.18 TEM

TEM allows visualizing objects by means of electron beams. Objects down to 0.2 nm (including viruses) can be identified with TEM.

3.2.8.19 Circular Dichroism

Purpose To analyse secondary structures of proteins.

Working Principle Circular dichroism is based on the physical phenomenon that light waves of a certain type are diffracted by optically active substances. This diffraction can be detected with the methods of UV and IR spectroscopy.

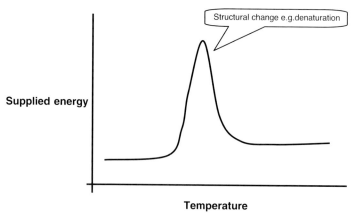

Figure 3.14 Typical plot of DSC.

3.2.8.20 Differential Scanning Calorimetry

Purpose To investigate protein structures. Differential scanning calorimetry (DSC) can analyze protein folding as well as aggregation.

Working Principle The physical foundation is that structural changes in molecules are characterized by the flow of energy to or from the molecule. For example, the clipping of bonds between aggregated molecules requires an energy flow to the molecule (endothermic), while the formation of bonds usually releases energy (exothermic).

In DSC the energy that is used to heat a sample is measured. Thereby, energy is supplied to the sample and compared to energy supplied to a reference sample without protein. In contrast to the sample, this reference should not undergo structural changes during heating in the contemplated temperature interval.

The structural changes in the protein sample will lead to the effect that it requires more or less energy than the reference when the temperatures of both samples are kept equal.

This difference in energy consumption is plotted versus the temperature and provides information about structural effects as well as the transition temperatures (Figure 3.14).

3.3
Process Analytics

Process analytics (in-process control, IPC) includes the fields of process control and the testing of intermediate steps.

Process control allows regulation of process parameters like temperature, pressure, pH value, conductivity, aeration, cell density, feed stream and so on. It ensures consistency of the process and therefore substantially contributes to product quality. The specifications for quality-relevant process parameters are part of the regulatory drug dossier and have to be met to release the product, just like the product specifications themselves. Process analytics is usually performed directly at the process equipment (fermenter or chromatography column), whereas product analytics (described in the previous section) relies on lab work. The control of the process by lab-based methods has its disadvantages since the time for sampling and testing can be relatively long, therefore a quick reaction to undesired conditions is aggravated. For unavoidable lab-based IPCs a so-called IPC-lab is usually established to enable operators to react faster to these data.

IPC laboratory

Process analytics is often based on relatively simple analytical principles; however, the analytical devices have to be suitable for the special conditions in the operational field (cleanability, sterilization, robustness, low calibration effort, in-line capability).

3.3.1
Fermentation

In the first instance, fermentation analytics serves to maintain consistent living conditions for the suspended cells or microorganisms. This includes monitoring of the physicochemical environment, such as pH value, conductivity and dissolved gases (oxygen, carbon dioxide). A sample port to support off-line measurements (product concentration, cell viability) is always available. The measuring program is supplemented by physical parameters like weight, temperature, filling level, stirrer rotational speed and gas flow. Sight glasses facilitate visual controls.

The possibilities for in-line analytics are restricted by the limited ability to automate testing methods as well as the special requirements with regard to cleanability and thermal sterilizability.

In routine production, samples are usually drawn only at the end of the fermentation process in order to avoid contamination by the sampling process. See Figure 3.15.

3.3.2
Purification

As an example of process control in purification, Figure 3.16 presents the analytical methods during chromatography. pH value and conductivity as well as temperature and flow are measured in the feed stream to the column. Some of these measurements can also be made in the buffer hold tank. A UV detector at the column outlet monitors the flow of proteins. Fractions can be separated according to preset flow, time or by means of the UV signal at the column outlet.

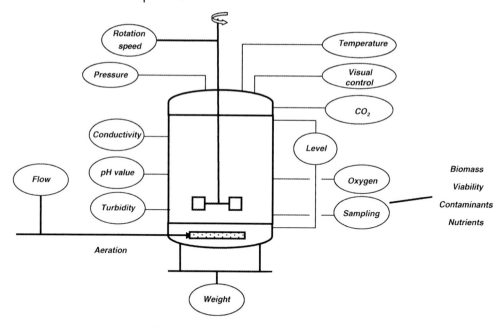

Figure 3.15 Analytics around the fermenter.

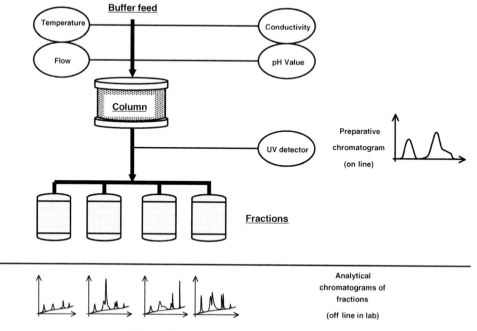

Figure 3.16 Process analytics in purification.

Analytical chromatograms of the fractions are generated in the lab. In this case the product fraction has to fulfill the specifications; however, the other fractions also give information about process consistency.

For filtration operations the most important parameters are the flow and the transmembrane pressure. The latter helps to recognize the operational condition of the membrane; when the membrane blocks, the pressure rises significantly while a decreasing pressure mostly indicates membrane damage. Retentate and permeate are mostly investigated by lab-based methods like chromatography.

3.3.3
Formulation and Packaging

Formulation and packaging are the last steps of the product on its way to the patient. Here, the requirements for manufacturing of sterile drugs apply as well as the pharmacopoeial test for excipients remaining in the finished product. Apart from making important biochemical tests of the product, parameters like filling level in the vial or the correct packaging and vial imprint are monitored on-line.

Figure 3.17 illustrates the most important on-line measurements during formulation, filling and packaging. The visual inspection checks whether the solution or lyophilization cake contains particles, or whether there are cracks in the glass or insufficiently closed vials. Often this is done by looking at the vials, since optical devices are not precise enough to differentiate between undesired and normal conditions. Since after this step the drug is released to the market, the primary and secondary package, as well as the drugs they contain, undergo rigorous (often 100%) visual testing.

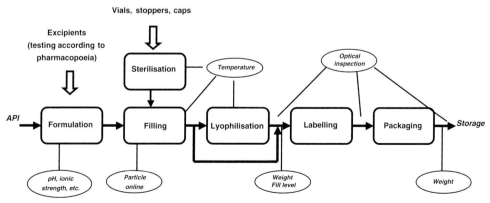

Figure 3.17 Process analytics in formulation and packaging.

3.4
Environmental Monitoring

Engineering measures for prevention of contamination (see Chapter 8)

Due to their generally 'life-sustaining' conditions, biological processes are particularly susceptible to microbial contaminations like bacteria or molds. A contamination in fermentation leads to the destruction of the product and is easily recognized from the deviating fermentation course. In contrast, in purification a certain level of microbial load can be tolerated, which is minimized at the process end by filtration. Once the product approaches the patient (i.e. during formulation and filling), microbial contaminations have to be avoided under any circumstances. Their pathogenic effect when administered parenterally is unacceptably high. Apart from microbial load, particle contamination generated by dust or abrasion plays a major role. The regulatory requirements for clean rooms therefore contain limits for particles and microbial contamination. It is the task of environmental monitoring to survey the purity of the ambient contamination sources and to minimize the risk for the product.

Figure 3.18 shows the main pathways of contaminations into the process or the process environment. Obviously a contamination of the ambient does not automatically result in product contamination. The more open the product solution is handled and the higher the resulting risk from a potential contamination would be, the higher the environmental requirements. Pharmaceutical water, buffers, chemicals, gases or clean steam come into direct contact with the product or with product-contact surfaces and can therefore contaminate the product.

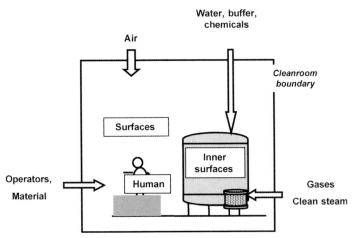

Figure 3.18 Sources of contamination of the process.

Moreover, particles can access the product through inner surfaces when gaskets decompose or cleaning has been insufficient. In the clean room itself the surfaces are contamination sources (e.g. abrasion of color particles from walls, accumulation of dust on horizontal or rough surfaces). Particles are carried in by the air supply as well as the personnel and material traffic; human beings are by far the dominant source for particle and microbial contaminations. *Sources of contamination*

Purity of water and clean steam is ensured by the testing of the following parameters: *Water monitoring*

- Total organic carbon (TOC).
- Microbial contamination (bioburden).
- pH value.
- Conductivity.
- Endotoxins.
- Particles.
- Inorganic components like chlorides or heavy metals.

Process gases and room air are mainly tested for particles, microbial contaminations, humidity and temperature. Inner and outer surfaces are tested by swab or wipe tests for attached microbial or other contaminations; for that purpose tabs are brought into contact with the surface to be tested and transferred to agar plates, which are incubated in the lab. The presence of microorganisms is apparent from their proliferation on the plate; one germ forms a circular colony. Once a microbial contamination is detected, the germ is identified by Gram staining and other pheno/genotypic methods. *Air monitoring*

3.5
Raw Material Testing

Raw material testing secures the quality of the deployed chemical and biological auxiliaries, like powders for buffer or media preparation and chemicals for the process (salts, acids, bases, antifoaming agents, solvents, etc.). *HSA risk*

Biological processing aids from human [human serum albumin (HSA) or animal [sera like fetal calf serum (FCS)] origin play a special role. HSA is used in liquid formulations to enhance protein stability; FCS serves as additive in cell culture media. The main risk associated with HSA is the potential contamination by viruses or prions.

Viruses can to a great extent be ruled out by comprehensive testing; however, a residual risk remains since the tests are very specific for the targeted test virus. Therefore, complete documentation of the donor anamnesis is part of the quality assurance concept that is associated with high organizational effort.

The main risk with FCS is its potential contamination with TSE. TSE is a generic term for different diseases occurring in animals and humans. Cows and calves suffer from BSE ('mad cow disease'); humans suffer from Creutzfeldt–Jakob disease (CJD) and Gerstmann–Straussler syndrome (GSS). This disease is activated by prions and transferable from animals to humans. The prions cannot be inactivated by thermal treatment, since this would also destroy the valuable and not harmful components. The disease has an incubation time of up to a decade. Analytical methods for the detection of prions are under development; however, up to now no robust method exists that would allow risk-free use of serum. Thus, the use of calf serum relies on certified origin; that means that production from definitely BSE-free cow herds needs to be attested. It is foreseeable that the growing biotech market as well as the progressing BSE issue will make bovine material an even scarcer raw material than it already is today. A remedy could be to develop reliable testing methods for prions; however, more promising in the long run would be to research further in serum-free media.

3.6 Product Comparability

If changes are made to the manufacturing process it has to be proven that the products before and after the change are comparable with respect to quality, safety and efficacy.

Process changes happen naturally during process development that often runs in parallel to the clinical development; however, even after market launch, process changes are common due to economical or regulatory reasons.

How can comparability be demonstrated? The specialty of biological products to be determined by the process and to be only limited analytically accessible makes this task particularly challenging for protein drugs. As a basic principle, each comparability exercise results in a case-by-case decision and depends on the extent of changes. Due to time and cost reasons, it is highly desirable to base comparability solely on analytical data; however, it can happen that these data are not sufficient and that additional pre-clinical or even clinical data need to be provided. In particular, questions around immunogenic properties of proteins and their degradation products or HCPs can practically only be shown in clinical tests.

It is challenging to define when a product can be considered comparable to its precursor. Which deviations from the known impurity profile on which process step can be tolerated and which ones jeopardize safe and efficacious use of the drug? A deep understanding of the biochemical mode of action and the physiological effects of impurities is of essence in this decision, since it is the basis for the establishment of a well-justified comparability specification. The more clinical experience with the product exists, the easier the evaluation.

If the biochemical mode of action is only hypothetically known and the influence of byproducts unclear, the product is highly 'empiric'. The more product properties are defined empirically, the higher the likelihood is to necessitate empirical studies for demonstration of comparability. Empirical in this sense means that properties are defined on the basis of pre-clinical or clinical data and that the predictive ability of analytical data is poor.

It is a major advantage for the comparability exercise if the previous process as well as the deployed analytical methods are known. This allows evaluating process changes by bridging from the old to the new process as shown in Figure 3.19. This aspect is also relevant for generic biologics. Since the information about analytical methods and the manufacturing process is proprietary to the original manufacturer and

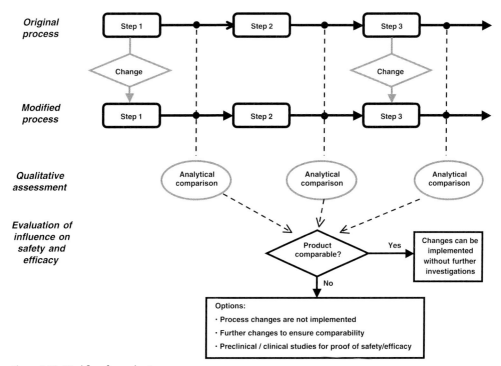

Figure 3.19 Workflow for evaluating comparability. Comparison of product quality in different process steps.

not public, copies of the biological drug are not easy to make. The establishment of biogenerics fundamentally depends on the quality of analytical characterization and the possibility to correlate the analytical data with physiological effects.

Risk analysis

A comparability study can be accompanied by a risk analysis that describes the process changes and their impact on measurable product properties (identity, purity, stability, activity, appearance), and provides an assessment regarding the safety and efficacy of the finished product.

Regulatory guidance for comparability questions can be obtained from the ICH guideline Q5E (Section 7.3.5).

Part Three
Pharmacy

4
Pharmacology and Drug Safety

The goal of pharmaceutical research and production is to provide a safe and efficacious drug to the patient. Safety and efficacy are demonstrated in clinical studies. Data collected in these trials are used to evaluate a risk–benefit ratio, which is ultimately decisive for the market approval of a drug. Side-effects of drugs can be easier to accept if the drug targets a severe disease, while the same side-effects would lead to rejection of the license application if the drug targeted a less severe disease.

Risk–benefit evaluation

Safety aspects encompass symptoms of poisoning, reactions of the immune system, tendencies to harm genetic material or the fetus, or cancer promotion. Efficacy is demonstrated by measuring a defined symptom improvement.

The relevance of these basic pharmaceutical contemplations for production is manifold:

- Clinical trials usually dominate the timelines of product development. Understanding the development events enables the production responsible to estimate the necessity for building up the supply chain for market launch.

  Clinical development determines the speed of product development

- In biotechnology, in particular, the requirements for process development and product manufacturing grow together in the final phase (phase III) of the clinical trial.

  Manufacturing process should be fixed before phase III

  This phase should definitely be supplied with material from the later launch process and the launch manufacturing site. A change after the clinical trials could result in an extended comparability study that may cause a significant delay of market entry. That is why the production process is practically fixed in early phases (before phase III, Section 4.3.2) of product development.

  Patient numbers in phase III can be significant; this may present a logistical challenge for development units and the manufacturing environment may be better positioned to satisfy the product demand.

  Clinical trial material has to be manufactured under GMP conditions (Chapter 6). A strong separation between 'free' development and the 'controlled' manufacturing environment no longer exists.

Structure of chapter 4

The following sections give an overview over the basic terms of pharmacy. We start with illustrating the administration and effect of drugs in humans (Section 4.1). One section describes how drug action can be influenced by the pharmaceutical formulation and route of administration (Section 4.2). The section addressing drug testing elaborates on pre-clinical and clinical testing (Section 4.3). It is followed by a section highlighting the path of the medicament once it leaves the controlled area of the manufacturer (Section 4.4). The last section is dedicated to the surveillance of drug safety for an established market product (Section 4.5).

4.1
Action of Drugs in Humans

In order to achieve a therapeutic effect in the human body, the active substance has to:

(i) Access the body.
(ii) Migrate to the desired location.
(iii) Display the desired effect at the target location.

Pharmacokinetics and pharmacodynamics

The different pathways into the body are called administration routes. The local and timely distribution is the subject of the pharmacokinetics (PK); it can be influenced by the route as well as the pharmaceutical formulation. The effect at the target location is investigated in the pharmacodynamics (PD).

Administration form, route, dose and frequency

Figure 4.1 shows a very simplified illustration of these correlations. The active substance is delivered in a certain formulation (solid, liquid), via a certain route (oral, intravasal, etc.), at a certain dose and frequency (single treatment, multiple treatment). The PK study investigates the bioavailability (i.e. how much and how long the active substance is effective at the target location, before it is eliminated by excretion or chemical conversion). The effect of the substance on the molecular as well as the phenomenological level is investigated in the PD study. Thereby, the desired effects (efficacy) and undesired effects (safety) are taken into consideration. The PD models result in a dose–response correlation; PK delivers a time–concentration curve (PK profile). The PK profile shows how long and at what intensity the substance stays in the patient. The results of PK and PD are combined into a dose recommendation for therapeutic treatment, assuming a certain time-dependent effect course in the patient.

Bioavailability

In this section, first, the basic PK and PD terms will be explained, and we subsequently highlight how the PK profile can be influenced by route and form.

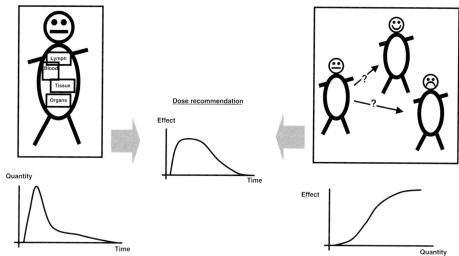

Figure 4.1 Correlation between drug quantity, drug effect, mechanism of action and residence time.

4.1.1
Pharmacokinetics

PK investigates the timely and local distribution of active substances in the body. The parts of the process that lead to the therapeutically effective drug level in the blood are contemplated in the liberation–absorption–distribution–metabolism–excretion (LADME) model:

LADME model

> *Liberation.* The liberation of the substance from the formulation stands at the beginning of drug transport in the body. Most proteins are injected directly; an example of a more controlled route may be the administration of polymeric microparticles in which proteins can be enclosed and released slowly from the depot.
>
> *Absorption.* When migrating to the desired compartment a drug usually has to pass biological membranes composed of biological cells. The intravenous route circumvents these membranes; the substance is directly injected into the blood. If the substance is needed in the blood – as in the case of hematological medicaments – further transport is not necessary. In the case of subcutaneous administration (i.e. the 'skin' membrane is

circumvented), migration into the blood or the lymphatic system is necessary. These processes are denoted as absorption. The molecular migration through the membranes can be slow and dominate the time to reach the effective drug level in the blood. Bioavailability is defined as the percentage of the total amount of delivered substance that can be found in the blood. Therefore, in the case of intravenous administration – the only method which on purpose circumvents absorption – bioavailability is always 100%.

Distribution. The distribution of the substance in the body after application is determined by the flow velocity of the blood, the different blood circulation (fat and bones have weak blood circulation), and the solubility of the substance in the aqueous blood and the hydrophobic tissue, respectively. Moreover, special biological barriers exist like the blood–brain barrier or the placenta that also influence distribution. The compartments are the plasma volume (blood), also denoted as the central compartment, the extracellular fluid, the total body water and special compartments like the brain. The substances can dissolve and be transported in all these fluids.

Metabolism. The substance is chemically degraded while it circulates in the body. This degradation can lead to inactivation of the drug; however, toxic functions can also be activated. In the excretion organs (liver and kidney), chemical conversion serves to improve the ability to excrete the substance fragments.

Elimination. Elimination denotes all processes that lead to decreasing concentration of the applied substance in the body. Apart from chemical conversion, other mechanisms are known like deposition or excretion to the outer milieu (renal excretion via the kidney).

Figure 4.2 illustrates the pathways of the drug in the body when administered systemically (i.e. into the blood stream). In contrast to that stands topical (local) administration in which the substance is brought directly to the target location (e.g. a tumor). Unfortunately the target locations are not always easily accessible, making topical administration an exception. Moreover, topical application would – due to the distribution mechanisms which let the drug arrive in the blood circuit – also trigger a systemic reaction.

Many proteins are administered intravenously so that they do not have to be absorbed in order to get into the blood. The distribution of the active substance into the different compartments is driven by blood flow and diffusion (i.e. by directed molecular movement). The substance strives to distribute itself in the surrounding environment until it reaches chemical equilibrium. As a consequence the substance ends up being in the whole body, which stands in contrast to the desire that

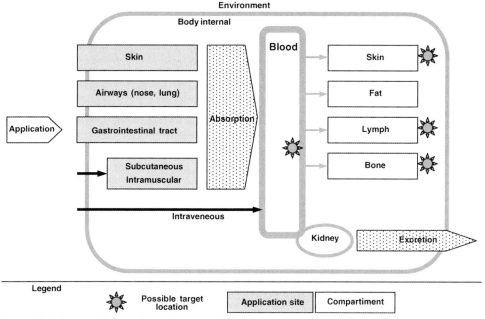

Figure 4.2 Pathways of drugs in the body at systemic application.

the drug act as specifically as possible at the target location. The latter can be achieved with highly specific antibodies, which in the first place also are distributed in the whole body, but ideally concentrate at the target location and unfold their activity there (targeting). The substances are partly degraded biochemically (metabolized) and partly destroyed by simple physicochemical mechanisms like oxidation or denaturation. Excretion is mainly through the kidney. All processes that eliminate the drug from the contemplated compartment are summarized in the term 'clearance'. It denotes the volume that is cleaned from the substance in a given time interval and is provided in units of milliliters of blood plasma per minute. The half-life is a measure of the speed of the clearance. It defines the duration in which the concentration drops by the half of a selected initial value.

Targeting

Clearance

Figure 4.3 shows the typical course of a time–concentration curve for single administration, as well as the essential PK parameters. After administration the concentration of the drug in the blood plasma increases, goes through a maximum (C_{max}, t_{max}) and thereafter decreases again due to elimination processes.

In this typical case the concentration curve lies well above the line of the minimal effective concentration – the concentration at which the

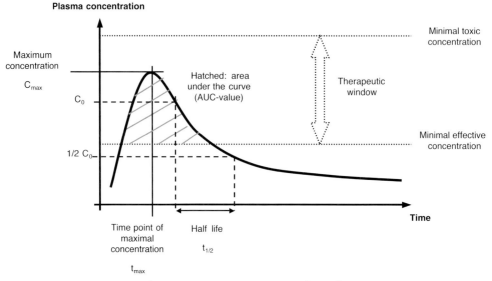

Figure 4.3 Time–concentration curve for single dosing. C_0 = concentration value at beginning of the measurement.

Area under the curve

desired drug effect is noted. The area between the concentration curve and this line is denoted as the area under the curve (AUC) and is an important measure of the therapeutic availability of the drug. One target of formulation development is to maximize the AUC. When doing so, one has to be careful that the plasma concentration always stays below the minimal toxic concentration – the concentration at which unacceptable side-affects occur. The distance between the minimal therapeutic and the minimal toxic concentration is called the 'therapeutic window'. It is the subject of PD to determine these limits.

Often a single treatment is not sufficient and the drug is administered repeatedly. The duration and the treatment scheme of the therapy determine the total dose per patient.

Figure 4.4 shows the plasma–concentration plot for repeated treatment. Elimination processes are compensated by multiple administration of the drug. This can lead to cumulative effects (i.e. a slow increase in concentration). The ideal situation is to keep up a constant mean concentration within the therapeutic window.

4.1.2
Pharmacodynamics

PD investigates the effects of the drug in the body. The result is a dose–response relationship and the definition of essential parameters

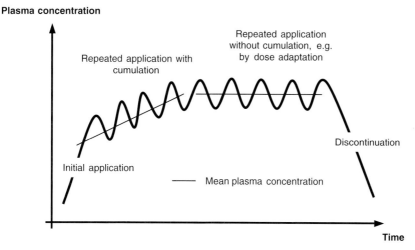

Figure 4.4 Time–concentration curve for multiple dosing.

of this relationship like minimal effective concentration or minimum toxic concentration.

4.1.2.1 Principles of Phenomenological Effects

Each drug has a diverse spectrum of activity in the body. This spectrum includes the desired effects as well as the undesired side-effects. The side-effects do not necessarily have to be harmful, yet they can lead to a product profile that is unacceptable for the market.

Drug safety requirements demand that the undesired side-effects have to be adequately balanced against the desired effects.

Qualitative side-effects do not depend on the quantity of the drug; they can occur at minute exposures, a typical example being drug allergy. Allergy develops from the repeated consumption of the drug, which generates an antigen–antibody reaction. If the body reacts at first contact with the drug this is denoted as an idiosyncratic side-effect. This can be caused by specific inherited enzyme defects or an immature enzymatic system (in the infant). *Qualitative and dose-dependent side-effects*

In contrast, dose-dependent side-effects are activated by a certain amount of the drug. The formation of tolerance, as well as addiction, belongs to this category.

If side-effects occur upon concomitant use of other drugs, they are characterized as drug interactions. They can result in both effect reduction as well as amplification. *Drug interactions*

The incidence of side-effects can lead to contraindications such that certain patient populations are excluded from getting the drug. This requires the physician to pay more attention when prescrib- *Contraindications* ing the drug, which can weaken the market profile of the drug significantly.

4.1.2.2 Parameters of Drug Effects

Drug effects are quantified by pre-clinical testing with animals and clinical testing with humans. Section 4.3.2 elaborates on the aspects of how such studies are conducted. Prior to that, this section explains the most important PD parameters that result from the studies.

Total spectrum of activity

When administering a drug, one will usually see that a higher dosing is accompanied by an increasing incidence of effects. Thereby the individual effects (desired effect and side-effects) can follow different dose dependencies. The quantification of such effects is the basis for evaluating the safety and efficacy of the drug, and consequently determines its dosing pattern.

Two questions are relevant for quantification:

- How do different doses impact the same individual? ('Analogous problem.') This method gives the strength of the effects in dependence of the dose.

- How does the same dose impact different individuals? ('Alternative problem'. This method defines a threshold value, above which an effect is considered to have occurred and below which it is considered to not have occurred. The method allows reduction of the obtained data to a rateable size and constitutes the basis for the definition of PD indicators.

Figure 4.5(a) illustrates how data from an analogous survey with multiple patients can be transformed into data for the alternative problem. The investigational drug is given to a larger patient group. Certain measurable effects are defined like the desired main effect and the undesired, but accepted, side-effects. Moreover, a threshold value is defined for which an effect is considered to have occurred (clinical endpoint).

Figure 4.5(b) shows an enlargement of the right diagram of Figure 4.5(a). The effective dose ED_{50} is defined as the dose at which 50% of the patients have experienced the desired effect. Figure 4.5(b)

Therapeutic window

also shows a tolerable side-effect that has to be labeled as such in the patient information leaflet. The red line is the cumulative frequency of a non-tolerable side-effect. It determines the upper limit of the therapeutic window [maximum tolerable dose (MTD)] that is closely related to the minimum toxic concentration measured in the blood. The width of the therapeutic window that spans between the minimum therapeutic dose and the MTD is of utmost importance for the safe use of the drug. The wider this window is, the lower the risk of the patient suffering from undesired side-effects when errors happen in administration or if

(a)

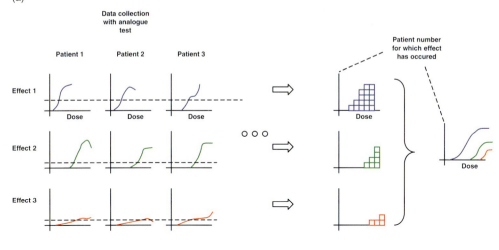

(b)

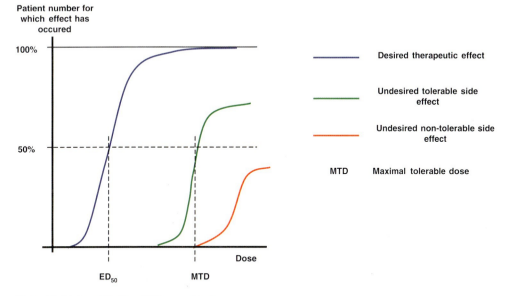

Figure 4.5 (a) Quantification of PD parameters by analog data collection and alternative evaluation. (b) Cumulative frequency of dose-dependent effects.

resorption characteristics vary. If the therapeutic window is very narrow, the drug can only be administered under control of the physician. A measure of the therapeutic window is the ratio between the effective dose (ED_{50}) and lethal dose (LD_{50}). One can also find the term lethal index (LD_5/ED_{95}). Naturally the lethal dose cannot be determined for humans – it has to be extrapolated from animal data. Therefore, the span is better defined by means of the MTD.

4.2
Routes and Forms of Administration

Route and form of administration have a significant impact on PK. Proteins are almost exclusively injected under the skin (subcutaneously), in the muscle (intramuscularly) or into the vein (intravenously).

Pulmonary administration

Alternative routes are taken with increasing success, like the lung (pulmonary, e.g. pulmonary insulin) or nose (nasal, e.g. Flumist®). Here, the relatively good absorption capacity of the nasal mucosa and the large surface area of the respiratory tract are used. However, these routes are associated with questions concerning immunogenicity (formation of antibodies in the tissue), degradation pathways and availability of the substance at the target location.

Drawbacks of parenteral administration

The gastrointestinal tract as well as oral absorption via the oral mucosa does not yet play a role in protein delivery. This is due to the poor stability of proteins and their susceptibility to proteases. Moreover, the migration velocity and general ability to permeate membranes is very limited because of the size of the proteins. As a consequence the drawbacks of parenteral administration have to be accepted:

- Quick elimination in the blood system, resulting in high dosing frequency.
- Reduced patient comfort by using this invasive method (e.g. pain at injection site, limited possibility for self-medication). This can result in low patient compliance (Section 4.4) and reduced quality of life.

Innovative approaches to formulation

Current strategies aim at clearing these disadvantages through innovative drug formulation which can, for example, reduce application frequency. In order to manipulate the release behavior, active substances are encapsulated in biodegradable microparticles. These capsules dissolve in the blood and release the protein. One can also think of a diffusion of the protein through the capsule, in which case the release is widely independent of the degradation velocity of the capsule. Other ways of immobilizing proteins are to transfer them into a stable crystal structure or to attach them to a carrier material from which they dissolve slowly. Some proteins are linked to PEG molecules, which results in higher stability and residence time in the blood system.

A prominent example for a pegylated product is Amgen's Neulasta®, which consists of a pegylated G-CSF (Filgrastim) at whose protein of 19 kDa a PEG-molecule of 20 kDa is conjugated. By this manipulation the molecule doubles its size. Modifications of the amino acid sequence can influence *in vivo* stability and therefore PK. The molecular chains of the insulin forms NPH and Lantus® differ in three amino acids out of a total of 53. While NPH insulin goes through a pronounced effect peak and loses its effect after around 15 h, the effect curve of Lantus® is significantly flatter and still at the same level after 24 h.

However, none of these modifications can eliminate the need for injection. Almost all of these products are injected as liquids. Usually these liquids consist of up to 5% protein, 5–10% excipients like sugars, amino acids, enzyme inhibitors or acids/bases and 90–95% water.

Many proteins are unstable in liquid solution and are therefore freeze-dried (lyophilized). They are supplied as powder or solid lyophilization cake and have to be dissolved in water (reconstituted) before injection.

Many proteins are lyophilized

4.3
Drug Study

A successful drug study is a prerequisite for market approval of a drug. The study is made in multiple steps. After a component has been identified as being potentially active in a first step, it is tested in animals or suitable model systems (cell cultures) with regard to its basic tolerability and toxic features. This so-called pre-clinical phase will be explained in Section 4.3.1. The pre-clinical phase is followed by clinical testing in humans, which is split up in three phases. The regulatory market approval can be obtained after the successful clinical study. Surveillance of the drug continues after market launch and is denoted as clinical phase VI. Section 4.3.2 is devoted to phases I–III of the clinical study; Section 4.5.2 on 'pharmacovigilance' elaborates on measures taken after market launch.

Effects of drugs in humans are characterized by strong variations. Not only relatively well-defined factors like age, gender and interactions with other drugs, alcohol or nicotine contribute to these fluctuations. Biological variability is also due to patient-specific factors like general constitution, type and timepoint of ingestion, and specific allergies.

The result of the drug study is a dose recommendation for the statistical mean of the investigated patient population. This dose recommendation is connected with information about the total spectrum of activity (desired effect and side-effects). It is ultimately left to the physician to judge individual cases and decide about how to use the drug.

Dose recommendation

The large number of influential factors on drug effects makes drug study a real challenge. In clinical trials it is intended to reach conditions that are controlled as much as possible in order to obtain data on the individual influence of these factors. The type of influence has to be defined anew for each medicament.

Clinical endpoint

A clinical study is successful if it proves that the investigational drug is superior with respect to safety and/or efficacy aspects or with respect to its risk–benefit profile than potential alternative therapies. The definition of this measure of success (clinical endpoint) is a crucial aspect of the study. First, a target profile for the investigational drug is developed. This profile is geared to existing therapies. Compared to those, hard criteria like mortality rate, recommittal rate, relapse rate, complications and disease-specific indices can be improved. However, less well-defined criteria like quality of life and patient comfort can also be part of the target profile. Thereby, the suggested improvement should be large enough to be noted by patients; however, the bar should not be raised too high not to put the success of the study at risk. In the clinical study, the achievement of the defined clinical endpoints is tested by suitable methods.

Study design

Consideration about significant influences and expected effects leads to a study design that has to be approved by regulatory authorities prior to execution.

The complete product development is mainly dominated by the clinical study. The study defines the timelines for process and analytical development, and the timepoint for commercial market launch. At the same time it goes along with a high risk of failure. Figure 4.6 shows a rough time course of drug development from research through development until market launch. The left axis shows the average spending in million US dollars, the arrows on the right side point to the probability of success with which a medicament gets market approval. It can be seen from Figure 4.6 that currently roughly US$450 million has to be spent during drug study, provided that the drug makes it the 7-year long way from entry into phase I until market approval.

The costs include the costs for abandoned projects. Obviously, only about 20% of the projects that enter clinical study are approved. Patents are filed during target identification in the research phase. The consequence of the long time between patent filing and commercial return is further detailed in Chapter 10. It should be noted that the cost figures in Figure 4.6 are subject to discussion. Some sources provide higher, some lower numbers.

A special aspect in the framework of drug studies is the question of biogenerics. Since product comparability is difficult to prove, it is currently unlikely that an imitator product gets through without a drug study to demonstrate bioequivalence; whether this study has to be as

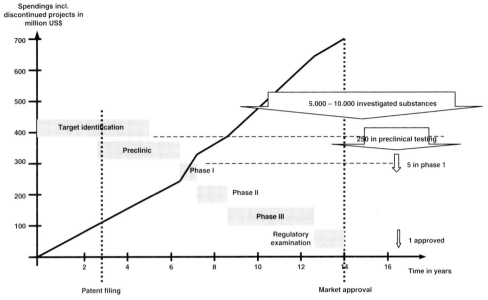

Figure 4.6 Times, costs and risks during drug development.

comprehensive as described in the subsequent sections for new entities has to be decided on a case-by-case basis.

Biogenerics
(see also Section 3.6)

4.3.1
Pre-Clinical Study

Successful completion of the pre-clinical study is the prerequisite for the start of the clinical study. The pre-clinical study is dedicated to proving the mechanism of action, the metabolism, and the recognition of dangerous side-effects and toxic reactions. Therefore, *in vitro* lab tests such as cell cultures as well as animal experiments are carried out.

Essential parameters investigated in pre-clinical studies include:

- Pharmacology, i.e. PK and PD *in vivo* (animal); effects on the cardiovascular system, respiratory tract and central nervous system.

- Toxicology.
  - Acute toxicity: effects occurring after single dose treatment; observation time 1–2 weeks; usually two species.
  - Subacute toxicity: effects occurring after repeated drug application over 2–4 weeks; investigation of other effects such as accumulation and tolerance formation.

- Chronic toxicity: observation time 3–24 months; testing of different doses thereof, at least one with guaranteed adverse effects. Two species: rodent and non-rodent.
- Reproductive toxicity: effects on reproduction, course of pregnancy. Male and female rats, also rabbits.
- Teratogenity: effects on descendants during gestation (fetal toxicity)
- Mutagenicity: effects on genetic information. Test with mice.
- Carcinogenicity: cancer-activating properties. Tests with mice and rats.
- Local tolerance: test for reactions in the area of the injection site (swelling, pain, etc.).

- Stability of the active substance *in vitro* and *in vivo*.

- Immunogenicity. The reaction of the immune system to the substance (e.g. allergic reactions or formation of neutralizing antibodies). Due to the specificity of the human immune system, these characteristics can only be inadequately investigated with animals. If reasons exist to believe that the protein is immunogenic (many proteins are indeed suspected of being immunogenic) then immunogenicity has to be investigated in humans.

Relevant animal models

In principle, some of these experiments can be made *in vitro* using cell cultures. It is, however, unavoidable to conduct animal testing to minimize the risk for the patients in the following clinical trial. Regulatory guidelines for biotechnological medicaments require that relevant animal models are selected. That means that the animals should represent the human organism as well as the disease as closely as possible. The animals in question are rodents (mouse, rat, rabbit, guinea pig) and other mammals (cat, dog, monkey, pig). They can be – if possible and necessary – lesioned or infected with the disease under investigation. These animal models allow us to make initial statements on the efficacy of the substance.

In addition to the ethical questions raised in connection with animal experiments, the financial and time effort for conducting animal trials is significant. Diligent selection of the animals is as important as the development of analytical methods to evaluate the experiments. Due to ethical, time and economic considerations, the experimental design should involve as few animals as possible.

4.3.2
Clinical Study

In the clinical study the medicament is delivered to humans for the first time. The risks of the application are foreseeable from results of the preclinical studies to a large extent; however, the trial is split in phases, each of which involves higher patient numbers than the preceding one in order to minimize the risk of severe unexpected side-effects even more. The following section describes these clinical phases. The subsequent section elaborates on the design of clinical studies, which should ensure that profound information regarding safety and efficacy necessary for market approval is obtained with reasonable effort.

4.3.2.1 Phases of Clinical Studies

The clinical trial preceding the regulatory marketing application consists of three phases (phase I–III). Studies conducted after market launch are denoted as phase IV studies. Table 4.1 shows the timeframe, usual participant numbers and targets of the individual phases.

The duration of the phases to a great extent depends on the disease and cannot be generalized. Especially for indications in which biological drugs play a role, the disease pattern can be complex and may allow evaluating drug effects only after a certain time after the application.

Table 4.1 Phases of clinical testing.

Duration	Participant	Target	Requirements for investigational drug
Phase I			
3–6 months	up to 10 healthy volunteers[a]	PD: comparison with animal data PK dosing framework dose dependent tolerance metabolism	EU: material manufactured under GMP
Phase II			
3–12 months	30–300 patients	proof of efficacy tolerance, spectrum of side-effects PD and PK dose–response relations important drug interactions	EU: material manufactured under GMP
Phase III			
1–3 years	100 to over 1000 patients	like phase II with larger population and higher real market relevance application under clinical and practice conditions long-term tolerability	material manufactured under GMP analytical methods validated validatable process final formulation

[a] Volunteers are healthy humans who receive small doses in first-in-man studies. Exceptions are made if it can be expected that the drug will show unacceptable side-effects; in this case patients are also treated in phase I.

Moreover, the high claims regarding analytical characterization and process definition to ensure consistent product quality can have an impact even on clinical timelines, so that the duration of the studies cannot simply be derived from small chemical entities.

Phase I

In phase I, the drug is tried for the first time in humans; it is applied to a few, mostly male healthy volunteers. Treatment starts with a fraction of the dose that has been identified as being well tolerated in animals and is increased until the first side-effects are noted. The studies are carried out in special labs or medicinal institutions under tight medical surveillance. In order to start this phase, analytical methods have to be in place for the analysis of the substance and its degradation products in the blood. Instead of healthy volunteers, patients can also be taken for phase I trials. This is done if the drug is expected to have severe side-effects, which is the case for most cancer drugs, for example.

Phase II

In phase II, the first therapeutic treatment in humans is made, hence this is conducted with ill patients. Dose-finding studies are also conducted here to compare desired effects and side-effects. The study ends with the important proof of concept (or proof of principle). The

Proof of concept

proof of concept is decisive for whether product development continues or not, since it gives first qualitative and quantitative data as to the future product profile. The end of this phase produces a dose recommendation. The phase II is often split into IIa and IIb. The differences between these sections are not always clear. They may relate to the homogeneity of the patient group with regard to disease pattern or stage, patient age or anamnesis (preliminary case history). The distinction can also mean that in phase IIa the focus lies on PK and PD, whereas phase IIb part focuses on efficacy. Due to the huge number of patients and the (compared to phase I trials) relatively low risk, phase II studies are conducted at several sites by several investigators (multicentric study).

Phase III

Phase III studies include many more patients than in phase II. Once a project has made it to phase III there is a good probability that it will reach the market. Conceptually in this phase the same questions have to be answered as in phase II; however, the drug is applied in a setting which is – although still tightly controlled – much closer to the intended commercial use, and used by many clinics or physicians. The therapeutic effect should be confirmed, the occurrence of rare side-effects recognized and data collected concerning the long-term use of the drug. At this stage, comparative studies with established medicaments as well as placebo-controlled studies (Section 4.3.2.2) are conducted. These studies finally allow making the final risk–benefit assessment. Phase III studies that are relevant for approval are called pivotal studies.

Pivotal studies

The drug supply of biotech studies cannot be directly compared to chemical drug investigations. Section 2.1 elaborated on the special meaning of the process for quality of biotechnological drugs. The transition from clinical supply to commercial supply requires significantly increased manufacturing capacities. In the case of chemical entities such a volume increase can be achieved by scaling up the reaction vessels. The comparability of the drug before and after the scale-up can mostly be demonstrated by analytical testing. For protein drugs this pathway is more difficult due to the limited analytical accessibility. A scale-up can result in variations to the product. For example, the living organism in fermentation can behave differently or the separation characteristics of the chromatography columns can be changed. It can happen that the changes are recognized (e.g. new byproduct), but their impact on humans is unknown or that the change is not directly measurable (e.g. new HCPs). As a consequence, after significant process changes, there is a risk that the previously gathered data for safety and efficacy can only be transferred to the new situation if the comparability of the drug has been demonstrated in humans. This is achieved by clinical comparability studies ('bridging study'). These studies are expensive and naturally comprise the risk of showing non-comparability. Two specialties for biotechnological products result: *Bridging study*

(i) The material for the pivotal studies of phase III should come from the finally defined process and the same site as the intended market supply.

(ii) Process optimization and scale-up projects can be associated with high effort. Consequently, one can find manufacturing processes that are not optimized and have been scaled-up by duplicating the established small-scale equipment.

Both aspects have a huge influence on manufacturing costs. Aspects of manufacturing should be considered early on in biotechnological product development to minimize the risk to defer costs of manufacturing or comparability studies. Moreover, the early fixing (before phase III) of a process and a manufacturing site limits the freedom of operation for supply questions.

4.3.2.2 Design and Conduct of Clinical Trials

Clinical trials are regulated by the rules of Good Clinical Practice (GCP) representing a directive for the balance between the legitimate interests for the research of new drugs versus the safety of patients. *Study conduct*

GCP rules formulate certain requirements regarding planning, execution and documentation of clinical studies. The examination follows a fixed trial protocol that has to be approved by regulatory authorities. The approval is subject to a positive vote of an independent ethics committee. The protocol details the course and the evaluation of the study. The study sponsor generates the plan and is responsible for compliance with the plan. Additionally, independent monitoring of the *Test plan* *Ethics committee* study is expected. The sponsor provides the principal investigator who coordinates the investigators in the field. The investigators should have sufficient experience with the disease and be able to react appropriately to unexpected drug reactions. The evaluation of the study has to be made by means of *a priori* defined measuring tools and against *a priori* fixed target values (clinical endpoints). Patient data are documented with so-called case report forms. The patients have to be informed about the potential risks of the drug application and consent in writing (informed consent) has to be obtained. They participate voluntarily, can recall their consent and drop out of the study at any time.

Study design

The design of clinical trials should ensure the statistical relevance and objectivity of the investigation result.

Statistical relevance

Statistical relevance is achieved by the number of patients for which a correlation between the application of the drug and its effect is measured. This effect can depend on many parameters (e.g. age, gender, other diseases, nicotine addiction, etc.). If necessary, it is possible to separate all these influences in the investigation. Depending on the number of parameters that need to be separated, a participant number results that makes the trial result statistically relevant (sample size calculation). Which patients qualify for the study is controlled by inclusion and exclusion criteria. A wide span between both criteria can result in a patient group that bears the risk of having so many disturbing factors that the statistical relevance between drug and effect cannot be shown. On the other hand, if the span is to narrow, the market potential of the drug is limited at the outset. The study design is characterized by certain features that we will look at in more detail:

Inclusion and exclusion criteria

Monocentric/Multicentric The study can be conducted at one site (monocentric) or at different centers (multicentric). The latter is mostly applied in phases II and III due to practical aspects such as the number of patients that have to be recruited. Phase I is mostly monocentric. Multicentric studies are challenging with respect to comparability of patient data and hence can require training of the investigating personnel in order to avoid the impact of non-drug-related effects (center effects).

Controlled/Non-Controlled Each study is well monitored. In the given context, the expression 'controlled' means that in a controlled study, the application of the investigational drug is evaluated against a control group. This control group can either receive an established competitive therapy or a non-active solution (placebo). As opposed to the placebo, the real drug is called the verum. Placebo-controlled studies discriminate non-specific psychogenic effects from the specific effect of the new drug. In principle, placebo treatment poses ethical questions since the patient is deprived of a potentially superior therapy. Therefore, investi-

Placebo

gation versus an established therapy standard is preferable. However, in such comparator studies the logistical challenge is that in a blinded trial no visible difference between the investigational and the reference drug may be present in order not to disclose the real identity of the applied drug to doctor or patient. The non-controlled study does without a reference group.

Blinded Study/Open Study In an open study, both the patient and physician know which product is administered. This design is chosen in phase I since here clear pharmacological parameters are measured that are not prone to psychogenic falsification. Moreover, in this phase, healthy volunteers are exposed which would not be able to register the main effect of the drug anyhow. In contrast, a blinded study design is taken in phases II and III. In one-sided blinding, the doctor knows which drug he/she delivers, but not the patient. In double-blinded studies, neither doctor nor patient know whether they are dealing with the verum or reference; hence, double-blinding is as far as one can get for objectifying the investigation. It is preferred to exclude psychogenic effects at the patient as well as the evaluating physician.

Parallel/Cross-Over Design In a parallel design, different patient cohorts are tested simultaneously with the different treatment methods (verum/placebo; verum/comparator; verum/verum with different dose). The evaluation allows for an interindividual comparison between the groups. In a cross-over design, patient cohorts are exposed subsequently and alternating to the different modes of therapy. This allows for an intraindividual comparison, which means that it can be seen how the individual reacts to the different modes. This method is only applicable for non-curative treatments where the patient is not cured by the treatment.

Randomized/Non-Randomized In a randomized study the participants are randomly allocated to the trial cohorts. This results in mixed cohorts as to age, gender, disease pattern and so on – all within the framework of the inclusion and exclusion criteria for the participants. Non-randomization is relevant if one is interested in the effects of the drug independent from side factors, which is often the case in phase IIa. Studies with such homogeneous patient populations are obviously less representative for the total target population.

Typical study designs for phase I are monocentric, non-controlled and open with 10 healthy volunteers. In phase IIa one can often find non-randomized, single-blinded, non-controlled studies with 10–100 patients. Phase IIb comes with designs between those of phase IIa and III with 100–200 patients. In phase III, typically randomized, double-blinded, controlled studies with over 1000 patients can be found.

Typical study designs

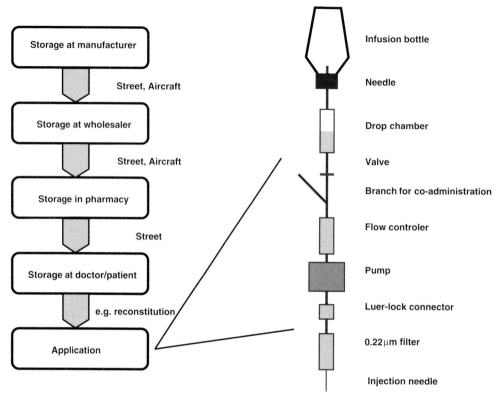

Figure 4.7 Path of the drug from the manufacturer to the patient (example for infusion in hospital).

4.4
Path of the Drug from the Manufacturer to Patients

The manufacturer holds the biggest knowhow about the properties of their product. They try to design the drug in a way that it still complies with all requirements regarding safety and efficacy when it leaves the area controlled by them. Figure 4.7 gives an impression which influences the medicament is exposed to in the clinical field. After storage at the manufacturer, the product starts its voyage to the wholesaler via street and air transport where it sits until being transferred to the pharmacy. Finally, it arrives at the patient or doctor where it is stored again. While transport and storage are pretty well controlled in an expert environment by doctors or pharmacies, it can be questioned whether storage by patients is adequately diligent.

The manufacturer is obliged to provide the administration and storage information of the product. Formally this is achieved by writing it on the packaging insert, which is mostly so crowded that concerns seem justified that patients may not be able to capture these instructions.

The right side of Figure 4.7 shows an example from the hospital. Different requirements for the drug can be derived from this typical infusion design. It has to be stable at room temperature during the infusion. It may not adsorb to the used materials. The physicochemical properties should be known as well as the biological effects when used concomitantly with other drugs. Last, but not least, it should not be too sensitive to mechanical stress. Typical application errors are wrong storage at too high or low temperatures. Also, hefty shaking during reconstitution of a lyophilisate can generate a solid precipitate in the vial and leave the product unusable. If the product cannot be designed in a way that these sensitivities are avoided, the only remedy is to train doctors or patients.

A further factor of uncertainty is to what extent patients adhere to the prescribing instructions of the drug – a varying behavior named patient compliance. Especially in the outpatient setting, a lack of patient compliance can be observed. This may be due to a lack of understanding or agreement with the 'decreed' therapy. Side-effects that are perceived as unacceptable can provoke denial of the treatment pattern. The better the mutual trust and communication between patient and doctor, the more patient compliance can be expected. Types of non-compliance include deliberate non-intake after experienced side-effects, unsteady intake and temporal omission. It is also observed that patients, motivated by the approaching doctors appointment, start to take the drug more regularly ('toothbrush effect') or they consume the previously skipped doses in too short frequency before the appointment ('parking lot effect'). The latter can lead to dangerous conditions. Simple communication, an easy treatment scheme as well as a comfortable application form reduce the risk of non-compliance.

Patient compliance

4.5
Drug Safety

Drug safety is a central theme in pharmaceutical industry. It is ensured by quality systems which reach from field surveillance (pharmacovigilance) through GMP-controlled drug manufacturing and documentation. Deficiencies in drug safety can result in sanitary consequences for patients; associated with this is an economic risk for the company and a personal liability risk for the responsible individual.

Guaranteeing drug safety starts with pre-clinical and clinical investigations. Naturally, these pre-marketing measures comprise a relatively small and homogeneous patient group. In the market place the drug experiences other challenges as to the age of patients, unforeseen comedication and effects of non-compliance. All in all, the use of the drug is much less controlled and involves many more individuals. Moreover, rare side-effects can only be detected if a large enough patient population is exposed to the drug; therefore, the activities of drug safety continue after market launch.

The following will introduce a classification of side-effects as to their incidence, relevance and causes. Then, the data sources and methods for pharmacovigilance are explained.

4.5.1
Causes and Classification of Side-Effects

Adverse event and adverse drug reaction

'Adverse events' can occur during the course of a drug application. In the first place it is unclear whether this event is linked to the drug. Only if an investigation reveals that there is a link is the adverse event classified as an 'adverse drug reaction' (ADR).

The reaction is considered to be 'serious' if it has one of the following consequences: death, life threat, remaining and/or significant disablement, congenital impairment, hospitalization required and medicinal significant action to avoid one of the above-mentioned consequences. Serious adverse reactions have to be reported to regulatory agencies.

The severity of the reaction is given in three steps: light (the patient feels disturbed), moderate (the patient feels hindered) and strong (the patient feels disabled).

The incidence of an ADR is described by the following categories: very frequent (above 10%), frequent (above 1 up to 10%), occasionally (above 0.1 up to 1%), rare (above 0.01 up to 0.1%) and very rare (0.01% or below).

Pharmacological classification of adverse events

Pharmacologically, different types of adverse events and ADRs can be distinguished:

- Type A reactions occur in the framework of the pharmacological mechanism of the drug. They are widely foreseeable.
- Type B reactions occur in the framework of hypersensitivity reactions (allergy, pseudo-allergy, etc.) to the drug. They are usually not foreseeable.
- Type C reactions describe long-term effects, addiction and discontinuation effects.
- Type D reactions denote toxic side-effects like carcinogenic, teratogenic, fetal and neonatal reactions.

This classification has a huge significance for the evaluation of adverse events. The assignment of a side reaction to a group of high incidence and/or severity can change the risk–benefit ratio to be unacceptably worse. As a consequence, the drug can either not be placed favorably on the market or even has to be withdrawn.

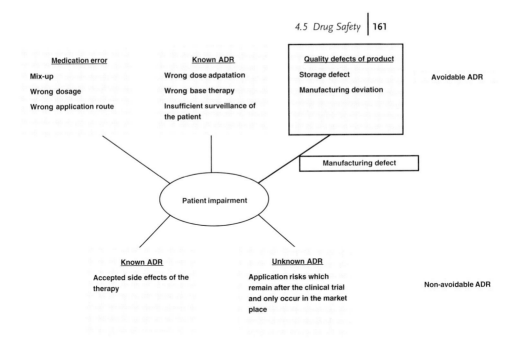

Figure 4.8 Avoidable and non-avoidable causes for detrimental side-effects. (Hartmann, 2003).

It is therefore of utmost relevance to find the root cause for the adverse event. Figure 4.8 schematically illustrates the potential causes for ADRs.

Non-avoidable ADRs include known and accepted side-effects that have already been taken into account in the risk–benefit assessment of the drug. The underlying data are of statistical nature and cannot be applied directly to the individual; hence, it can happen that the use of the drug shows higher side-effects in the individual case without questioning the overall drug profile. In these cases the physician has to decide whether or not the drug can be administered to this patient. Equally non-avoidable are rare, hitherto not observed side-effects, since they only show up in a larger patient population. These can be allergic effects or drug interactions that have not been investigated in the clinical trial.

Avoidable reactions include medication errors like drug mix-up, wrong dosage or wrong application. Also avoidable are known ADRs that can be caused by wrong dose adaptation for elderly patients or in cases of dysfunctional digestion; also a lack of scrutiny on the part of the treating doctor can contribute here. Most relevant from the production perspective are the avoidable ADRs due to quality defects of the product. Extensive measures of GMP (Section 6.1) and tight product control from the manufacturer through to the client have made these rare events; however, when a serious adverse event occurs, the question of product quality is always raised. In these cases it is of utmost importance for pharmacovigilance that the manufacturing process for the relevant dose can be traced back on the basis of the manufacturing documents. This documentation is explained in Section 6.1.6.

Once an adverse event occurs, it is important to identify the cause. The basic assumption is that an event that has occurred in close time vicinity to the drug administration is ascribed to the drug unless the reporting person explicitly abnegates this causality. In a subsequent investigation, other potential causes for the event have to be evaluated with differential diagnosis and a pharmacological plausibility check. If the investigation concludes that the event is associated with the drug, action by regulatory authorities can – depending on the incidence and severity of the event – reach from a notification in the patient and prescription information up to a marketing prohibition.

When thinking about adverse events it is of interest how the manufacturer or the authorities obtain the information about these field-related cases and how they react to these reports. This will be explained in the next section.

4.5.2
Methods of Supervising Drug Safety (Pharmacovigilance)

There are a various systems for surveillance of drug safety after market launch of a drug. The most important are listed in Table 4.2. The systems can be roughly classified into analytical and descriptive systems. The analytical trials serve to check a specific hypothesis (e.g. that a drug interacts with another in an unfavorable manner). The descriptive pathway serves to identify and formulate a hypothesis.

Analytical studies are often lengthy and – with the exception of clinical phase IV studies – prone to greater statistical uncertainties. Descriptive methods, on the other hand, allow us to react much faster to adverse events. They therefore constitute the cornerstones of pharmacovigilance. The most important instrument is the registration of suspected adverse reactions, which is a system in which the healthcare professionals have volunteered to report observations regarding adverse events to registration centers. The centers are integrated into the health authority system and build the link to regulatory authorities.

4.5.3
Measures at Incidence of Adverse Reactions

What happens after an adverse event has been reported by the registration system?

Table 4.2 Sources for post-marketing data collection of safety-relevant information.

Study type	Conduct	Type
Clinical study phase IV	experimental, randomized study; target often is indication expansion, therefore not suitable for safety data, but statistically highly relevant	analytical
Case-control study	a patient population that has shown a certain syndrome is compared to a population that does not show this symptom; afterwards it is determined retrospectively which risk factors are specific to the groups	analytical
Cohort study (longitudinal study)	a selected patient group is observed over a longer period of time over certain intervals; it can also be made as a controlled cohort study with control group	analytical
Cross-sectional study	evaluation of a patient group at a certain timepoint; it is checked whether a hypothesis regarding a side-effect can be confirmed for larger patient populations	analytical
Correlation studies	correlation of unspecifically collected data, e.g. cancer cases in the population versus the amount of cigarettes sold	descriptive
Registration of suspected adverse reactions	spontaneous, voluntary report by healthcare professionals to registration centers; most important part of pharmacovigilance	descriptive
Field observations	post-marketing surveillance studies; after market launch observations are noted without study-specific pre-settings and then evaluated; number of patients and duration of the observation can be chosen arbitrarily	descriptive

The physician observes an adverse side-effect that he/she deems so severe that he/she reports it. The report goes to the professional association or directly to regulatory authorities. At the authority the observation gets registered and preliminarily classified as to the risk. There are three risk classes that are equal for the US and Europe: *Risk classes*

- *Class 1*: the defect is potentially live threatening or could cause severe health damage.
  - Wrong product (content deviates from declaration).
  - Right product, but wrong substance strength with severe medicinal consequences.
  - Microbial contamination of sterile injectable or ophthalmologic product.
  - Contamination with other products in relevant quantities.

- *Class 2*: the defect can cause disease or wrong treatments and does not fall under Class 1.
  - Wrong or missing text or numerical value.
  - Wrong or missing product information.
  - Microbial contamination of non-injectable, non-ophthalmologic sterile products with medicinal consequences.
  - Chemical or physical contamination.
  - Deviation from specifications.
  - Insufficient closure with severe medicinal consequences.

- *Class 3*: the defect is not a significant risk for health.
  - Defect packaging (e.g. wrong batch declaration or wrong expiry date).
  - Closure defect.
  - Contamination, foreign matter.

Rapid alert process

If the case falls under Class 1 or 2, the marketing authorization holder has to provide a statement within 24 h. If they do not recall the product, the authorities can mandate a recall. In this case the so-called 'rapid alert system' is used, under which the license holder and/or manufacturer and/or authorities form a task force and evaluate the case in a short time. A complete and traceable manufacturing documentation is of utmost importance to be able to preclude product quality defects that could harm other patients and to protect the pharmaceutical company from significant damage.

The recall of a product batch is the most immediate measure for ensuring patient safety. After having determined the root cause of the deviation – if possible at all – other activities can start, such as:

- Change in the patient and prescription information.
- Implementation of warning notices.
- Change or withdrawal of the market authorization.

Trust in the healthcare system is very important and at the same time fragile; therefore, recalls or changes in the product characteristics have to be communicated carefully. As far as possible, communication is routed via an expert group of physicians or pharmacists. The responsible pharmaceutical company is obliged to inform the expert group about the changed product properties ('Dear Doctor Letter'). Only in special well-founded occasions is public media used, which usually leads to enduring damage to the reputation of the drug and its manufacturer.

Consequently, the quality of drugs is very important under the aspects of patient safety and personal liability, and for company survival.

Part Four
Quality Assurance

The assurance of product quality plays a major role in the pharmaceutical industry. Quality defects can be life-threatening to patients and cause significant economical damage for the company.

It is important to register warning signs early on and take steps for risk mitigation. This requires that instruments are in place for risk recognition, communication, initiation and implementation of preventive and corrective actions, and evaluation of such actions; in short, instruments to practice quality assurance.

Quality assurance is often understood as an abstract, hardly tangible construct which in part is composed of self-evident activities or even of activities that are perceived as being superfluous. The most established measure for quality assurance – the testing of the final product – is the most understood and accepted one.

Quality management defines measures beyond that and reaches much further into the organizational environment of product manufacturing. It includes all sectors of the company since only the common and shared understanding of quality requirements generates the necessary support by management and operational personnel for the mentioned provisions. The following chapters will explain the sense and concrete implementation of these provisions.

Chapter 5 is dedicated to the basic concepts of quality assurance. Chapter 6 explains the design of the quality system for product manufacturing and the rules of GMP that cover a large part of quality assurance.

5
Fundamentals of Quality Assurance

5.1
Basic Principles

The terms 'quality assurance' and 'GMP' are often used in the same context; however, they should be discriminated from each other.

Quality assurance signifies the entirety of measures to secure quality. It encompasses company-wide provisions for product development and corporate organization. These measures are set forth and aligned in the quality management system. GMP signifies the portion of quality assurance dedicated to product manufacturing and testing. The GMP guideline of the EU stipulates:

Good Manufacturing Practice (GMP)

Quality assurance

> 'Good Manufacturing Practice is that part of quality assurance which ensures that products are consistently produced and controlled to the quality standards appropriate to their intended use and as required by the Marketing Authorization or product specification. Good Manufacturing Practice is concerned with both production and quality control'.

Apart from those of manufacturing, more and more 'Good Practice' rules are being developed explaining the concrete realization of the general rules of quality assurance in the different areas of industrial practice. Table 5.1 shows examples for these so-called GxP rules and their background.

GxP

5.2
Benefit of Quality Assurance Activities

The benefit of quality assurance is understood intuitively when considering patient risk upon consumption of adulterated drugs; however, it is difficult to put a number on the added value that a quality system contributes to the value chain of the company. Non-pharmaceutical companies voluntarily adhere to quality assurance measures ('ISO 9000 certified') due to marketing considerations. On the other

---

Manufacturing of Pharmaceutical Proteins: From Technology to Economy. Stefan Behme
Copyright © 2009 WILEY-VCH Verlag GmbH & Co. KGaA, Weinheim
ISBN: 978-3-527-32444-6

Table 5.1 Examples of 'Good Practice' rules.

	Good ___ Practice	Content
GMP	Manufacturing	manufacturing and control
GEP	Engineering	design and document control for technical planning
GTP	Tissue	handling of human or animal derived tissue for therapeutic applications
GCP	Clinical	clinical development
GSP	Storage	storage
GLP	Laboratory	lab work
GAMP	Automation	automation of manufacturing processes

hand, pharmaceutical companies are bound to comprehensive regulatory guidance dictating the implementation of a quality management system and adequate measures for GMP-compliant manufacturing. In this case the voluntary and the legal motivation overlap. The following arguments support the implementation of a quality assurance system:

- Assurance of optimal patient safety (ethical motivation).

- Legal obligations. In the pharmaceutical field the implementation of a quality management system is a legal obligation.

- Minimization of the economic risk for the company. Risks arise from the liability for the product and the significant market disadvantages that a company experiences after loss of trust in its products.

- Cost savings. The costs can be kept lower with efficient quality assurance than without. The categories for quality costs are described in Table 5.2. A concrete cost–benefit analysis has to balance the failure prevention costs against the potentially avoided failure costs. It is difficult to find discriminations between the provisions that would have been done anyway and the ones that are triggered by the quality assurance system. Therefore, no established method exists for a tangible cost–benefit analysis.

- Liability reasons. If serious adverse effects occur that can be attributed to quality defects of the product, the individuals responsible for product release are liable in person. In such cases, complete and traceable documentation can relieve the signatory from suspicion. Moreover, it can be demonstrated that the manufacturer has complied with their due diligence for the production and distribution of the product, and that they have fulfilled their legal obligations.

Table 5.2 Categories of quality costs (Schneppe and Müller, 2003).

Failure prevention costs	Testing costs	Failure costs
Quality planning	testing of incoming goods	rejects
Validation	in process control	reprocessing
Quality capability analysis	final product testing	recall
Supplier evaluation	environmental monitoring	quantity deviations
Test planning	testing equipment and aids	depletion
Audits	maintenance of testing equipment	screening inspection
Management of the quality system	quality certifications	re-examination
Quality control	laboratory investigations	disposal
Quality assurance training	testing documentation	failure investigation
Quality enhancement programs	other measures and investments into quality testing	quality-related downtime
Quality benchmark		warranty
Other measures of failure prevention		product liability image loss

5.3
Quality Management According to ISO 9000

The state of the art as to quality systems is described in the guideline series ISO 9000 (9001–9004). This series of rules is not specific for the pharmaceutical industry, but describes the basic elements of quality assurance that can be applied in every company that wishes to manufacture high-quality products. The rules allow approaching quality assurance systematically; 20 fields of activity are described that have to be considered in a quality system.

5.3.1
Fields of Activity

Figure 5.1 shows the fields of activity of quality assurance and their position in the organizational workflow of the company. The numbers of the fields of activity correspond to those in ISO 9000.

The core process of manufacturing, here shown in gray, begins with the supply of starting materials. These are brought in by another company for further processing (7) and have been selected as a result of a purchasing process (6). It is followed by the actual manufacture (9) after which the products are tested (10), for which again suitable testing

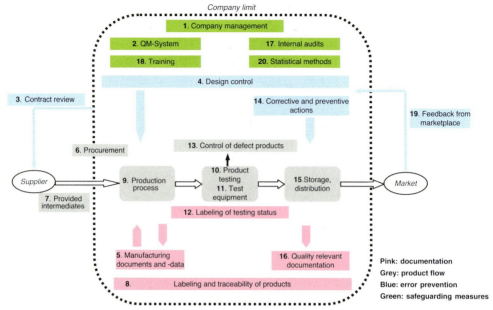

Figure 5.1 Fields of activity of quality assurance according to ISO 9001.

equipment (11) has to be in place. After testing, the goods are transferred to storage and dispatching (15).

At this stage the goods are delivered to the market and at latest here quality has to be real. The control of defective products (13) ensures that no goods with quality defects are distributed to the market; moreover, the analysis of the defects should lead to corrective actions. Indeed, quality to a great extent depends on the ability to evaluate defects and identify suitable corrective actions to address these defects (14). Further information as to defective product is obtained as feedback from the market via the sales organization (19).

Documentation

Information from activities 13 and 19 can only be evaluated if the manufacturing process is accurately documented. The time gap between manufacturing and client feedback allows identifying a failure on the basis of the descriptions of manufacture and testing only long after the error has been made. Therefore, a main part of quality assurance is documentation, which essentially has two tasks: data are collected to ensure traceability of products (8), and the production process is standardized, unified and documented by written protocols and reports (5). Special consideration is given to quality-relevant documentation (16), but it is not always easy to differentiate this from non-quality-relevant parts. As a rule of thumb one could regard at least the data laid down in the marketing authorization as being quality relevant; all other data necessary for optimized control of process and environment should be checked case by case for their quality relevance. Clear labeling of the testing status of equipment (12), raw materials or intermediates ensures that only suitable and prepared equipment is used.

The activities marked in blue serve for failure prevention. The 'ear at the client' (19) belongs to that as well as the diligent review of supply contracts (3). The latter should be associated with external audits in order to ensure that the supplier of materials can indeed fulfill their contractual obligations. *Failure prevention*

Steering and control of failure correction and prevention measures is an essential portion of quality assurance; this aspect will be explained in detail in Section 6.2 where typical operational workflows are shown. The design control (4) is a very challenging aspect of failure prevention. It tries to steer the development process in a way that potential quality defects are anticipated and eliminated by suitable provisions. Procedures known as 'risk analysis' fall into this category. They are applied for product development as well as for the process and testing program. Finally, the technical facility should also be constructed to be 'quality safe'. Since at the timepoint of planning no real-world experience with the product exists, design control to a large extent depends on the background and experience of the people doing it. The earlier an error is detected, the cheaper it is to circumvent it. Therefore, design control should receive the highest attention possible. *Design control*

Finally, the green area constitutes the safeguarding measures, without which quality management cannot function. Company management is responsible for the introduction and implementation of quality management (1), one part of it being a quality system (2) in which the quality-relevant flows are written down and which integrates the different provisions for quality assurance. Frequent training of personnel (18) is a pivotal column of quality assurance and has to be documented. Internal audits (17) check the compliance with quality-relevant rules. The 'statistical methods' (20) field of activity denotes a collection of mathematical instruments and evaluation procedures for the interpretation of the documented results. Only these methods guarantee that quality data are read as objectively as possible and at the same time are representative of the entity of samples or activities. *Safeguarding measures*

The quality concept reaches far beyond the final testing of the product. Only the combination of the different fields of activity yields the desired system-inherent quality that builds upon avoidance of errors in product design and manufacturing.

5.4
Structure of Quality Management Systems

The quality management system constitutes the organizational framework for documentation, application and surveillance of quality assur-

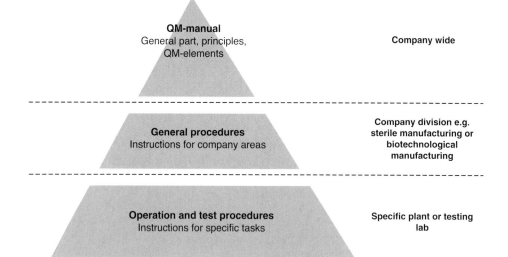

Figure 5.2 Documentation inside a quality management (QM) system. Modified from Brauer (1996).

Quality management manual

ance measures. This is accomplished by writing down the provisions of the system, providing a set of binding rules for quality-relevant activities.

The system therefore can be considered as the 'law book of quality assurance'. The 'laws' are set forth as working protocols that have to be followed. The protocols are subject to change-control procedures; moreover, individuals are named who are responsible for the compliance with the listed measures.

Figure 5.2 shows the basic structure of a quality management system. The more general procedures and rules that are described in the quality management manual are at the peak of the pyramid. The quality management manual can consist of a structured collection of individual procedures. The basic fields of activity, such as personnel training or change management (Section 6.2.2), are fixed here; their requirements are roughly drafted.

The underlying level contains procedures with significance for the whole company division; they can address the execution of analytical or process validation (Section 6.1.5). A further example for this level is the gowning rules for certain clean-room classes.

Manufacturing and testing procedures regulating the operational work in detail can be found in the lowest category. Here, the execution of a product-specific analytical method or how to perform a certain manufacturing process step can be described.

The procedures are denoted as standard operating procedures (SOPs). The SOPs of the upper level are sometimes called quality management SOPs or directives, and the ones of the middle and lower level simply SOPs. The discrimination into 'general SOPs' and 'specific SOPs' may be more meaningful.

SOPs

All procedures of the quality management system are subject to change management. That means that a change is only possible if a certain change-control procedure is performed.

Change management

In general, the procedures contain the following information:

(i) Distribution list.
(ii) Scope.
(iii) Applicability.
(iv) Definition of used terminology.
(v) Roles and responsibilities.
(vi) Description of the subject matter.
(vii) Reference and accompanying procedures.
(viii) Attachments.

It becomes apparent that the actual scope is only described in parts (ii) and (vi). The other parts regulate the interfaces to other systems and name individuals to be informed or being responsible.

While procedures define how to perform activities, execution documentation reports the actual course and outcome of performed activities. Thus, plans for the execution are considered to be part of the execution documentation. They are generated according to the procedures, but address singular, process- and product-specific activities (e.g. stability investigations, process validation), and hence are not put into a SOP format. These plans are denoted as the 'protocol'. After execution of the protocol, the outcome is recorded in 'reports' (e.g. validation report, stability report).

5.5
Quality Management System Components in the Pharmaceutical Area

5.5.1
Documentation

In the GMP-controlled environment it holds true that: 'What hasn't been documented hasn't been done!'. This sentence is sometimes perceived as being provocative; however, at the first regulatory inspection its deep truth becomes apparent. The only way for external authorities to get an overview of manufacturing quality is by looking

Table 5.3 Examples of documentation in the pharmaceutical area.

Section	Rule	Execution plan	Execution report
Training	'Training has to be undertaken frequently'	topics, dates and participants of the training	training documents and demonstration of success
Manufacturing[a]	'All parameters have to be noted'	master batch record	executed batch record
Control of defective products	'No defective product may enter the market'	deviation investigation[b]	deviation report
Change of SOPs	'Changes have to be reviewed and approved by responsible experts'	signature page, distribution and review list	signature page signed
Audits	'The compliance with GMP rules has to be checked regularly'	audit plan	audit report with observations

[a]The important field of manufacturing documentation is covered in detail in Section 6.1.6.
[b]See Section 6.2.1.

at the facility, talking to personnel and reviewing documents. Thus, the documentation plays an essential role since the quality status of the production processes is manifested here. Consistency of documentation from the quality management SOP over process procedures down to the detailed protocol is key. It is, for example, inappropriate not to report information in the execution documents that has been requested in the quality management SOP. See Table 5.3.

Documentation encompasses all fields of activity of quality assurance. In general, documentation consists of procedures, protocols and reports.

Apart from describing what to do, the systematic approach of compiling a protocol is a good exercise to obtain to a thorough understanding of the topic. The plans describe the execution of the activity, expected outcomes and as well proceedings when deviations from the expected results occur.

Finally, the reports contain the process of execution, the obtained data, a statistical evaluation and consequences from the activity.

5.5.2
Failure Prevention and Correction

Even the best quality system cannot preclude failures; however, a good quality system should be able to handle three important aspects when dealing with failures:

(i) Defective products should be recognized and rejected before they are delivered to the market.

(ii) If the defect is detected only after delivery to the market, the risk for the consumer should be evaluated quickly.
(iii) The analysis of the root cause of the failure should result in a learning experience on how to prevent such failures in the future.

The first point refers to the control of defective products. They have to be recognized, sorted out, marked and analyzed. Depending on the outcome of the failure analysis, the product can be further used or has to be rejected.

Control of defective products

Figure 5.3 shows a flow diagram for the control of defective products. The testing of the product gives a result which either fulfils the specifications or not. If the test result meets specifications, the product is released for distribution or – in the case of intermediates – for further processing. If the test results do not fall within the specified ranges, either the testing method has delivered misleading data or the product is indeed out of specification. This decision is not always straightforward, especially

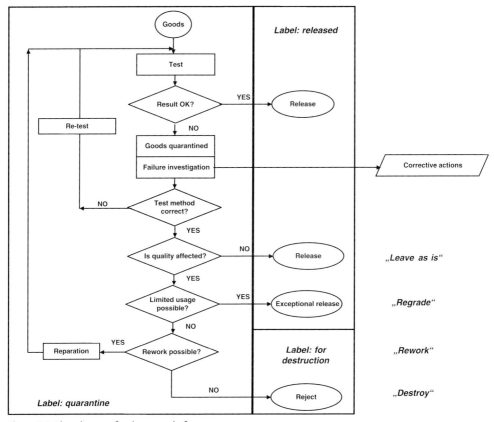

Figure 5.3 Flow diagram for the control of defective goods prior to delivery. The goods can be intermediates or final product.

when heterogeneous mixtures with limited analytical accessibility are tested. The basic assumption is that a diligent implementation of analytical methods (method validation, use of reference materials, calibration) should widely exclude analytical artifacts (wrong measurements) in product testing. For this reason, product release based on retesting can only be done on very rare and well-justified occasions.

Once testing has been completed correctly, four options remain to handle the defective product. The first is to check whether the failure to meet product specifications limits product quality. A typical example of this in pharmaceutical manufacturing can be given from the area of packaging. In cases where the text on the packaging does not conform with the licensed text, the authorities can agree to an exceptional release in order to satisfy urgent medical needs of patients. On the other hand, an out-of specification result in the content of the vial will usually lead to rejection of the product.

The term of testing can also be extended in the pharmaceutical sense: the release of the product is not only subject to successful testing and certification of the final product quality (Certificate of Analysis), but it is also necessary that the manufacturing process has been performed in conformance with the registered procedures certified by a Certificate of Compliance. In routine production, deviations from the described manufacturing process can happen (e.g. irregularities of fermenter aeration, blocking of filters or membranes, irregular flow through a chromatography column) that do not result in measurable quality deficiencies. In order to release the product despite such deviation, the failure investigation has to demonstrate that the product itself has not experienced any adverse influence. In these cases, 'leave as is' is applied.

Limited use

The second option is limited use despite detected quality defects (regrade). This option does not have practical relevance for biotechnological pharmaceuticals since the prediction of potential severe adverse effects is simply too vague. Again, the example of non-conforming packaging and the supply shortage which has to be avoided can be considered. It could, for example, be possible that market distribution is only approved in countries where urgent medical needs cannot be satisfied by other ways, which would resemble a limited usability.

Reparation

The third option (i.e. reparation or reprocessing) is possible in pharmaceutical manufacturing, provided that – if safety and efficacy could be affected – the measures and the process of the rework have been fixed, investigated, validated, submitted with the registration dossier and approved by the regulatory authorities. An unforeseen rework in the stages of API manufacturing or formulation is impossible. A foreseen rework can, for example, occur for the blocking of filters or chromatography gels. In the packaging area, non-anticipated rework can occur when, for example, the glue for the labels does not stick and the labeling process has to be repeated.

If all this does not help, the product has to be destroyed. This is usually done upon the occurrence of slightest deviations from specifications of the final product. However, significant deviations from the specified storage conditions can also lead to rejection and destruction of the product. *Destruction*

Control of defective goods has to make sure that released, quarantined and rejected goods are not confused. For that purpose, unambiguous labeling systems and physically separated storage areas need to be in place.

Unfortunately, not all product defects are detected before market distribution. Once a defective product enters the market, the measures of drug safety have to work. In the first place it has to be clear where the point of information for these technical complaints is. This customer interface has to communicate effectively with the client, the manufacturer and the relevant authorities. Subsequently, the report on the technical defect has to be evaluated as to the potential risk for patients and by investigating the root cause, it has to be clarified whether the defect is a singular event or whether the whole batch or even other batches are also affected. If necessary, the affected goods have to be recalled. It is of utmost importance that both the manufacturing process and distribution path are well documented in order to take a qualified decision about the significance of the defect.

One of the essentials of quality management is to learn from errors. Quality assurance does not mean to provide a short-term solution for issues, but to find a failure root cause and eliminate it. Identification of the root cause sometimes requires comprehensive scientific investigations for which production needs technical development resources.

One measure for failure prevention is to perform risk analysis in which a manufacturing process is scrutinized step by step for potential quality risks. Systematic methods for this approach are the 'Failure Mode and Effect Analysis' or the Pareto analysis and the Ishikawa diagram. The use of the risk analysis is also required more and more by regulatory authorities at all steps of drug development. The ICH Q10 (Quality Systems) and ICH Q8 (Pharmaceutical Development) guidelines refer to the methods for systematic risk treatment described in ICH Q9 (Quality Risk Management). *Risk assessment*

The risk management process is performed in three steps:

- *Risk identification.* The risk is described and assigned a probability of occurrence, a severity upon occurrence and an index that expresses how easy or difficult it is to detect the risk.
- *Risk control.* It is decided whether and how the risk shall be reduced.
- *Risk communication.* The risk is communicated to management and relevant personnel, and the result of the risk management process reviewed.

Risk management processes have a high value for the so-called risk-based approach from which the regulatory authorities expect that *Risk-based approach* industry will get a better understanding and thereby control of the risk potentials of drug development and manufacturing.

Change control

Change control for intended process changes is another measure for prevention of failures that could potentially result from insufficient evaluation of consequences of a change (Section 6.2.2).

Annual reports
US: Annual Product Review
EU: Product Quality Review

Trend analysis (statistical process control) and annual overview reports help to discover mistakes early on. For data trend evaluation, selected process or product parameters are recorded and assessed. Examples for slowly drifting processes are the loading capacity of a chromatography gel, which can be recognized by the shift of a peak from batch to batch, or the slowly proceeding fouling of a filter membrane, which becomes apparent by increasing processing time for filtration.

Figure 5.4 shows examples of the most important failure sources and preventive measures.

5.5.3
Responsibility of Management and Training of Personnel

The implementation of quality assurance measures finally is up to the company's personnel. In this context, management carries a special

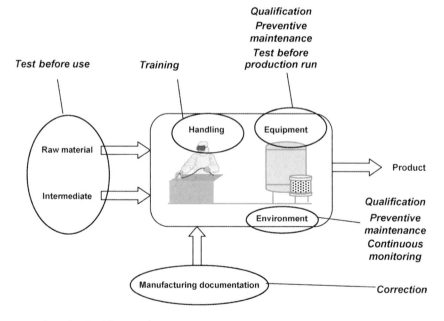

Figure 5.4 Sources of errors and preventive measures in manufacturing.

responsibility, since it has to support implementation and actively communicate the quality policy. Moreover, it allocates appropriate resources and the organizational framework, such as:

- Existence of documents describing the quality system.
- A clear delegation of personal responsibility inside the company documented by organizational charts or responsibility matrices.
- The implementation of an independent organizational unit which controls the compliance with quality assurance policies.

Management is obliged to provide the necessary financial and human resources to ensure that quality-relevant tasks can be accomplished. It nominates a quality responsible person within the upper management and it has to support the regular review and improvement of the quality system. *Management*

The personnel of the company that have influence on quality-relevant activities have to be trained frequently. The training program has to be fixed as to content and participants, and sufficient trainers be provided. The training has to be documented and it needs to be demonstrated that it was successful, which in practice means that employees are tested for their GMP knowledge afterwards. *Training*

5.5.4
Audits

A quality audit is a review of the quality status of an organization. The audit can have different key aspects, such as the quality system, manufacturing documentation, production facilities or personnel training.

Internal audits are performed by the company's own personnel (consultants also are taken); in external audits, the client reviews the quality status of the supplier. *Internal audits*

Both internal and external audits are obligatory since the company has to continuously update its quality assurance system and the client has to make sure that the supplier can provide the required quality. Audits can be made 'for cause' if quality defects have given rise to concerns regarding quality readiness. Moreover, a client is obliged to perform frequent routine audits. Supplier audits typically are performed every 2 years. *External audits*

The result of an audit is a list of observations classified as to their criticality. The classification often is geared to the one used by regulatory inspections. 'Critical observations' are directly relevant to quality and have to be corrected immediately. They can be reason enough to reject a product; insufficient cleaning of a vessel could fall into this category. 'Major observations' are at least as relevant in that they have to be corrected in order to guarantee a safe product in future manufacturing.

Examples for this could be insufficient documentation of manufacturing steps or non-conforming sampling timepoints for environmental monitoring. 'Minor observations' have to be corrected with lower priority. Smaller mismatches in the documentation could be given as an example. The audit result is documented and discussed between the parties. The auditing side may provide recommendations for improvements. The audited side reacts with a list of corrective actions, which again are discussed with the auditing party. After agreement has been achieved over the necessary corrective actions, these actions can then be implemented.

Audit result

5.5.5
External Suppliers

Often intermediate products and raw materials are procured from external suppliers. The receiving party has to ensure that these goods comply with its own quality requirements; however, while a contract including quality provisions is a good and necessary step in this direction, it is not considered to be sufficient. The client has to make sure that the supplier provides acceptable quality by the following measures:

- The company defines specifications for the product to be supplied.

- Interesting candidates for the supply task are qualified (i.e. by means of a quality audit, visit of manufacturing facilities, review of organizational processes and discussions with key personnel). It can also be interesting to evaluate an existing client–supplier relationship. Usually it is not deemed acceptable on the client's part to rely on the supplier's certification according to ISO 9000 only, since the client has to make sure that the quality provisions are implemented for its specific product.

- Upon contract closure, the final product and testing specifications have to be provided, and the procedures fixed for the case of deviations from expected values. A quality assurance agreement has to be closed in which *inter alia* the procedures for changes of the specifications or the process are laid down.

Quality assurance agreement

- After beginning routine supply, quality has to be continuously monitored. This is done by testing of incoming material on the client's part and yearly statistical evaluations.

- If the quality is insufficient, adequate corrective actions have to be initiated.

Supplier interfaces are regulated by legal agreements. The next section elaborates on the basic principles of contract review.

5.5.6
Contract Review

Contracts define the scope, service and compensation between contract partners; they provide an understanding and serve to avoid conflicts at the interface between the companies. Contracts between pharmaceutical companies and suppliers often are very comprehensive and affect different functions in the companies (Sections 9.4.3 and 12.3.2). Contract review should ensure that both companies can fulfill their contractual obligations. In detail this means that:

- All relevant functions have to be included in the review and the making of the contract.
- The requirements are communicated to all departments affected at both the client and the supplier.

A typical example of the obligations of the pharmaceutical company to the supplier of a biological ingredient is the regulatory coverage of process changes. The supplier wanting to change the manufacturing process has to solicit approval by authorities; however, the authorities will only accept communicating with the responsible marketing authorization holder (i.e. the pharmaceutical company). It falls to the latter to contact authorities and file change requests. Ultimately, the customer has to build up their own in-depth know how enabling them to lead an adequate communication with regulatory authorities. This obligation is especially expressed for biotechnological products since regulatory dossiers are particularly detailed.

During negotiations, both parties should share the same understanding of the agreement. Therefore, practical examples for certain situations can be played through and attached as contract appendices (e.g. price calculation, forecast scheme). The contract review should be documented. This can also be relevant in later legal disputes.

The basic principles of contract review should also be applied to the submission dossiers of a drug. The dossier constitutes a written obligation to the authority to manufacture the product in the described manner. Affected functions in the company (e.g. production, quality control and quality assurance) should check whether they can fulfill these obligations. *Marketing authorization dossier*

5.6
Quality Assurance in Development

During development of a product or the design of a technical facility, design control should ensure that the result does not bypass the client's requirements. In innovative projects, in particular, planning and development teams tend to focus on their own solution strategies. While the innovative power of the team is indeed desired, it is important to scrutinize whether these solution strategies lead to a satisfaction of client (e.g. marketing or production) desires and are aligned with regulatory requirements. In this process, quality assurance is a team approach in which ideally the person responsible for quality assurance leads the organizational effort for quality reviews on certain steps or the control of documents.

Design control can be applied to process and analytical development, clinical development, and technical planning of a production plant. All these activities need certain milestones at which a quality review can be performed.

Activities of design control

The main activities of design control are:

- Clear definition of design requirements. Define the procedures for treatment of design changes.

- Compilation of a design process plan that defines the design milestones.

- Assignment of personal responsibility for work portions.

- Fixing of organizational interfaces of functions or companies involved in the design process.

- Documentation of design results. Procedures for the documentation and approval of design changes.

- Procedure for the release of results for further use.

- Fixing of methods for result testing. This can be a use test, a special testing method or any other acceptance test.

- Design verification: ensure accordance between design result and design requirements. This can be made in regular design reviews in which the design team reflects the original requirements. These can be safety requirements (safety review), quality requirements (GMP review) or client specifications. In the planning process for technical plants, qualification activities are included in these reviews (design, installation, operation, etc., see Section 6.1.4).

- Design validation. Proof that the product is in accordance with the requirements. In the construction of technical plants, qualification of performance as well as validation activities are included here.

6
Quality Assurance in Manufacturing

The differentiation between general quality systems and GMP rules was explained at the beginning of Chapter 5. Indeed, GMP rules have been developed in parallel to the more generic standards of ISO 9000 and are not a focused application of these rules. Initially, GMP rules only covered the manufacturing area; lately, however, more and more areas are being added which – according to the original abbreviation – are denoted as GLP, GSP and so on (Table 5.1). They describe the implementation of the generic rules of ISO 9000 in operational practice.

For reasons of clarity, it is noted that this chapter focuses on GMP.

6.1
GMP

GMP rules should ensure the quality of the medicinal product. Compliance with GMP is a legal requirement in all countries of the world, and is controlled and enforced by health authorities. *GMP compliance*

The rules are continuously modified; in the vast majority of cases an intensification of the regulations can be observed. What had not been an issue in regulatory inspections 5 years ago could today cause a notice of defect. The reasons for this development are, on the one hand, adaptation to technical progress and, on the other hand, increased demand as to interpretation of regulatory guidelines by authorities. The actual interpretation of GMP is therefore called current GMP (cGMP). Due to *Current GMP* this dynamic it seems fair to say that cGMP compliance is not a state, but a goal, since the requirements may have changed right after having reached the alleged status. Consequently, regulatory inspections in most cases do not end with an immediate closure of the manufacturing site even when major observations have been registered, but the manufacturer is given the opportunity to correct the deficiencies.

GMP rules are published in regulatory guidelines that are compiled in collaboration with industry. The guidelines themselves are in most cases not directly legally binding; however, they define the state of the art of science and technology, and serve as a standard for the regulatory grant and maintenance of manufacturing permissions (Section 7.2).

EU GMP guideline

The 'EU Guidelines for Good Manufacturing Practice', issued by the EU, gives a good overview over the regulatory scope of such guidance. This guideline is Part 4 of the 'Rule Governing Medicinal Products in the European Union' and details the requirements of the Directive 2003/94/EEC laying down the principles of GMP. It is made up of Part I, in which the rules for pharmaceutical manufacturing are laid down, Part II, which contains provisions for active pharmaceutical ingredients, and the Annexes, in which the rules are defined more precisely for the relevant application. Table 6.1 provides the index of contents of the very comprehensive and easily understandable guideline. The scope of the corresponding US guidelines is comparable.

GMP rules aim at reducing the risk for patients as far as possible. Therefore, they stipulate a comprehensive documentation and control of changes to process, analytics, environment and documentation, which is time consuming and expensive. One could try to eliminate even the smallest risk for the patient by intensifying these efforts to a discretionary extent. This would come at the price that the development times for new products and the costs for existing drugs could become so high that access to these drugs could be jeopardized. As a result of this risk–benefit contemplation, certain alleviations from the full implementation of GMP rules addressing the special requirements of development processes as well as early manufacturing steps have been established. It holds generally true that the closer the manufacture is to the patient, the more restrictive the regulations. This is valid for the temporally distance of development products as well as for the technical distance in commercial manufacturing.

Stringency of GMP rules

Figure 6.1 illustrates this relationship. During product development the quality of the product can be guaranteed by intensive testing; moreover, only a small, well-monitored patient population is exposed to the drug. At this stage, certain alleviations from the complete set of GMP regulations are allowed. The product for pre-clinical testing can by all means be made in a research laboratory without GMP status. The first-time-in-man application, however, remains a critical point in development at which for the first time manufacturing under GMP is required. The US rules in this area allow higher flexibility than the European 'Clinical Trials Directive' (Directive 2001/20/EC), hence less resistance on the development path. One difference between 'full GMP' and the transitional area is, for example, that investigational drugs have indeed to be manufactured under GMP; however, contrary to commercial material, process validation (Section 6.1.4) does not have to be completed for clinical use.

Taking a closer look at the manufacturing flow, the borders become clear. The MCB is made under GLP rules. All process steps that lead to

Table 6.1 Structure of the EU GMP Guideline (EudraLex, Vol. 4).

Chapter	Title
Part I: Basic Requirements for Medicinal Products	
1	Quality Management
2	Personnel
3	Premises and Equipment
4	Documentation
5	Production
6	Quality Control
7	Contract Manufacture and Analysis
8	Complaints and Product Recall
9	Self Inspection
Part II: Basic Requirements for Active Ingredients Used as Starting Materials	
1	Introduction
2	Quality Management
3	Personnel
4	Buildings and facilities
5	Process Equipment
6	Documentation and Records
7	Materials Management
8	Production and In-Process Controls
9	Packaging and Identification Labeling of APIs and Intermediates
10	Storage and Distribution
11	Laboratory Controls
12	Validation
13	Change Control
14	Rejection and Reuse of Materials
15	Complaints and Recalls
16	Contract Manufacturers (Including Laboratories)
17	Agents, Brokers, Traders, Distributors, and Relabelers
18	Specific Guidance for APIs Manufactured by Cell Culture/Fermentation
19	APIs for Use in Clinical Trials
Annexes	
1	Manufacture of Sterile Medicinal Products
2	Manufacture of Biological Medicinal Products for Human Use
3	Manufacture of Radiopharmaceuticals
4	Manufacture of Veterinary Products other than Immunological Veterinary Products
5	Manufacture of Immunological Veterinary Products
6	Manufacture of Medicinal Gases
7	Manufacture of Herbal Medicinal Products
8	Sampling of Starting and Packaging Materials
9	Manufacture of Liquids, Creams, and Ointments

(Continued)

Table 6.1 (Continued)

Chapter	Title
10	Manufacture of Pressurized Metered Dose Aerosol Preparations for Inhalation
11	Computerized Systems
12	Use of Ionizing Radiation in the Manufacturing of Medicinal Products
13	Manufacture of Investigational Medicinal Products
14	Manufacture of Medicinal Products Derived from Human Blood or Plasma
15	Qualification and Validation
16	Certification by a Qualified Person and Batch Release
17	Parametric Release
18	Good Manufacturing Practice for Active Pharmaceutical Ingredients (since October 2005 as Part II, see above)
19	Reference and Retention Samples
20	Quality Risk Management

the API or the WCB are subject to GMP rules for APIs (e.g. EudraLex, Vol. 4, Part II).

For biotechnological drugs, due to the special significance of the manufacturing process, the stringency of GMP requirements for the API and final product is basically equal. In pharmaceutical manufacturing of parenteral products, the requirements for sterile production have to be considered additionally.

Many materials are used in the manufacturing process so that the question arises: 'If and if yes, to what extent these have to be

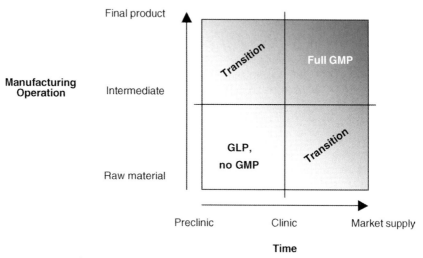

Figure 6.1 GMP requirements increase with increasing market supply proximity.

manufactured under GMP?'. These materials can be classified in four types:

- Starting materials are precursors to the product. The MCB is an example. It must not necessarily be produced under GMP; however, other rules like GLP and rules for analytical testing apply. *Starting materials*

- Intermediates are also precursors for the products that are generated in the different process stages and are forwarded for further processing. In biotechnology the intermediate always contains the active molecule since it originates from the beginning of the process in fermentation. An intermediate therefore always has to be manufactured under the relevant GMP rules for active ingredients. *Intermediate products*

- Pharmaceutical excipients are used for formulation and remain in the final product. They have to be manufactured under GMP; the water for the final formulation is an example for this class. *Excipients*

- Raw materials are deployed on every step of the process. They should not appear in the final product; however, remainders or leachables of them can sometimes be found in the drug in trace amounts. Water belongs to that group, as do chromatography gels, fermentation media, salts and chemicals. Raw materials used in API production do not have to be manufactured under GMP, yet it has to be proven that they can support the process in the framework of their specifications. *Raw materials*

There is a trend to expand GMP rules to further upstream process steps. This tendency seems to be justified in the light of the integrated view of quality systems. However, too strong regulation hinders the improvement process on the way to quality optimization because what is GMP and what not to a great extent is up to the opinion of regulatory authorities. They decide about a license for manufacturing and have to be able to understand the manufacturer's measures for risk control. The competent individuals in the authorities carry a high personal responsibility, which understandably expresses itself in that the safe and established way is preferred over the innovative way. The companies, for their part, do not want to take the risk of delaying approval of their license; therefore, in order not to provide the authority with a reason for finding faults, a small safety margin is willingly added to what is considered to be the standard solution. The results so presented (e.g. for clean-room finishing or documentation quickly) establish themselves as the new doable state of technology and find their way into the repertoire of regulatory requirements. Thus, the GMP spiral is driven up more and more. The question remains how high the effort can be increased while still increasing benefit. Obviously, entry into clinical research or even commercial production becomes increasingly challenging for smaller companies lacking the financial power or GMP knowhow. *GMP spiral*

The following sections highlight some focuses and typical implementations of GMP requirements. For manufacturing facilities, only the GMP framework rules are mentioned here; further realization of technical solutions will be described in Chapter 8.

6.1.1
Personnel

GMP stipulates comprehensive obligations for personnel at the management and operational level.

Head of Production and Head of Quality Control

Two individuals hold key positions: the Head of Production and the Head of Quality Control. Both are responsible for compliance with GMP rules, and guarantee in person that each batch has been manufactured in accordance with regulatory requirements and the approved drug dossier. The Head of Production is responsible for the GMP-compliant manufacture of the product and the generation of the batch record. The Head of Quality Control is responsible for the review of the batch record as well as all testing of the final product, the raw materials and the intermediates. In the EU, responsibility for the release of the product lies with the so-called 'Qualified Person'.

Personnel training

Personnel involved in manufacture and testing have to be trained regularly. Not only operators and laboratory technicians belong to this group, but also personnel deployed in the cleaning of manufacturing areas. The training and the success have to be verified. Personnel are bound to a clear hygiene regime regulating health checks, behavior and gowning procedures. For example, eating, drinking, chewing and smoking are forbidden in the clean-room areas. Table 6.2 lists the essential gowning rules.

Table 6.2 Clean-room gowning for different clean-room classes.

EU GMP Annex 1	US 209 E	ISO 14644	Gowning
D	NA	9	hair and beard covered, gown, overshoes, street clothing below allowed
C	100000	8	hair and beard covered, completely closed jump suit, gloves, overshoes street clothing below not allowed, particle-poor materials
B	10000	7	head cover closed over hair and beard, face mask, completely closed jump suit, gloves, clean-room boots, sterilized overshoes, underwear street clothing below not allowed, particle-poor materials
A	100	5	like B, no permanent work place, only necessary interventions allowed

[a]The US standard FS 209E (Federal Standard 209E: 'Airborne Particulate Cleanliness Classes in Clean Rooms and Clean Zones') was replaced by ISO 14644 in 2001. Nevertheless, the terminology of standard 209E is widely used. The numbers (100; 10 000; 100 000) relate to the particles per cubic foot counted during the operation of the facility ('in operation'). Since the European standard stipulates the particle count during both the non-operating state ('at rest') and 'in operation', the comparison between the two standards is prone to confusion.

In particular, spontaneous creativity of the personnel is limited. Personnel are held to operate strictly according to operating procedures; deviations of these are only possible under special circumstances, and have to be recorded and notified to supervisors. Changes to the manufacturing procedures may only be made by authorized persons and are subject to change-management procedures. Arbitrary changes in the production flow or the analytical methods are not allowed; deviations from the expected or specified values are to be recorded and notified.

6.1.2
Premises and Equipment

Requirements for premises and equipment are mainly driven by considerations of external contamination, cross-contamination and product confusion.

External contamination denotes impurities (Section 2.5.1.1) that are accidentally introduced into the process from outside sources and that have the potential to jeopardize the safety of the product:

- Microbial or non-process-related viral contamination.
- Contamination with particles of any kind.
- Leaching of materials of construction (e.g. gasket softeners, metal ions).

Apart from using appropriate materials, the best counter-measure against external contamination is to create a process that is completely closed to the environment. Section 6.1.2.1 explains how this can be achieved.

Cross-contamination denotes impurities that have been created by the product itself or other products. Cross-contaminations are particularly dangerous if they relate to potentially pharmacologically active substances. Typical representatives are:

Cross-contamination

- Vitiation of the purer process stage with intermediates from earlier process stages (carry-over).
- Other products that have been produced earlier in the same facility or are simultaneously produced in adjacent facilities.
- Wrong raw materials that may not be used in the current campaign.

Carry-over

Measures against cross-contaminations will be discussed in Section 6.1.2.2.

6.1.2.1 Measures to Avoid External Contamination
Due to their life-supporting conditions, biotechnological processes are particularly prone to microbial contamination. Moreover, the

limited ability to analyze them does not always allow detection of a contamination. Since many parts of the process are not completely closed, the product faces certain risks when being exposed to the environment. For example, the process is partly opened to the environment if chromatography skids (Section 2.5.3.1) or hoses are connected to tanks.

Therefore, the process environment is of utmost significance for purity and safety of the produced drugs. Biotechnological manufacturing either happens in completely closed equipment in a controlled environment or in a partly open process in classified rooms. Thus, a difference is made between 'controlled' and 'classified'. Classified rooms conform to the clean-room conditions as defined by relevant regulations; controlled space fulfils company internal standards for cleanliness, monitoring, air conditioning and access control. The existing clean-room standards of the US (Federal Standard 209E) and the EU (cGMP Guideline, Annex 1) are not completely harmonized so that despite having a common ISO guideline (14644), misunderstandings occur pretty frequently (compare Table 6.2).

Controlled non-classified

Figure 6.2 shows the prevailing clean-room cascade for biotechnological manufacturing. The cell bank is manufactured under high safety precautions since potential contaminations would proliferate

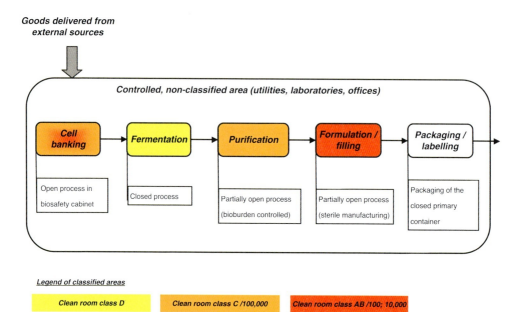

Figure 6.2 Clean-rooms in the process chain of biotechnological manufacturing. Red = laminar flow hood in B environment.

to all batches derived from this bank. Cell banking is usually done in a separate laboratory of clean-room class ISO 8 under a biological workbench of class ISO 5 (laminar flow hood). In fermentation, the process is enclosed in the equipment, therefore class ISO 8 or even lower is sufficient here; media and inoculum are fed via sterile connections or get sterile filtered prior to use. In purification, often at least partly open processes are performed. In this area the goal is to control the bioburden as much as possible; sterility is not required. At the transition to sterile manufacturing, the solution is led through a 0.2-μm filter, capable of removing germs. Formulation takes place in a very clean environment (ISO 7) and filling into the open vial under sterile conditions (ISO 5) since the protein cannot be terminally sterilized.

The described clean-room configuration is not mandatory, but common for many processes. It is important to understand that if all processes would be closed, then the entire manufacturing could be made in class ISO 8 or lower.

Since most processes are at least partly open, a battery of protecting measures against external contamination is required. They relate to the cleaning and monitoring of the personnel and material entering the manufacturing areas, including:

- Filtration of the air in the production rooms.
- Monitoring of particulate and microbial burden.
- Good cleanability of clean-rooms, frequent cleaning and sanitization.
- Air pressure cascade between rooms.
- Airlock systems for personnel and material traffic.
- Personal hygiene and clean-room gowning.
- Cleaning of all materials in a material airlock.
- Testing of materials of construction for the emission of particles or dissolution of molecules.
- Ensuring purity of process auxiliaries (water, gases, chemicals).

These measures form the primary barrier against contaminations. The inner barrier is made up by the seclusion of the process as requested by GMP guidelines. Even if all the primary measures are taken, the process has to be designed as enclosed as possible and the remaining partly open steps have to be particularly protected.

The classification of clean-rooms goes along with regulatory requirements for gowning (Section 6.1.1), and limits for particle count and measurable microbial counts in the air, on outer surfaces, and equipment. The upper allowed limits for viable and non-viable particles are specified in the guidelines. Moreover, the rooms have to fulfill certain criteria as to temperature, humidity, pressure gradients and airflow between the rooms. It is the manufacturer's responsibility to define suitable conditions. They should provide technical solutions

Specifications for viable and particle counts

'suitable for the intended purpose'. Crossing from one clean-room class into another is only allowed via airlocks that guarantee that no impurities enter into the cleaner environment. Section 8.6 elaborates on the technical standard to realize this.

Equipment requirements

All equipment (i.e. apparatus and analytical devices) is subject to similar rules as the rooms. As opposed to clean-room walls, the apparatus often has contact with the product, therefore such equipment needs to be especially clean and the material of construction may not interact with or contaminate the product. Typical threats for products are dissolution of particles or molecules from materials (bleaching), disintegration of gaskets and precipitation or folding of product components at surface-active substances. All in all, the equipment has to be:

- Easily cleanable.
- Qualified.
- Inspected and maintained (and analytical devices have to be calibrated regularly).
- Documented in an equipment log.

6.1.2.2 Measures to Avoid Cross-Contamination and Product Confusion

The measures to avoid cross-contamination decisively influence the design and utilization possibilities of a technical production plant. A general rule applies:

> *Processes for different products are to be separated either physically or temporally. Different process stages of the same product are to be separated either physically or temporally if it is expected that the carry-over of impurities can jeopardize product quality.*

For physical (spatial) separation the question arises where the process streams have to be separated. This is particularly interesting for supporting functions like quality control, equipment cleaning, storage, and media and buffer preparation, but also the supply of water, air and other utilities. Special attention has to be dedicated to areas in which a flow towards the direction of the manufacturing process is possible (e.g. ventilation of rooms with recirculating air, personnel, cleaning of equipment in conjointly used cleaning areas and stationary cleaning of larger vessels).

For these processing materials another general rule applies:

> *Non-product-containing substances that flow towards the manufacturing process can be handled in jointly used areas. Potentially product-containing substances may not flow in the direction of other product manufacturing processes.*

According to this rule, media and buffers for two products can be prepared in one single room. Likewise, storage can be shared among different products. In contrast, recirculation of air between areas with different products is not allowed. Also, personnel taking care of the products have to be separated. The common use of the quality control laboratory is allowed if no sample returns from the lab to the manufacturing area.

The risk of cross-contamination to a great extent determines the broadness of usability of a production plant. Already the one-product facility requires that flows of product and impure waste streams should not cross-over. In multi-product plants, the risk of cross-contamination by carry-over is supplemented by the risk of contamination between the products. A result of this claim is the segregation of flows of equipment, personnel, product, raw materials and waste, either spatially or temporally (Section 8.2.3), as far as possible.

Figure 6.3 shows potential configurations of a production plant. The single-product plant consists of a sequence of rooms – here fermentation and purification – that have to be segregated from each other. Between the process batches, so-called batch-to-batch cleaning is *Batch-to-batch cleaning*

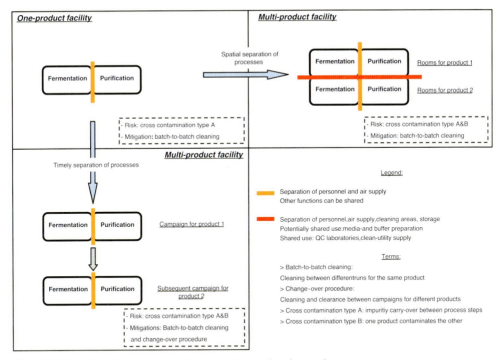

Figure 6.3 Prevention of cross-contamination in different plant concepts. Type A = contamination of the purer product step with intermediates from upstream production steps; type B = contamination by other products that have been produced earlier in the same plant or which are manufactured simultaneously in adjacent facilities. QC = quality control.

Change-over procedure

performed. The facility can be used for multiple substances if a comprehensive change-over procedure is performed between the runs for different products. For this purpose mobile equipment has to be removed from the rooms, and cleaning validated for all parts of the equipment and rooms.

Parallel processing of different substances in one facility requires an almost complete separation of production and supporting areas. Only far remote support functions like the generation of clean media and the quality control laboratories can be shared.

In addition, avoidance of cross-contamination results in very concrete requirements for the rooms:

- Production, storage and quality control areas may not be used as traffic areas. Vice versa, traffic areas such as passage or hallways may not be used for any of these upper purposes.

- The order of the rooms should mirror the logical order of the process flow, entailing rising clean-room classification. In order to attest and assure these attributes, so-called 'GMP flows' are designed in which the flow of personnel, product, mobile equipment, raw materials and waste is documented (Section 8.2).

- Sufficient space should be provided to avoid product confusion.

- In storage areas, the substances have to be segregated depending on their status, intended use and storage conditions (Table 6.3).

- Quality control laboratories as well as social rooms have to be separated from production rooms.

- Only one product may be produced at the time in a room.

Table 6.3 Different storage areas have to be set up for different substance classes and testing statuses – the table shows storage areas without consideration of different temperatures (each cross denotes a separate area).

	Raw material	Intermediate	API	Final product
Quarantine	×	×	×	×
Released	×	×	×	×
For destruction[a]	×	×	×	×
Recalled[b]	NA	NA	×	×
Returned[c]	×	NA	NA	NA

[a] Goods that do not meet specifications and are not dispatched.
[b] Goods that have been recalled.
[c] Goods that have been returned after complaint.
NA = not applicable.

6.1.3
Equipment Qualification

When producing and testing drugs, it has to be demonstrated that the described process and testing methods fulfill the intended use and lead to consistent quality. This is proven in two steps: in a first step the equipment and methods are qualified, and subsequently the processes are validated.

Qualification is the documented confirmation that all relevant systems are designed, installed, tested and function in accordance with GMP requirements and their specifications. Validation is the documented confirmation that methods, processes, equipment, materials, workflows and systems indeed, reproducibly, and in accordance with GMP requirements, lead to the expected results. *Definition of validation and qualification*

The following Section 6.1.4 is dedicated to validation; here, qualification will first be explained. A closer look at the above-mentioned definition of qualification helps to understand its purpose:

- *...documented confirmation...* the qualification activities and results have to be noted in qualification protocols and reports. *Documented evidence*

- *...relevant systems...* only relevant systems have to be qualified. These systems are characterized by the fact that they can have an influence on product quality, such as the water-supply system. Other systems are analytical instruments, process equipment, clean-room environment, supply and utility systems, and automation and computer systems. For non-critical systems like cooling systems typically commissioning is sufficient. Relevant systems are also denoted as 'critical systems'. They can vary from process to process. For example, the temperature control of a fermenter can be critical for a temperature-sensitive product, whereas it is irrelevant for a robust product. Consequently, identification of critical systems requires knowledge of critical process parameters. *Relevant systems*

- *... in accordance with GMP requirements and their specifications...* implies that GMP rules and specifications exist for these systems. GMP rules are known for clean-rooms and utility systems with product contact such as water and gas supply. Specifications exist for all systems since they are needed for ordering. Yet it has to be assured that the equipment ordered with those specifications is capable of supporting the process adequately.

- *...designed, installed, tested, and function...* means that qualification is a multi-step process that starts with the design and ends with the proof of functionality of the equipment.

Figure 6.4 illustrates the staggered approach for equipment and facility qualification. Reading the V-model of qualification from the left, it *V-model of qualification*

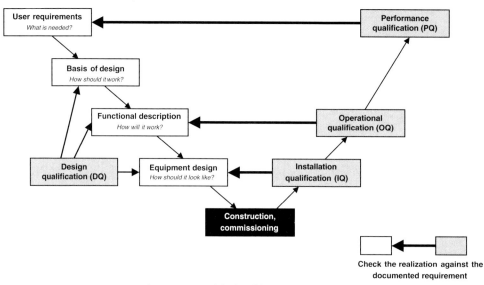

Figure 6.4 V-model of qualification.

User requirement specification

shows the essential qualification documents that are generated during facility design. In the subsequent construction phase, the technical realization is checked against these documents on the different steps. Design activities start with identifying and documenting requirements in a so-called 'user requirements specification' document. Based on this document, the proposed technical solutions are more and more refined in the engineering process, resulting in another document, the 'basis of design', which provides a first answer as to how the tasks are going to be fulfilled. The functional descriptions as well as the equipment design already include fairly detailed execution plans. The progress of the design activities is accompanied by the design qualification. This is a process in which the plant and equipment design is scrutinized under quality aspects – a check that can be made in a dedicated team session

Design qualification

and/or be supported by external consultants. Design qualification is the documented confirmation that the design of process equipment, control systems and facility is in accordance with the requirements of the user and regulatory authorities.

After construction and commissioning follows the installation qualification. In essence this is a check whether the relevant systems indeed have been constructed as designed. For example, for a clean-steam system the slope of the piping, necessary for self-draining, would be

Installation qualification

checked following a pre-set schematic. Installation qualification is the documented confirmation that process equipment, control systems and facility have been installed in accordance with the approved design documents, and that the finish complies with regulatory requirements.

After the installation qualification follows the operational qualification, which is the documented confirmation that the process equipment and the critical utilities function in accordance with the specifications. It includes the check of quality-relevant operation conditions (and minimal and maximal load) as well as safety interlocks and alarms. Operational qualification assures functionality of the plant. After the operational qualification the system can be used for commercial manufacturing of drugs. However, during the start-up phase the system has to be closely monitored in order to demonstrate reproducibility of quality.

Operational qualification

Therefore, operational qualification is followed by performance qualification, which is the documented confirmation that a critical system can constantly and reproducibly deliver a product which fulfils the pre-defined quality attributes. The term performance qualification is used for utility systems and clean-room installations since process equipment delivers this confirmation by process validation.

Performance qualification

After having completed qualification, the testing effort can be significantly reduced. An unsuccessful qualification can substantially delay the start-up of a production plant since it has to be successfully completed for GMP production.

Section 9.1.3 shows the course of a qualification after construction of a new plant and Section 8.4.1.1 shows as an example the effort for the qualification of a water system.

6.1.4
Process Validation

Validation is the documented confirmation that methods, processes, equipment, materials, workflows and systems indeed (and reproducibly) lead to the expected results. Hence, in validation the focus lies in the result generated by the coactions of the prior qualified systems.

Successful validation is a prerequisite for regulatory approval of drug manufacturing. Non-validated or non-validatable processes are unacceptable for drug production. This section exemplifies process, method, cleaning and transport validation.

Virus validation
(Section 2.5.3.3)

In order to validate the process it is run in intended routine operation. The validation is called successful if it can be shown that the required quantity and quality of the product can be yielded reproducibly. Usually this is done with three consecutive successful process runs (so-called 'conformance lots' or 'consistency batches'). The analytical effort is higher during validation than in routine production, since the process course should be monitored and characterized as thoroughly as possible. Once consistency has been proven, the monitoring program is reduced to the necessary level. The validation reports form part of the drug dossier and therefore have to be completed before submission to

Process validation

the authorities. The product that has been produced during a successful validation can be marketed after obtaining getting regulatory approval.

Cleaning validation

Cleaning validation follows rules similar to those of process validation. It can be integrated into process validation; however, in order to minimize potential detrimental influences on process validation it should be completed beforehand. In cleaning validation it is shown that a desired cleaning success is achieved when applying a pre-defined cleaning procedure to a certain fouling. The cleanliness is confirmed by swab tests or analysis of the last flushing step with water (final rinse).

Analytical validation

Validation of analytical methods is performed by checking the general criteria of analytical methods described in Section 3.7.1. These are accuracy, precision, range, linearity, specificity, LOD and LOQ. The optimal setting is to measure against a known reference substance; however, such a substance rarely exists so that a comparison with different testing methods, which allow a similar detection, provides the necessary security for the method. In method validation, the reproducibility of an analytical procedure according to the developed protocol stands in the foreground, be it in different laboratories or by different staff in the same laboratory.

Transport validation

Transport needs to be validated if it can be assumed that the product is particularly jeopardized by shipping conditions. For example, temperature-sensitive products are shipped together with a temperature logger to demonstrate the reliability of the cold chain. Shipping validation for special temperature requirements includes demonstrating that the used packaging can indeed maintain the temperature over the whole shipping period. Therefore, the boxes are exposed to representative temperature profiles in climate chambers. The boxes are also tested for their resistance to mechanical stress.

Validatability

A process or an analytical method is called non-validatable if it is unable to deliver reproducibly the same result. For example, a chromatography process at which end different results are observed at each batch is not validatable. A non-validatable process can ultimately be ascribed to a lack of process understanding, because if the relevant parameters responsible for the scattering are known, they could be changed in a way such that the scattering is reduced. By this mechanism the validation requirement forces the process owner to accurately investigate and control the process, hence guaranteeing consistent product quality.

Likewise, transport for which a too narrow temperature window has to be guaranteed would likely not be validatable. In practice, this means that the product specifications have to be made so broad that the process remains validatable.

Re-validation

After a certain time period a re-validation can become necessary. This means that in routine production the comprehensive monitoring program of validation is applied. Also, major changes in the process have to be verified by validation studies. Already the replacement of a filter by one of a different type requires re-validation.

For complex validation projects a validation master plan lays out the necessary activities. It describes the systems to be validated and the rough workflow of activities. The activities for validation of a single unit have to be fixed in a validation protocol; the validation result is described in a validation report.

Validation master plan

Usually validation is performed prospectively – a protocol is agreed and executed before the process is used for drug manufacturing. In addition there is retrospective validation – an exceptional option that is used for already existing processes for which a validation requirement has been identified. Those can be validated retrospectively by evaluating existing process data.

Prospective and retrospective validation

6.1.5
Computer Validation

Automation systems are often quality-relevant, and have to be qualified and validated like process equipment. The qualification of hardware is completed by the validation of the software. Both activities are summarized under the term 'computer validation'. Computer-based systems are, for example:

- Process-control systems for operating process equipment (e.g. chromatography columns, fermenters and CIP processes).
- Electronic systems for batch documentation ('electronic batch record').
- Control of analytical devices (e.g. HPLC equipment).
- Laboratory information systems for the processing of batch-relevant data ('laboratory information management system').
- Building automation system.
- Warehouse management systems.
- Archiving systems for GMP-relevant data.

In this case validation has to demonstrate that the computerized systems and the implemented software fulfill regulatory requirements. It has to be assured that the electronic data management is on the same safety level as the manual one. Hence, in computer validation special consideration is given to:

- *Electronic signature*: the electronic individual identifier of the authorized signatory.
- *Data safety* – protection against unauthorized or unnoticed manipulation of the stored data as well as the program code.
- *Tracking of changes and accesses to data and code*: audit trail.

- *Access control*: assignment of individual rights for data reading, filing and manipulation.
- *Data accessibility and readability*: for the entire archiving period, the data have to be stored in a form which allows reading them.

6.1.6
Documentation

GMP regulations require that documented evidence is provided for 'almost everything', this includes proving that manufacturing complies with legal regulations as well as the approved marketing authorization. Hence, all influences on and around the manufacturing process need to be documented, such as:

- Clean-rooms.
- Raw materials.
- Deployed clean utilities such as water or gases.
- Cleanliness and testing status of equipment.
- Process flow and course.
- Course and result of the analytical testing.

Figure 6.5 schematically shows when various documents are generated during manufacturing. Equipment cleaning documentation is not shown here for the sake of clarity.

The base document for the manufacturing process is the 'master batch record'. It constitutes the generic manufacturing protocol. Based on the master batch record, a 'batch record' is created for each batch. The batch record is issued for a specific batch number and filled in by operating personnel with process data during manufacturing. Hence, the batch record is both at the same time: an operating standard and a report form in which, for example, weight or process time data are entered. All observations are documented here. The completed batch record after manufacturing is denoted as the 'executed batch record'.

Raw materials

Quality control tests the raw materials prior to use and issues release certificates. Clean utilities and clean environments are subject to continuous monitoring. At least while processing, the measured data have to be inside the specified ranges. Testing and monitoring of raw materials, clean utilities and the environment is regulated in SOPs. The permitted ranges are stipulated in the specifications.

Analytical report

After manufacturing the product is tested by quality control. Here, an analytical report is generated based on analytical SOPs and product specifications.

Deviation investigation

The performance documentation is reviewed by quality assurance. If there are no objections, the product batch can be released. If deviations from the normal process course occur, a deviation investigation has to be opened and a deviation report compiled, which is

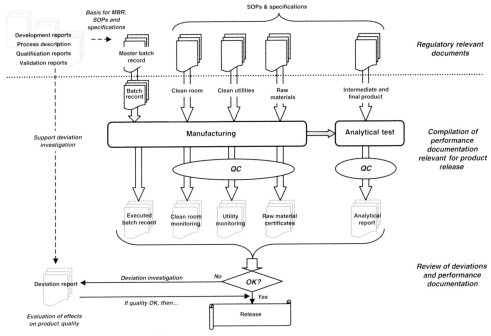

Figure 6.5 Documentation in manufacturing and testing. MBR = master batch record.

part of the release certification. In many cases the assessment of the deviation depends on the knowhow generated during product and process development. Therefore, Figure 6.5 also shows development reports, the process description, and the qualification and validation documentation as core documents for process evaluation.

Documentation can either be electronically or paper based.

6.2 Operative Workflows under GMP Conditions

This section outlines typical organizational flows in GMP-controlled environments and explains the roles of different departments (e.g. quality assurance, quality control, operations, and regulatory affairs) in these contexts.

6.2.1 Product Release and Deviation Management

The release of the product to the market is accompanied by numerous quality assurance activities. Figure 6.6 illustrates the typical course of a

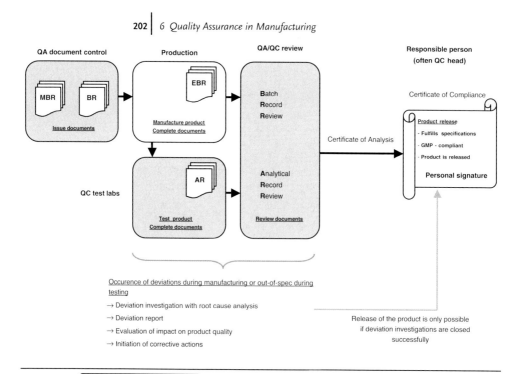

Figure 6.6 Process of product release under GMP. QC = quality control; QA = quality assurance.

product release and the functional groups involved. How to release a product is geared to country-specific legislation and is therefore a little different from case to case, yet an underlying basic scheme is common.

The manufacturing documents are released by quality assurance and issued with a batch number. The process is run according to these documents, and the process data, IPC results (Section 3.3) and any other observations are entered into the manufacturing reports. The completed executed batch records are combined to the manufacturing documentation, which is reviewed by the quality department (batch record review). If everything has been performed in agreement with the applicable procedures and rules, the Certificate of Compliance can be issued. With the Certificate of Compliance, the manufacturer certifies that the path of manufacturing has been in compliance with company directives.

The quality control laboratory checks whether this path has led to the desired result. Here, an analytical report is compiled that is reviewed in

Certificate of Analysis

an analytical report review process. If positive, the result is a Certificate of Analysis in which the results of the release tests are summarized.

If both reviews are successful, the individual responsible for market release can release the product to the market. This individual is personally liable for product quality.

In routine production some deviations from the expected operation can occur. Examples are: *Deviations*

- Process deviations (blocked filter in purification, insufficient gas flow in fermentation, briefly exceeding the storage temperature, a deviating value in pH adjustment, etc.)
- Deviations in environment or clean utilities (partial outage of HVAC units or demoisturizer, interruption of water flow, interrupted vessel heating, decrease of room temperature, etc.)

Not every deviation must lead to a quality-deficient product. However, after a deviation has occurred, a failure investigation has to be initiated to analyze the root cause (root cause analysis) and to identify the potential impact on product quality. Moreover, corrective actions have to be implemented if the event is not classified as being singular. The results of the investigation are summarized in a deviation report and have to be presented for product release. If the analysis shows that the deviation indeed has caused a quality defect, the relevant batch usually cannot be released for the market. In biotechnological production compliant processing is an important part of product quality, therefore deviations can quickly lead to batch rejects. The process of improving the process after having detected deficiencies is known as the CAPA (corrective action–preventive action) process. *CAPA*

In analytical testing, the result can fail to meet the specified criteria. Such an out-of-specification result is subject to investigation as well. First, it has to be determined whether the value was correct or measured incorrectly. If a faulty measurement can be excluded, the failure root cause has to be identified and again be evaluated in terms of product quality.

Only if all investigations relating to out-of-specification results and process deviations are successfully completed can the product be released.

6.2.2
Changes in the Manufacturing Process

In biotechnological manufacturing any changes to the process are usually associated with much higher effort than for small-chemical pharmaceuticals. As mentioned previously, the main reasons are that

(i) the process defines the product properties, (ii) the possibilities for analytical characterization are limited and (iii) clinical consequences of quality changes are difficult to project.

Voluntary change

Changes in the analytical program or the process can be desired for process optimization or quality improvement. Typical examples include:

- Use of a new auxiliary material.
- Replacement of a process step by another one.
- Yield optimization.
- Change of equipment or manufacturing site.

Change obligation

Apart from these voluntary changes, manufacturers can be obliged to make changes, for example:

- Regulations require adapting methods and equipment to the current state of technology.
- The obligation to obey other laws like occupational health and environmental protection.
- The mandatory reaction to observations after having been inspected.

The main effort emerges from the studies ensuring comparability of the product before and after the change, which for biotechnological products cannot always be based on analytical data only. Moreover, due to the higher degree of detail in biotech dossiers, the regulatory effort is significantly higher than that for chemical pharmaceuticals.

Thus, the effort is driven by the risk potential for safe and efficacious use which is created by the change. Unfortunately the terminology for process changes is different between the US and the European guidelines, often creating confusion, yet for both regions the following common principles apply:

Major change/variation

- Changes that have a potential for compromising safety and efficacy (EU: major variation; USA: major change) need regulatory approval prior to implementation. There are fixed timelines for the change processing time by authorities; however, requests for additional information can lead to a clock stop so that the duration for change approval on the part of the authorities ranges from 3 to 9 months. Examples of this change category are change of chromatography steps, changes to fermentation, transfer of production to different equipment, scale-up and extension of product specifications.

Moderate change/minor variation

- Changes that have a low potential to compromise safety and efficacy (EU: minor variation, US: moderate change) do not need explicit approval prior to implementation; however, they have to be notified to the authorities before being implemented. The authority then has a certain time period to react to the change submission (EU and US: 30 days). If the authority fails to do so, the change can be implemented. Examples of this change category are: duplication of a manufacturing train without changing the technology and transfer of an analytical lab.

- Changes that definitely do not have a potential to compromise safety and efficacy (EU: minor variation; US: minor change), but which relate to processes or equipment mentioned in the regulatory dossiers, do not need explicit approval. In this category, the European and the US regulations are different. In the US, such a change can be implemented and notified to the agency afterwards in the yearly compiled 'Annual Report'. In Europe, the change has to be notified before implementation, but the agency only has 14 days to object. The notification of these changes serves to align the regulatory approval dossier with the operational reality. An example of this change category is the addition (neither change nor deletion) of a release test.

Minor change/minor variation

- The fourth and 'lowest' category of changes is made up of those changes which do not need regulatory notification (non-reportable). All the changes that do not mean a risk for the drug and do not relate to circumstances mentioned in the dossier fall into this group. They include, for example, editorial changes to manufacturing documents that have not been submitted with the drug dossier.

Examples for the classification of changes can be found in the EU guideline 1085/2003/EC and the US Food and Drug Administration (FDA) guideline 'Guidance for Industry: Changes to an Approved Application for Specified Biotechnology and Specified Synthetic Biological Products'. Table 6.4 gives an overview of the regulatory classifications of manufacturing changes. It should be noted that in Europe the options to file a Type I variation for biotech products is very limited.

Table 6.4 Regulatory classification of changes to the manufacturing process.

	US (Food and Drug Administration)	Europe (EMEA)
Only internal documentation	– not reportable	– not reportable
Notification requirement, retroactive report to authority	– Annual Report (minor change)	– –
Approval or non-rejection by authority required[a]	– –	– Variation Type Ia (minor variation), notification
	– Changes Being Effected, CBE 30	– Variation Type Ib (minor variation)
	– Pre Approval Supplement, (major change)	– Variation Type II (major variation)

[a]CBE30 and Type Ia/b can be implemented if the authority does not object within a certain time period. Pre Approval Supplement and Type II are subject to explicit authority approval. EMEA = European Medicines Agency.

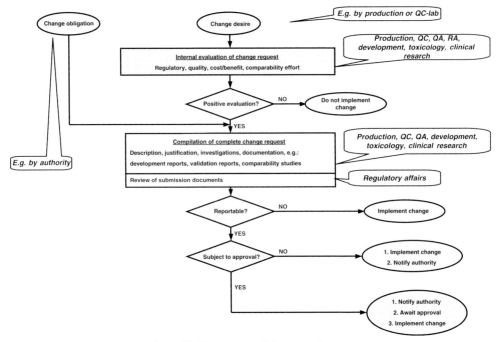

Figure 6.7 Management of changes to the manufacturing process. QC = quality control; QA = quality assessment; RA = regulatory affairs.

Practically each change affecting the manufacturing process or the analytics is considered to be a Type II variation.

The typical flow for the implementation of a change to the manufacturing process is outlined in Figure 6.7. A change can either be desired by production for the process or by quality control for analytics. A first cost–benefit analysis should identify the essential impact on product quality, the possible saving potential, regulatory implications and other consequences for product comparability. If the suggestion looks beneficial after this first check, the compilation of the change request can start. It consists of the description and motivation of the change, and includes an evaluation of the quality impact. Regulatory affairs review the request against the approved drug dossier and determine the regulatory strategy. Depending on the quality of the change it is either notified to the agency or implemented.

The demonstration of product comparability after larger process changes can be associated with pre-clinical or clinical studies. The final decision about this is taken by the regulatory authority. This means that the pharmaceutical company carries a risk since the perception of the regulatory body as to this specific question cannot always be predicted easily. In order to increase the safety regarding regulatory requirements it is possible to inform the authority about the intended comparability study and obtain regulatory input on the projected plan. In the US, the output of such a strategy can be a 'comparability protocol' that outlines the scientific pathway to follow for the comparability study. This protocol does not anticipate the study outcome; only if the found data support the study hypothesis can comparability be confirmed. In addition, in both Europe and the US, regulatory bodies offer to provide 'scientific advice' in which the authorities' perspective can be discussed before making investments.

Comparability protocol/ scientific advice

In hardly any other field of biotechnological production do the limitations to creativity become as apparent as in change processes; however, patient safety and experience of unpredicted cases of product alteration after changes seem to justify the conservative approach. It is not the deployment of scientific staff for investigations alone, but also the long regulatory approval times which mean a high effort for the pharmaceutical company. The high effort for risk mitigation resulting from uncertainty makes production structures inflexible.

6.3
Production of Investigational Drugs

The production of drugs for clinical trials has a special position within the GMP framework. Drugs in development change by nature of the early stage, leading to a conflict of interests with the rigorous change-control procedures requested by GMP. In the past, a non-GMP environment had been accepted for investigational drugs; quality was guaranteed by the rules of GLP. GMP rules in a strict sense were only applied to commercial production (Section 4.3.2.1).

The directive 2001/20/EG of the EU, effective May 2004, stipulates that also investigational medicinal products have to be manufactured under GMP regulations. These rules are further detailed in the GMP guideline of the EU, especially Annex 13. In the US, a requirement as strong as this does not yet exist. However, it cannot be expected that a clinical trial is supplied with the final product ready for market supply. This would mean to work sequentially through process and clinical development, which does not make sense either from the clinical or from the scientific perspective. Therefore, the manufacture of clinical trial material is different from that of commercial drugs in essential aspects:

- Development processes are not validated.
- Specifications change during development.
- Personnel are recruited from the scientifically oriented development environment.

- Batches are so small that manual filling of sterile drugs may be necessary.

The increased risk potential resulting from these conditions is compensated for by comprehensive quality control testing and by the increased level of patient monitoring in clinical trials.

However, the claim to establish as much GMP as possible has to be fulfilled. Naturally this can be done rather for the static facility environment than for the dynamically changing core manufacturing process and analytics. The environment, monitoring, personnel training, cleaning protocols, separation of quality control and manufacturing, the diligence of documentation, and finally the quality system have to live up to GMP standards. All this is done recognizing the fact that changes to process and analytics during development are needed and desired.

Part Five
Pharmaceutical Law

7
Pharmaceutical Law and Regulatory Authorities

This section describes the legal rules of pharmaceutical law that have to be followed when producing biotechnological drugs. Pharmaceutical law regulates everything that is relevant for safety, efficacy, quality and marketing of drugs; the GMP rules are anchored here. Based on an overview of the fields of pharmaceutical law, the subsequent sections are focused on the aspects relevant for manufacturing. Authorities and institutions and the most important rules issued by these bodies are described as well as official enforcement measures. A short look at the regulations for drug approval completes the picture.

In the production of biopharmaceuticals it is not only pharmaceutical law that has to be obeyed. A second large part of regulatory guidance relates to permits for construction and operation of technical facilities. The main aspects of this are environmental, occupational health and building laws; this field is covered in Section 9.6.

7.1
Fields of Pharmaceutical Law

The rules for controlling drug approval, manufacture and distribution primarily aim at providing optimal medical care for patients, and protecting them from unacceptable quality and safety risks. The fields of law are similar in all legislations. They include the following aspects, of which only some (i–iv) are directly relevant for production:

(i) Quality and manufacture of drugs. Foundation of GMP.
(ii) Quality control and quality assurance in manufacture and distribution.
(iii) Liability for drug injuries.
(iv) Placing on the market of drugs. Here the rules of pharmacy-only products are stipulated.
(v) General requirements as to the safety and efficacy of the final drug product.

Manufacturing of Pharmaceutical Proteins: From Technology to Economy. Stefan Behme
Copyright © 2009 WILEY-VCH Verlag GmbH & Co. KGaA, Weinheim
ISBN: 978-3-527-32444-6

(vi) Market approval of drugs. Prerequisites and procedures.
(vii) Design and conduct of clinical trials.
(viii) Observation, collection, and evaluation of drug risks. Responsibility of manufacturer and regulatory authorities.
(ix) Import and export of drugs.
(x) Price regulation.

In Germany, these topics are regulated in the 'Arzneimittelgesetz' (AMG), in the US in the 'Federal Food, Drug and Cosmetic Act' and the 'Public Health Service Act', and in Japan in the 'Pharmaceutical Affairs Law'.

7.2
Bindingness of Regulations

The legislative bodies in the pharmaceutical arena are manifold. They are characterized by differences in legal obligations, level of detail and publishing body; Figure 7.1 provides an overview with examples.

Laws that are adopted on the level of national legislation comprise the legal framework for drug surveillance. These legal texts are further detailed and interpreted as to their technical/scientific realization in regulations issued by responsible agencies. In the US, the 'Code of Federal Regulations' constitutes such a body of legislation. Laws and regulations are published in the official registers of the legislative institutions (US: *Federal Register*, EMEA: *Official Journal of the European Union*).

Specifics of European legislation (Section 7.3.2)

Scientific/technical-oriented institutions and authorities provide further practical guidance in standards and rules. These documents

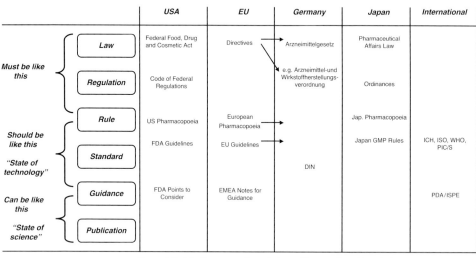

Figure 7.1 Bindingness of regulations. See text for abbreviations.

are in a strict sense non-binding if they are not referenced in the law or the regulation; however, they constitute the recognized state of science and technology ('state of the art'), and serve the authorities as a yardstick. These rules should be complied with unconditionally. As an example of such referenced rules, the pharmacopoeias can be mentioned. Although not being referenced, the GMP guidelines of the FDA and EMEA and other national authorities in practice have a clearly binding character. However, an observation found in regulatory inspections by the FDA may only relate to the Code of Federal Regulations, not to the agency's guidelines.

State of science and technology

Pharmacopoeia

Guidelines

The bodies of rules published by international organizations like the ISO, ICH, World Health Organization (WHO) or Pharmaceutical Inspection Convention and Pharmaceutical Inspection Cooperation Scheme (PIC/S) have a comparably binding character.

Less binding – yet providing direction – are the scientifically oriented publications of authorities that are published as guidance for expected trends ('Points to Consider' and 'Note for Guidance'). When planning or developing a process, these documents provide a valuable future perspective and should be considered. They are not necessarily relevant for already operating processes.

Apart from these official rules, published by national and international institutions, there are others issued by professional associations like the International Society of Pharmaceutical Engineering (ISPE) or Parenteral Drug Association (PDA) (Section 7.3.5). These associations cooperate closely with the agencies and try to provide technically detailed guidance attempting to clarify which technology suits both regulatory and economical criteria. The focus of these implementation recommendations is to balance the targets of 'GMP compliance' and 'cost optimization'.

7.3
Authorities, Institutions and their Regulations

At first sight the structure of bodies of legislation, authorities and institutions appears somewhat confusing. Indeed, there is no unified GMP book, but a multitude of documents that are nationally or internationally relevant. However, a second view reveals that the different rules are identical in their basic principles. There are numerous activities aiming at harmonizing GMP rules. The most important ones are the ICH (Section 7.3.5) body of legislation and the efforts for common standards for regulatory inspections (PIC/S).

In Europe, at least, the official competencies appear likewise heterogeneous as the rules. The parallel existence of European, national and regional authorities has created a complex structure for which the characterization 'regulatory landscape' seems well justified. In the US, the FDA represents a strong central authority bundling most of the functions of pharmaceutical surveillance.

The present section will provide an overview of the regulatory landscapes in the US, the EU, Germany as an EU member state and Japan.

When trying to understand the authorities' roles, it is helpful to discriminate their main tasks:

- Approval of drugs and related clinical trials.
- Surveillance of the safety of marketed drugs (pharmacovigilance).
- Surveillance of the quality of manufacture and distribution (GMP aspects).

The focus of this work is on the GMP aspects; when reading the following sections it should be clear that this is only one part of the regulatory scope.

7.3.1
FDA

In the US there is one authority for the surveillance of drugs: the FDA.

The FDA is a part of the Department of Health and Human Services of the US and consists of eight so-called 'Centers' and 'Offices'; it has about 9000 coworkers, which for the major part are deployed in food control. The central sections that are responsible for drug surveillance are located around Washington, DC [Center for Drug Evaluation and Research (CDER) and Center for Biologics Evaluation and Research (CBER): Rockville, MD]. The focus lies on the areas drug approval and regulatory tasks; the 'Field Offices' that take care for inspection, control and laboratory work, as well as public relations are distributed over the US.

Figure 7.2 shows the organizational structure of FDA with its eight departments having the following roles:

- The CDER is responsible for approval and surveillance of all prescription and non-prescription drugs sold in the US, with the exception of those which are under CBER's responsibility (see below). Since 2003, in the biotechnological field CDER is responsible for:
  –Monoclonal antibodies for *in vivo* use.
  –Proteins intended for therapeutic use, including cytokines (e.g. IFNs), enzymes (e.g. thrombolytics) and other novel proteins, except for those that are specifically assigned to the CBER (e.g. vaccines and blood products). This category includes therapeutic proteins derived from plants, animals or microorganisms, and recombinant versions of these products, therapeutic proteins like cytokines, enzymes,

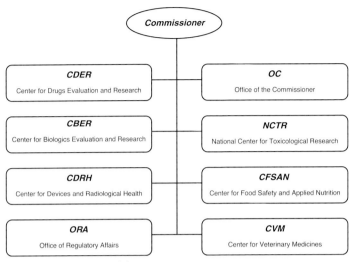

Figure 7.2 Organization of the FDA.

growth factors and all therapeutic proteins made from plant, animal or human cells as well as their recombinant versions.

–Immunomodulators (non-vaccine and non-allergenic products intended to treat disease by inhibiting or modifying a pre-existing immune response).

–Growth factors, cytokines and monoclonal antibodies intended to mobilize, stimulate, decrease or otherwise alter the production of hematopoietic cells *in vivo*.

- The CBER is responsible for products generated from human or animal-derived materials and vaccines: CBER
  –Products that consist of human or animal cells (cell therapeutics), or are composed of fragments of these cells (e.g. cell fragments as therapeutic vaccines).
  –Vaccines (independent of their method of manufacture).
  –Allergenic extracts used for the diagnosis and treatment of allergic diseases and allergen patch tests, toxins and anti-toxins.
  –Blood, blood components, plasma products and blood substitutes.

- The Center for Devices and Radiological Health (CDRH) is responsible for medical devices like pace makers, injection pens, hearing aids and so on. CDRH

- The Office of Regulatory Affairs (ORA) is the headquarters of the 'Field Offices'. The ORA performs inspections and uses its own laboratories for quality testing. Moreover, the ORA coordinates public relations ensuring that FDA regulations are publicly known. ORA

OC
- The Office of the Commissioner (OC) is the staff of the FDA Head (Commissioner) taking care of the efficiency of the agency.

NCTR
- The National Center for Toxicological Research (NCTR) performs scientific work especially in the field of biological mechanisms for the toxicological effects of substances. This work aims at recognizing early the risk potential in food and drugs.

CFSAN
- The Center for Food Safety and Applied Nutrition (CFSAN) is responsible for the safety of all food on the American market, with exception of meat, poultry and some egg products for which the Department of Agriculture has responsibility.

CVM
- The Center for Veterinary Medicines (CVM) is responsible for veterinary drugs and nutrition.

The CDER and CBER have issued many rule documents ('Guidance for Industry', 'Question and Answer Documents on cGMP') that are relevant for biotechnological products. They can be accessed via the FDA internet page. Worth reading are the 'Guides to Inspection' published by the ORA. The 'Freedom of Information Act' (FOIA) allows American authorities to publish regulatory admonishments; these 'Warning Letters' in the 'FOI Reading Room' are a valuable source for those trying to understand the FDA's attitude and inspection philosophy.

The FDA controls the implementation of GMP rules stipulated in the Code of Federal Regulations (CFR) and issues widely respected rules for the realization of those requirements. Important GMP documents include:

- Pharmaceuticals in general (21 CFR 210-211).
- Biological products (21 CFR 600-800).
- Electronic Records, Electronic Signatures (21 CFR 11).

Many GMP guidelines covering the framework of all drugs which are marketed on the US market can be retrieved via www.fda.gov.

7.3.2
EMEA

The EMEA is the drug-regulatory authority of the EU. In order to understand the role and responsibility of the EMEA, it makes sense to take a look at the European unification process. Growing together, the EU step by step transfers responsibilities of drug legislation and control away from the national to the European institutions.

European legislation
A general rule of European legislation is that the European Council of Ministers and the European Parliament issue directives that have to be implemented in national law in the individual member states to become legally binding.

Therefore, in Europe national law exist besides European law. Not all official processes are centralized as in the US. However, the European member states strive for harmonization and central control of the drug surveillance systems. In the case of biotechnological products, for example, the application for clinical trials has to be submitted to national authorities, yet the application for market approval has to be submitted to the central authority – the EMEA. This diversity is confusing at first glance and, moreover, it is in continuous transition; however, a closer view of the regulatory landscape reveals that the responsibilities are pretty clearly defined and justifies that the EMEA's reputation is ever growing, approaching that of the FDA.

Currently, European authorities primarily play a role as coordinator of national authorities. EMEA works in common working groups and committees that aim at implementing the European regulations in the member states consistently.

In 2007, the EMEA employed 440 staff primarily covering drug-approval processes. To fulfill its role the agency reverts to a network of about 3500 experts based in national authorities. It is directly responsible for approvals of biotechnological products, which have to be submitted in the centralized procedure. After having received the marketing application dossier, the EMEA announces the so-called 'rapporteurs' from national authorities that are responsible for assessing the dossier. The rapporteurs lead the review process and report to the relevant committee. The committee gives a vote that is forwarded as a recommendation to the deciding committee – the European Commission. The Commission grants approval if the committee's vote has been positive.

The EMEA has four committees, each of which consists of two members of each member state:

- The Committee for Medicinal Products for Human Use (CHMP) is responsible for all human drugs that do not have 'orphan drug status' (Section 7.5) or are regulated by HMPC. In 2004, the CHMP was renamed from the former Committee for Proprietary Medicinal Products (CPMP) and, therefore some of the guidance still can be found under this abbreviation. *CHMP/CPMP*

- The Committee for Medicinal Products for Veterinary Use (CVMP) is responsible for veterinary drugs. *CVMP*

- The Committee for Orphan Medicinal Products (COMP) is responsible for drugs for rare diseases (orphan drugs, Section 7.5). *COMP*

- The Herbal Medicinal Products Committee (HMPC) is responsible for the approval of herbal drugs. *HMPC*

In addition to the committees, the EMEA organizes working groups that are also composed of participants of the member states. These

Working Parties — working groups aim at developing a common European perspective on pharmaceutical questions and to work out specific guidelines addressing these questions. The groups include the Biotech Working Party, the Quality Working Party, the Safety Working Party and the Immunologicals Working Party.

EDQM — Apart from the EMEA, which primarily takes care of drug-approval questions, the 'European Directorate for the Quality of Medicines' (EDQM) gains increasing relevance for quality assurance and quality control of European drugs. The EDQM is an authority of the European Council that works independently of the EMEA, but cooperates in questions of quality assurance and control. It originates from the editorial office of the *European Pharmacopoeia* (PhEur) and continues to function as its editor. Additionally it has established itself as the coordinating body of European laboratories for drug quality control. The EQDM is responsible for certifying European laboratories and leads the union of European quality control laboratories 'Official Medicines Control Laboratories' (OMCL).

OMCL

When manufacturing drugs in a European state, the national authority is responsible for inspections and approving manufacturing licenses. According to the existing mutual recognition procedures between European member states, the inspections of one state are recognized in all other states.

The basic European directive regulating drug manufacture is:

- Commission Directive 91/356/EEC; laying down the principles and guidelines of good manufacturing practice for medicinal products for human use.

EudraLex — In order to clarify this directive, European authorities have issued a remarkable body of legislation named EudraLex. Here, all rules for control of drugs in the EU can be found. Table 7.1 provides an overview over this collection.

7.3.3
German Authorities

BfArM/PEI — The principal drug-regulatory authorities in Germany are the Federal Institute for Drugs and Medicinal Products [Bundesinstitut für Arzneimittel und Medizinprodukte (BfArM)] and the Paul-Ehrlich-Institut (PEI). The BfArM is responsible for approval and surveillance of drugs that do not fall under PEI's responsibility. The PEI is responsible for sera, vaccines, test allergens, test sera, test antigens and blood preparations. Both institutions are principal federal authorities and are subordinated to the Federal Ministry of Health and Social Security. The division of responsibilities resembles that between the CDER and CBER of the FDA.

Table 7.1 EudraLex – the European comprehensive body of legislation for the control of medicinal products in the EU (http://ec.Europe.eu/enterprise/pharmaceuticals/ eudralex; for the content of Vol. 4, see Table 6.1).

Vol. 1	Medicinal Products for Human Use, Pharmaceutical Legislation Directives Regulations Miscellaneous
Vol. 2	Medicinal Products for Human Use, Notice to Applicants Volume 2A – Procedures for Marketing Authorization Volume 2B – Presentation and Content of the Dossier Volume 2C – Regulatory Guidelines
Vol. 3	Medicinal Products for Human Use, Guidelines Volume 3A – Quality and Biotechnology Volume 3B – Safety, Environment and Information Volume 3C – Efficacy
Vol. 4	Medicinal Products for Human and Veterinary Use, Good Manufacturing Practices
Vol. 5	Veterinary Medicinal Products, Pharmaceutical Legislation
Vol. 6	Veterinary Medicinal Products, Notice to Applicants
Vol. 7	Veterinary Medicinal Products, Guidelines
Vol. 8	Veterinary Medicinal Products, Maximum Residue Limits
Vol. 9	Medicinal Products for Human and Veterinary Use, Pharmacovigilance Part I – Pharmacovigilance of Medicinal Products for Human Use Part II – Pharmacovigilance of Veterinary Medicinal Products Part III – General Information on EU Electronic Exchange of Pharmacovigilance Data Part IV – General Reference to Administrative and Legislative Information Relevant to Both Medicinal Products for Human use and Veterinary Medicinal Products

The control of quality and manufacture lies in the hands of regional authorities of the 'Länder', like the Office for Occupational Health, Sanity and Social Affairs in Berlin (Landesamt für Arbeitsschutz, Gesundheitsschutz und Soziales) or the Regional Board of Darmstadt for Hesse (Regierungspräsidium Darmstadt). In total, there are 34 regional authorities ('Inspektorate') that are coordinated by the Central Office for Use of Drugs and Medicinal Products [Zentralstelle für Gesundheitsschutz bei Arzneimitteln und Medizinprodukten (ZLG)]. The ZLG has similar functions as the EDQM in Europe, but on a national German level. It is responsible for coordinating regional activities and certifying testing laboratories. Coordination is done in

Inspektorate

expert groups in which regional representatives work on harmonization of different topics.

The principal bodies of legislation in Germany are the Drug Law (AMG) and the Regulations for Drug and API Manufacturing [Arzneimittel- und Wirkstoffherstellungsverordnung (AMWHV); formerly the Pharmabetriebsverordnung].

ZLG

AMG/AMWHV

7.3.4
Japanese Authorities

The Japanese regulatory landscape has been characterized by multiple reorganization efforts in recent years.

The healthcare administration is spearheaded by the Ministry of Health, Labor and Welfare (MHLW; Kosei-roudou-sho), the former Ministry of Health and Welfare. It consists of 11 Bureaus, out of which the Pharmaceutical and Food Safety Bureau (PFBS; former Pharmaceutical Affairs Bureau) is primarily relevant for pharmaceutical manufacturing. It has five Divisions, out of which one is responsible for the approval of manufacturing and import licenses (Evaluation and Licensing Division). This division only supervises the procedures of licensing; the actual evaluation (e.g. by audits) is performed by the Japanese prefectural authorities. There are 46 prefectural administrations that are responsible for the local granting of manufacturing licenses.

MHLW

PFBS

Prefectural administration

The principal institution for drug approval is the Pharmaceuticals and Medical Devices Agency (PMDA; KIKO). Although the PMDA is not subordinated to the MHLW it is funded by national payments. Applicants have to submit their dossiers to the PMDA, which coordinates the cooperation with the MHLW and the prefectorates. The approval itself is finally granted by the MHLW.

PMDA (KIKO)

The principal drug law in Japan is the 'Pharmaceutical Affairs Law'. As in other legislations, authorities issue regulations called 'Ordinance' and 'Notification'. The three most important ordinances for manufacture and quality assurance of pharmaceutical products are:

- Standards for Quality Assurance of Drugs, Quasi-drugs, Cosmetics and Medical Devices [GCP Ministerial Ordinance].
- Regulations for Buildings and Facilities for Pharmacies and so on [Buildings and Facilities Regulations].
- Standards for Manufacturing Control and Quality Control of Drugs and Quasi-drugs [GMP Ministerial Ordinance on Drugs and Quasi-drugs].

In addition to these ordinances, the ICH guidelines are adopted in Japan.

A comprehensive and understandable overview over Japanese laws and authorities can be found on the internet page of the 'Japan

Pharmaceutical Manufacturers Organization' (www.jpma.or.jp), whose 'English Regulatory Information Task Force' issues the brochure *Pharmaceutical Regulations and Administration in Japan* in English.

7.3.5
Other Important Institutions

US Pharmacopoeia
The US Pharmacopoeia is an independent scientific health organization. It aims at fixing quality standards for pharmaceuticals and is recognized as an official body. Amongst others it finances itself by selling the *US Pharmacopoeia* (USP). Its many honorary contributors come from healthcare professions and public as well as private organizations such as healthcare authorities, academia, the pharmaceutical industry and consumer associations.

ICH
The ICH is a common activity of the European, US and Japanese health authorities. It was founded in 1990 with the target to harmonize the ever-growing and diverging regulations for pharmaceutical manufacturing and control between these regions. The ICH guidelines are of utmost importance since they practically constitute a world standard. The ICH has issued guidelines for the fields quality (analytics, 'Q'), safety (preclinical tests, 'S'), efficacy (clinical trials, 'E') and some multidisciplinary guidelines ('M'). The latter category includes the guideline M4 (Common Technical Document, CTD) mandating the *CTD* format for the submission of manufacturing and control documentation. In the GMP field the main ICH guidelines are to be found in the Q series, which is listed in Table 7.2.

ISO
The ISO is a network of international standardization institutes of 153 countries and aims at harmonizing technical standards of any kind worldwide. ISO is a non-public organization, yet it occupies a place between the official and the private organizations since many members come from governmental bodies. In the GMP field, the ISO norms have contributed to formation of quality systems. Since 2001, the classification of clean rooms is based on ISO standards. GMP documents include:

- ISO 9000 series: Quality Management Systems.
- ISO 14644: Clean Rooms and Associated Controlled Environments.

WHO
The WHO is an organization of the United Nations. It has issued a series of GMP guidelines in its 'Technical Report Series' (Table 7.3).

Table 7.2 List of the ICH Q series (quality documents).

Stability	
Q1A	Stability Testing of New Drug Substances and Products
Q1B	Stability Testing: Photo-Stability Testing of New Drug Substances and Products
Q1C	Stability Testing for New Dosage Forms
Q1D	Bracketing and Matrixing Designs for Stability Testing of Drug Substances and Products
Q1E	Evaluation of Stability Data
Q1F	Stability Data Package for Registration Applications in Climatic Zones III and IV
Analytical validation	
Q2A	Validation of Analytical Procedures: Text and Methodology
Impurities	
Q3A	Impurities in New Drug Substances
Q3B	Impurities in New Drug Products
Q3C	Impurities: Guideline for Residual Solvents
Pharmacopoeias	
Q4	Pharmacopoeias
Q4A	Pharmacopoeial Harmonization
Q4B	Regulatory Acceptance of Pharmacopoeial Interchangeability
Quality of biotechnological products	
Q5A	Viral Safety Evaluation of Biotechnology Products Derived from Cell Lines of Human or Animal Origin
Q5B	Quality of Biotechnological Products: Analysis of the Expression Construct in Cells Used for Production of r-DNA Derived Protein Products
Q5C	Quality of Biotechnological Products: Stability Testing of Biotechnological/Biological Products
Q5D	Derivation and Characterization of Cell Substrates Used for Production of Biotechnological/Biological Products
Q5E	Comparability of Biotechnological/Biological Products Subject to Changes in their Manufacturing Process
Specifications	
Q6A	Specifications: Test Procedures and Acceptance Criteria for New Drug Substances and New Drug Products: Chemical Substances (including Decision Trees)
Q6B	Specifications: Test Procedures and Acceptance Criteria for Biotechnological/Biological Products
GMP	
Q7	Good Manufacturing Practice Guide for Active Pharmaceutical Ingredients
Pharmaceutical development	
Q8	Pharmaceutical Development
Risk management	
Q9	Quality Risk Management
Q10	Quality Management

Table 7.3 WHO GMP guidelines.

1. WHO GMP: Main Principles for Pharmaceutical Products
Good Manufacturing Practices for Pharmaceutical Products: Main Principles (TRS 908, Annex 4)
2. Good Manufacturing Principles: Starting Materials
Active Pharmaceutical Ingredients (Bulk Drug Substances) Pharmaceutical Excipients
3. Good Manufacturing Principles: Specific Pharmaceutical Products
Sterile Pharmaceutical Products (TRS 902, Annex 6) Biological Products Investigational Pharmaceutical Products for Clinical Trials in Humans Herbal Medicinal Products Radiopharmaceutical Products
4. Inspection
Pre-Approval Inspection (TRS 902, Annex 7) Inspection of Pharmaceutical Manufacturers Inspection of Drug Distribution Channels (TRS 902, Annex 8) Guidance on GMP Inspection (TRS 908, Annex6) Model Certificate of GMP (TRS 908, Annex 5)
5. Hazard and Risk Analysis in Pharmaceutical Products
Application of Hazard Analysis and Critical Control Point (HACCP) Methodology to Pharmaceuticals

PIC/S

The PIC and PIC-scheme are two separate yet cooperating organizations that are denoted together as PIC/S. In total, 28 public health authorities work together in PIC/S in order to harmonize regulatory inspection procedures. This is done by publishing GMP standards and inspection guidance as well as by common training of inspectors and networking activities between the agencies. The principal PIC/S document is the 'Aide Memoire: Inspection of Biotechnology Manufacturers'.

ISPE

The ISPE is an international professional association with the aim to provide guidance for technical implementation of GMP guidelines. The approach focuses on the identification of practical, economically meaningful solutions. At the same time, the widely recognized 'Baseline Pharmaceutical Engineering Guides' publications describe the current technical standard of industrial practice (Table 7.4).

PDA

The PDA is an association for professionals from pharmaceutical industry. The objective is similar to that of the ISPE. The PDA closely

Table 7.4 ISPE baseline pharmaceutical engineering guides.

Vol. 1	Bulk Pharmaceutical Chemicals
Vol. 2	Oral Solid Dosage Forms
Vol. 3	Sterile Manufacturing Facilities
Vol. 4	Water and Steam Systems
Vol. 5	Commissioning and Qualification
Vol. 6	Biopharmaceutical Manufacturing Facilities

cooperates with the FDA and therefore has high relevance for the formation of US guidelines.

7.4
Official Enforcement of Regulations

Governmental authorities enforce GMP rules. The FDA takes care of drugs approved in the US, while the EMEA takes care of products approved in the EU. The latter delegates its responsibility to national health authorities. In the member states of the EU, questions around manufacturing licenses in particular are delegated further down to regional authorities.

The most important instrument of regulatory control is the facility inspection which is performed:

- Prior to market approval.
- For special causes (e.g. apparent quality deficiencies) and routinely (two years).

An inspection can last between a couple of days up to several weeks. Typical foci are the condition and control of manufacturing facilities, quality systems, type and management of process deviations, personnel training, validation, and conformance of the actual process with the approved documentation.

At the end of the inspection the official representative provides a feedback, an official report and (if applicable) necessary recommended measures. The observations are classified according to their product quality risk potential:

After inspections, the FDA summarizes the major and critical observations on a form with the number 483; therefore, such observations are also called '483s'

- A critical observation can be corrected with interventional measures and immediate action.
- Major observations can be prosecuted with regulatory sanctions and monetary fine.
- Minor observations require voluntary measures which can be discussed with the authority.

The FDA has published two guidelines outlining the basics and execution of inspections:

- The *Investigations Operations Manual* (IOM) provides information over the course of inspections, cooperation with the inspected company and recall procedures. *IOM*
- The *Regulatory Procedures Manual* (RPM) contains precise allegations and forms (e.g. for administrative actions like issuing of warning letters, recalls and the treatment of other incidents). *RPM*

Other sources worth reading regarding inspections are the WHO guidelines (Table 7.3) or the 'Aide Memoire: Inspection of Biotechnology Manufacturers' published by PIC/S.

If patient safety permits, the pharmaceutical company gets the chance to correct the major and critical observations while continuing drug manufacturing. For this purpose the FDA issues 'Warning Letters' in which the company is made aware of sanctions that become effective if no corrective action is implemented. Regulatory sanctions can include: *Regulatory sanctions*

- Public admonishment with possible image damage.
- Request for recall of the potentially adulterated batches of product.
- Confiscation of the product.
- Revocation of the manufacturing license.
- Penal prosecution of individuals in case of wantonly negligent or deliberate misconduct such as concealment of serious quality defects with severe consequences.

7.5
Drug Approval

This section gives a very short introduction to the basic terms of drug approval since these processes play an essential role when preparing for market launch. It is beyond the scope of this work to explain all the regulations relevant for approval in detail.

In the US, there are two principal documents that need to be provided for drug development. For the initiation of a clinical study (phase I), an Investigational New Drug Application (IND) has to be filed. The actual market application is filed, the Biological License Application (BLA) or New Drug Application (NDA), at the end of the clinical trial. Market distribution can start after the regulatory review of these documents and a positive vote by the authority. *IND*

The BLA license is regulated under the Public Health Service Act; the NDA license is regulated under the Food, Drug and Cosmetic Act

Since May 2004, a directive exists in Europe which regulates the conditions and execution of clinical trials [Clinical Trials Directive (CTD); European Directive 2001/20/EC]. The European pendant to the American IND is the IMPD (Investigational Medicinal Product *IMPD*

Dossier). For each clinical trial involving biotechnological products, an IMPD has to be filed with the national authority in the country in which the trial will be conducted. After successful completion of the clinical trials, the 'Drug Dossier', which resembles the American BLA/NDA, can be filed with the EMEA. The centralized procedure with the EMEA is obligatory for biotechnological products.

Both the American and the European applications contain manufacturing data in the form of the CTD, which is described in the ICH guideline M4 (Section 7.3.5). Investigational drugs have to be manufactured under GMP; in Europe, the manufacturer requires an official manufacturing license.

In addition to these general rules there are two concepts for special product categories that are equal in Europe and the US:

Orphan drug status

- Drugs for rare diseases (orphan drugs, US: patient population less than 200 000; EU: patient population less than approximately 246 000) have special conditions for approval and marketing, including financial benefits and regulatory simplifications for conducting clinical trials and a market exclusivity for certain time (US: 7 years) regardless of the patent expiration. These simplifications were granted to motivate the researching pharmaceutical companies to invest into these therapies, despite the small return (Section 10.1).

Fast-track status

- Drugs for life-threatening diseases can be led through the regulatory review in the expedited procedure (fast track, priority review). This includes that the company can interact with the authority more often than usual and gets an official opinion on the path forward. It is not enough to target a life-threatening disease to qualify for this group; moreover; the drug has to give a reasonable expectation that it is going to be efficacious and that it provides a unique therapy option to patients.

Scientific Advice

In the normal procedure authorities offer 'Scientific Advice' by which companies can discuss further planned steps of the development program with the authority. This procedure increases the safety for future activities, should there be doubts that the authority coincides with the suggested path forward (e.g. for the design of a clinical trial or the relevance of a planned manufacturing process change). It is important to note that a Scientific Advice is never binding, yet it signals agreement with the proposed path forward; the actual assessment always is based on the data gathered along this path.

Part Six
Production Facilities

This part describes design configurations, the design process and engineering projects for planning and construction of biotechnological production plants. Chapter 8 is dedicated to design criteria, and Chapter 9 to the process of planning, construction and commissioning.

8
Facility Design

The design of a biotechnological production facility has to take into account regulatory, technical and economic aspects at the same time. This chapter begins in Section 8.1 with an overview over the necessary functional units belonging to a production plant. Section 8.2 gives an example that illustrates how GMP criteria are considered in plant design. Section 8.3 is dedicated to more general questions that arise when designing a plant, such as the question whether a plant should be tailor-made for a single product or be multi-functional. Further basic concepts such as disposable technology, flexible piping and fractal versus integrated construction are discussed here.

A large area in production plants is dedicated to clean media like pharmaceutical water, sterilization steam or clean gases. Section 8.4 describes this area of clean media and other utilities necessary for facility operation. Equipment cleaning is not only a quality-relevant factor, it is also one of the biggest users of water and energy. Section 8.5 describes typical cleaning protocols and installations. Section 8.6 provides an overview over the main design criteria for clean-rooms. Section 8.7 outlines the challenges and general approaches to process automation for biotech production. Quality control laboratories are the most important support function for the production plant; when designing a facility they have to be considered and are briefly highlighted in Section 8.8. Last, but not least, Section 8.9 discusses criteria for the selection of a site for a manufacturing facility.

8.1
Basic Principles

The manufacturing process is the core of a production facility; it is surrounded by supporting functions. Figure 8.1 shows a schematic arrangement of these functions. The four shells contain functional

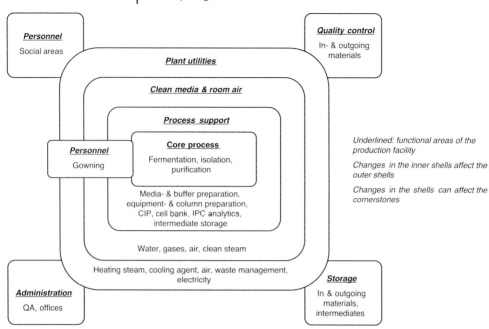

Figure 8.1 Functional areas of a biotechnological production facility.

Process description

areas that have to be designed as parts of the production facility. The shells are arranged in a way that changes of the inner shells affect the adjacent outer shell. As a consequence, when planning a facility one should start with the process and progress to the outer support functions. In practice, this means that a process description and cleaning protocols should exist before the conceptual design of the facility is made. Experience shows that the long design and permitting periods allow doing that with only limited accuracy. Therefore, when realizing a plant, a grasp of the technically necessary and economically possible is helpful to ensure that when the final specifications come out, the plant fits the purpose.

Reliability of support functions

At the same time the outer shells are indispensable for the functionality of the inner ones. Consequently, there is an interdependence that has to be considered in a way that the outer shells have to be designed with sufficient reliability and flexibility to accommodate potential changes of the inner shells.

The four corner stones are as important for the operation of the plant as the shells; however, their dependence on the core areas is less obvious and these functions can also be serviced by outside facilities (with the exception of personnel). Storage areas, for example, often are shared between different manufacturing facilities, yet these functions have to be provided around the core. The model shows four shells:

- *Core process.* It includes all areas in which the product is manufactured directly, like fermentation, isolation and purification. These functions constitute the clean-room core of a facility. — *Core process*

- *Process support.* It comprises the media and buffer preparation, equipment and column preparation, equipment cleaning, cell seeding/bank storage and IPC analytics. These functions do impact the quality of the product, but serve as support to the core functions. Changes to the core functions affect the process support functions. — *Process support*

- *Clean media and air.* This third shell of the model supports both inner ones. Clean media and room air come into contact with the product, and thus are quality critical. Again, a change in the inner shells affects this shell. — *Clean media and room air*

- *Utilities.* The outer shell contains the utilities needed for generating clean media and room air quality. — *Utilities*

The corner stones of functionality are:

- *Personnel.* Especially in small-scale poorly automated multi-purpose plants, like facilities for clinical supply, many production staff are deployed. This personnel needs adequate gowning zones and personnel airlocks in the clean-room area. Outside this clean space, social rooms need to be provided. — *Personnel*

- *Quality control.* The duties of quality control comprise product testing, microbiological testing and clean media monitoring, raw-material release, and IPC testing (e.g. chromatography fractions) as well as stability tests. Due to the special significance of analytics for biotechnological drugs, quality control labs are essential parts of manufacturing and should be considered early on when designing a manufacturing plant. — *Quality control*

- *Storage.* GMP requirements for storage become particularly apparent in the separation of different materials as to their release status (Section 6.1.2.2). A well-controlled receiving and dispatching section is an important part of a production plant. After all, inadequate storage and distribution routes can adversely affect product quality. Often production facilities have smaller receiving and dispatching areas, and share storage areas with other facilities. — *Storage*

- *Administration.* Each plant has a number of office places that host administrative functions like management, logistics and quality assurance. These areas can be located outside the manufacturing plant. — *Administration*

The described functional areas have to be considered when designing a plant. The process core itself is only a small part, yet it should be defined as well as possible in order to be able to design the support functions most optimally.

Figure 8.2 shows how the essential parts of the functional areas interact. The product flow is drawn in red; the green color shows the path of the critical raw materials on their way to the product. The water and CIP systems (Section 8.5) are shown in blue and are exemplary here also for other clean media. The cleaning of small equipment parts is an important supporting function besides the CIP installations. The flow is completed by the waste management systems shown in brown.

It would be meaningful also to show clean-room ventilation since apart from being relevant to quality, it is a dominant factor for calculating building size and energy installations. However, for the sake of clarity this part is omitted here and the illustration focuses on the sections in immediate contact with the product.

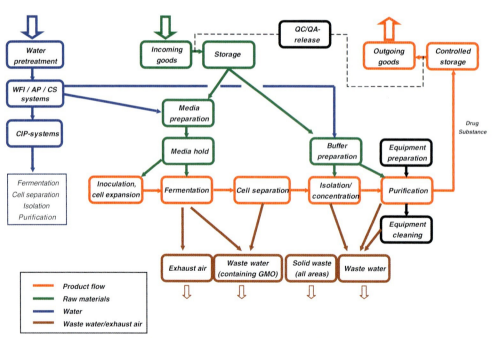

Figure 8.2 Principal functional areas and flows of the process core and support of an API plant using mobile equipment in purification. CS = clean steam; GMO = genetically modified organisms; WFI = water for injection; AP = aqua purificata.

8.2
GMP-Compliant Plant Design

The requirements for plant design resulting from GMP rules have been discussed in Section 6.1.2. The present section ties up to these thoughts and illustrates the implementation of the criteria by means of a layout example. It highlights the aspects of GMP flows and the zoning concept.

The main goal of GMP requirements is to achieve good and consistent product quality. One important part of that is avoiding contamination and confusion of materials and products. As a consequence, processes have to be separated from each other and conducted in a clean atmosphere. Segregation of products or process stages of the same product can be achieved in three ways:

- Temporal separation means processing different stages one after the other. In between the stages, the equipment and rooms used have to be cleaned. *Temporal segregation*

- In spatial separation, parallel processing in one building is allowed; however it has be assured that there is no link between the processing areas (e.g. via shared air ventilation or operating personnel). *Spatial segregation*

- Procedural separation is in practice only used to prevent crossing-over of material or personnel flows. Based on suitable procedures, it assures that different materials are not handled in parallel in the same area. A typical example would be a procedure in a multi-product plant that allows moving waste bins exclusively in the morning between 10.00 and 12.00 in the morning, product A in the afternoon between 13.00 and 15.00, and product B from 16.00 to 18.00. Procedural separation is a makeshift and should not be implemented in new plants; however, it cannot always be avoided. *Procedural segregation*

The consequences for facility design resulting from these segregation requirements are enormous. Ultimately they lead to the facts that manufacturing rooms are aligned according to the natural process flow, that stored materials are separated amongst each other, and that product, material and personnel flows in the plant are not only scrutinized from the logistical standpoint, but also from the perspective of product safety. Moreover, adequate space is requested for all operations, and the multiple use of rooms as manufacturing, traffic and storage space is precluded. *Segregation requirements*

Cleanliness requirements relate to everything that can potentially come into contact with the product, such as personnel, equipment, clean media and room air. The requirements are staggered depending on the seclusion of the process, the proximity to the final product and subsequent contamination reduction steps (e.g. filtration). Again, the *Cleanliness requirements* implementation of these requirements has a huge impact on facility design. Different classes of clean-rooms are necessary that have different standards as to finishing and construction type as well as clean-air ventilation. The different areas are separated by personnel and material airlocks (Section 8.6). In addition, qualified personnel are needed for cleaning and monitoring.

Conflict of interest between GMP and occupational health

GMP guidelines are not the only bodies of legislation that have to be followed for plant design. There are binding regulations of occupational health and environmental protection that partly contradict GMP rules. A typical example is the realization of a biological safety class aiming at preventing the release of substances to the environment as much as possible. This can be achieved by generating low air pressure to the ambient in the process room; in turn, GMP regulations stipulate that outer contaminations should be avoided which typically would be achieved by generating an over-pressure in the process room. Occupational safety rules require escape doors or sprinkler systems, which can become problematic in clean-room designs. The fire code can limit the use of certain materials or influence the design of air ventilation systems significantly.

Conflict of interest between GMP and maintenance

A further challenge for GMP-compliant design is the technical maintenance work for installations in clean-rooms. Maintenance and repair inside the clean-room always comes with a risk of contamination of the classified space. Hence, it is an emerging trend that high-maintenance equipment is accessed from the back side of the clean-room wall, again hugely impacting facility design.

There is no patent medicine for the GMP plant. Apart from GMP rules, there are many other design criteria that are revealed while working on the concrete design. Ultimately, however, GMP criteria are the largest determining factor of plant design.

The following sections will demonstrate how GMP requirements can be incorporated systematically in the facility design. Based on a flow scheme of the process, an exemplary conceptual facility layout is suggested. Subsequently, this concept is checked for its suitability to avoid cross-contamination and product confusion by drafting the principal personnel and material flows.

8.2.1
Production Flow Diagram

Facility design begins by generating a detailed production flow diagram, showing all steps of the core process. Figure 8.3 is an example of such a diagram for purification. An assessment of the contamination risk of the individual steps leads to the clean-room classification. Although no regulatory binding specifications exist, there are established standards that have emerged from experience with the process flows. The process

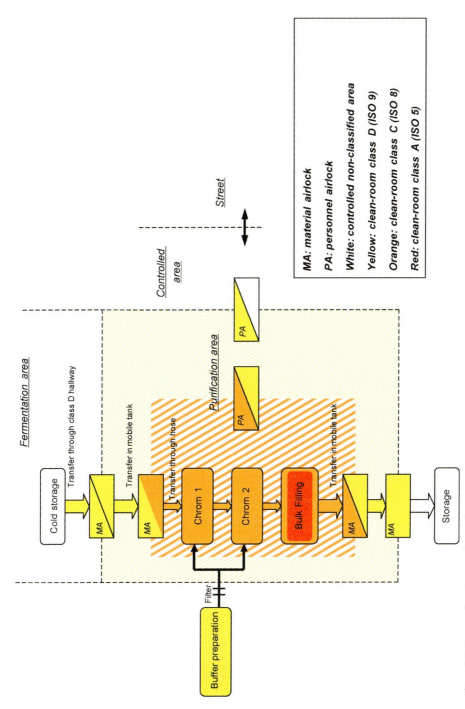

Figure 8.3 Process flow diagram for purification. Chrom = chromatography step; full areas = process steps; hatched areas = classified space.

flow diagram indicates which transitions for staff and material are necessary between the clean-room classes. The diagram is independent of constructional constraints; it defines the conditions solely based on process needs. Figure 8.3, for example, shows the passages between clean-room areas C and D for personnel and material. Substances which cross the clean-room limit via pipes are usually filtered, which is exemplified here by the chromatography buffers.

8.2.2
Conceptual Plant Layout

The actual architectural layout of the plant can be started after having analyzed the necessary segregation steps and clean-room conditions. Mostly in a first step, the specifications of the rooms are defined and converted into conceptual floor plans, giving a first idea of the required floor space. Subsequently, constructional constraints like shape of the building ground, construction regulations and existing infrastructure are included, and a complete conceptual draft of the building can be compiled.

Figure 8.4 gives a possible configuration of a conceptual design. Obviously, vertical concepts are also possible in which the functional areas are located one upon the other; however, for the sake of clarity, this example works with a strictly horizontal layout.

The example shows a goods receiving and storage area for incoming goods on the left side. Goods from this storage area can be transferred into the fermentation and the purification areas via an airlock. The fermentation area can be found in the upper section, the purification area in the lower. Both areas are completely separated from each other. The product is transferred from fermentation to purification via a material airlock. Each section has its own personnel with dedicated gowning space. In the middle, between these areas, support functions are located, which again are not shared. Both areas contain capacities for cold storage of product intermediates or chromatography columns. Goods dispatching is close to the purification area. The accesses to the production rooms are shown as airlocks in this example. It has to be noted that there is no general rule saying that purification rooms have to be separated from the hallway; however, for the chosen example, the flexibility of the plant is significantly enhanced by this design. Moreover, and this will become apparent from Figure 8.6, the rooms in this particular example are classified higher than the hallway, which makes the airlocks obligatory.

To the right, administrative functions are located as well as the quality control laboratories; the building entrance is also here. The core area of the building is surrounded by technical areas containing heating and cooling systems, clean utility supply, and waste treatment installations.

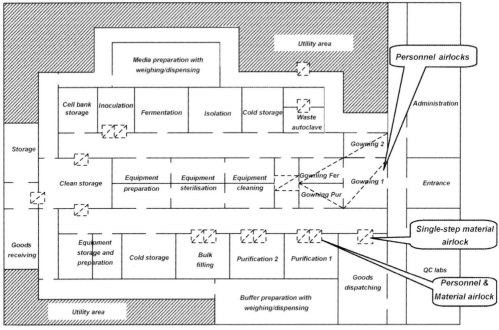

Example for multipurpose API plant with mobile equipment

Figure 8.4 Example of a conceptual design for a plant with mobile equipment. The airlocks separate areas of different room classifications. QC = quality control.

The example shows a multi-purpose plant for small production scales in which intermediates are transferred from one step to the next in mobile tanks. Typically, such facilities are used for the production of small amounts of clinical or commercial product. The design does not allow parallel processing of different products since the risk of cross-contamination would be too high, yet campaigning is possible. The layout accommodates both microbial fermentation and mammalian cell culture.

There are many ways to design a facility optimally for the intended purpose. Apart from being tied to process requirements, the final design always depends on existing boundaries like the building ground or existing infrastructure connections. There are, however, some basic design considerations that should be taken into account for each planning process.

Is the Facility Fit for the Intended Purpose?
This is the core of the planning process. Despite seeming self-evident to follow that goal, design processes often have their own *Fit for purpose* dynamics, carrying away the focus from the original target. Regular

supervision of this process is recommended. It is also for this reason that the planning process is structured by a series of milestone documents that enable focusing stepwise on the actual design target (Section 9.1).

Is the Facility cGMP Compliant?

GMP compliance

The impact of cGMP rules on plant design is huge. The design has to take into account the guidelines of different regulatory authorities of countries to which the product will be delivered. The principal impact comes from the physical separation of process stages, the separation of process and support functions, and the zoning concept with gowning areas and air-conditioning systems.

Is the Facility Flexible?

Flexibility

Building facilities means a long-lasting and significant capital commitment. Often the plant has to be built before official market approval, accepting the risk of not getting approval. Therefore, it is often requested to build in sufficient flexibility to enable the manufacture of other products. Technology platforms that differ only slightly from product to product provide the best prerequisite for this strategy. Flexibility is influenced by many factors, including equipment, floor plan, capacity of utility systems, and the ability to adapt process control and documentation systems.

Usually, facilities are classified and used according to their production technology (microbial fermentation versus mammalian cell culture) and their scale (laboratory, pilot, large-scale production). In the subsequent sections, addressing basic design concepts, some thoughts concerning flexibility are picked up again (Section 8.3).

Can the Facility be Expanded?

Expandability

Expansion can mean to increase capacity for the original product, but it can also denote expanding into manufacturing of other products. Planning for expansion is based on even further forward-looking scenarios than the actual purpose-focused planning. Often financial assets for the advance scope cannot be released; often even the basis of design simply is not clear. Expansion planning means investing upfront in larger facilities and in generous infrastructure feeds. The fractal concept that is discussed in Section 8.3.2 is a typical example of expansion design.

Is it Possible to Separate the Core Process from the Support Functions?

Functional segmentation

The core process is performed in clean-rooms, and relies on supporting systems like water, steam, refrigeration and electricity. These technical systems need to be repaired and maintained, and do not have to be designed in clean-room environments. The technical systems should be located as close as possible to the clean-rooms; on the other hand, it should be possible to maintain the systems without compromising the process space. Since it is not easy to cross from a non-clean-room to a clean-room, the operational flows for process and technical systems are conflicting. The resolution of this conflict has a tremendous impact on facility design. Typical configurations are to feed the supply pipes through the ceiling or through a side wall. The top access is associated with significant issues since the space is usually needed for ventilation systems and ceiling maintenance works; side penetration only works if a technical corridor splits up the clean-room space. The example in Figure 8.4 shows the side-wall transfer in the media and buffer preparation areas. In the areas of equipment cleaning, the only possibility is the ceiling path from the floor above.

Is the Plant Capacity Optimized and are Synergies with Existing Facilities Used?
This question addresses aspects as different as shared use of a buffer preparation tank for multiple chromatography processes, the common use of an autoclave or the selection of standardized dimensions of process equipment. It also includes questions as to the usage of existing infrastructure in storage, administration or utilities. *Optimized usage*

The ultimate goal of all these topics is to optimally allocate the invested capital. Some of the measures (usage, purpose focusing, synergies) work with cost savings, others like flexibility and expandability promise a long efficient service time at the cost of more initial spending.

Obviously each design project has its own emphasis within the framework of these requirements. In recent years, at least the GMP rules for biopharmaceutical facilities have widely consolidated, so that no big surprises can be expected in this area. Nowadays it is possible to buy plants 'of the shelf' that comply with regulatory claims.

The next section will show how to proceed if a conceptual plant layout needs to be checked for GMP compliance.

8.2.3
GMP Flow Analysis

Once the conceptual layout of the building is completed, it has to be checked whether it complies with the segregation requirements (Section 6.1.2.2). They manifest in the material and personnel flows and the clean-room concept, and are checked by drawing the flows into the layout plan. The paths of the different components are drawn into that plan independently of when the movement takes place. The diagrams indicate where the flows potentially cross and where risk potentials result that have to be minimized.

The product flow (Figure 8.5a) shows the path of the product from the cell bank to the dispatching area. Obviously, the order of the rooms follows the process flow. Only the transport of the cooled product runs contrary to the other flows. This transfer would be made in a protected container considering that the product is most probably transferred to a sterile filling and lyophilization area.

The personnel flow (Figure 8.5b) illustrates the movement of the personnel for product manufacturing and equipment preparation; they should be free of crossings. The drawing shows the separation of clean-room and non-clean-room staff, and fermentation and purification staff. The fact that the operators meet in the hallways of the different areas can be accepted since the purification steps are accessible via airlocks with additional gowning requirements. It should, however, be precluded procedurally that an operator engaged in equipment clean-

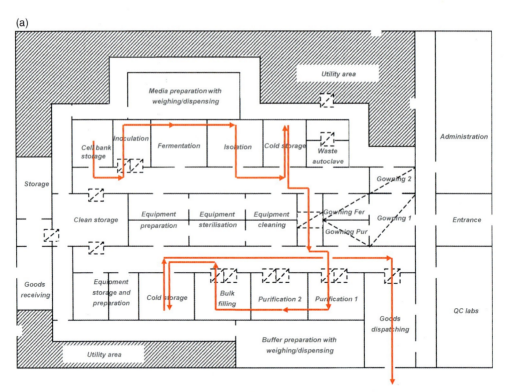

Figure 8.5 (a) Product flow in the plant with mobile equipment. (b) Personnel flow. Dashed line = non-clean-room staff; solid line = clean-room staff. (c) Flow of materials and equipment. (d) Flow of solid and liquid waste. Inactivation of waste which potentially contains genetically modified organisms.

8.2 GMP-Compliant Plant Design | 241

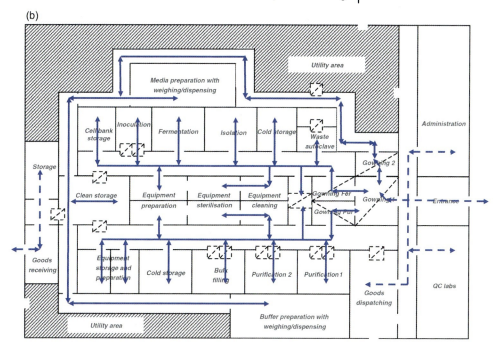

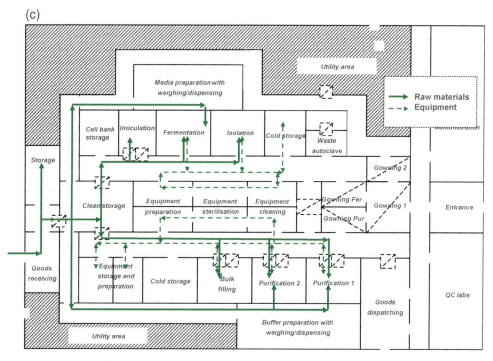

Figure 8.5 (continued)

(d)

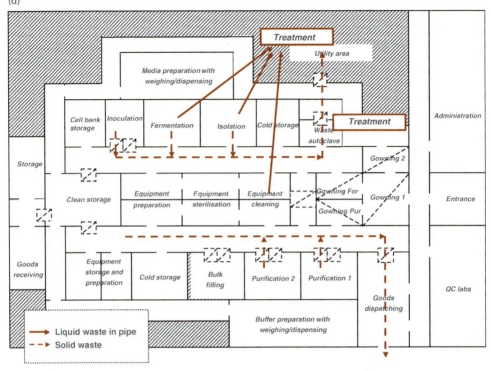

Figure 8.5 *(continued).*

ing enters the equipment-preparation area, in order to avoid that the cleaned equipment gets contaminated.

The different plant sections share the first stage of the gowning area in which street clothing is replaced by plant clothing; afterwards the paths separate. The gowning area is for entering and exiting the facility; unidirectional flow is not realized in the plant. Unidirectional configurations can be found in multi-purpose facilities with parallel processing; however, this example facility is designed for one product at a time, therefore unidirectional flow is not necessary.

The material flows (Figure 8.5c) indicate movements of raw materials and consumables. All materials are transferred into the plant via the storage area and further into the production space via airlocks. Raw materials for media and buffer preparation are not brought into the clean-room core area, but directly from the storage area into the preparation rooms. Solid consumables like hoses or filters are also brought in via the material airlock. The dashed lines show the paths taken by mobile equipment like tanks or chromatography columns. It becomes apparent that the cleaning areas for up- and downstream are separated from each other, and that there is unidirectional flow from dirty to clean, autoclaved equipment. The cleaning areas of a facility are particularly prone to carry-over contaminations, since equipment from different process stages meets here, is opened and washed. Therefore, a strict separation of clean and dirty equipment as well as unidirectional flow is obligatory.

The waste flow (Figure 8.5d) displays the pathways of liquid and solid waste. Waste comes with a high contamination risk. In that respect it has to be treated as diligently as the product itself. Moreover, especially the management of waste containing genetically modified organisms is regulated by environmental law. Prior to dumping it into the usual disposal routes (waste water, incineration) and before leaving the building, it must be biologically inactivated (Section 8.4.3). Since genetically modified organisms are exclusively handled in the upstream area, waste inactivation is indicated in Figure 8.5(d) for this area only.

8.2.4
Zoning Concept

Parallel to the analysis of the GMP flows, the building is segmented in different clean-room zones (Sections 6.1.1 and 6.1.2). Section 8.6 describes how the cleanliness of the environment is assured by measures of gowning, interior room finishing, personnel hygiene, air ventilation, material traffic and so on. The implementation in facility design is detailed in Figure 8.6.

The white areas are controlled, but non-classified areas (i.e. the access is controlled and cleaning requirements are increased over those of office space). However, there are no particle or microbial limits mandated by regulatory guidelines, but in-house specification is set for these rooms. The segment of production in which the product is handled in a closed manner is one cleanliness stage higher (here yellow) and accessible via an airlock (Section 8.6) only. This space is classified, and therefore regularly checked as to temperature, humidity, pressure gradient, particles and microbial contamination. The orange section characterizes the next higher step in which the product, or the substances and equipment with immediate product contact, are handled openly. Typically, this is the case in the purification rooms and the cabinet in which filling into the last container prior to formulation is done. This open filling process is performed in a laminar-flow hood to avoid contaminations. After this step, and mostly in a different building, final formulation and terminal sterile filtration follow. The sensitive and open step of inoculation is also done under very clean atmosphere to prevent microbial contamination. The laminar-flow hoods are indicated in Figure 8.6 by the red color.

Controlled non-classified areas

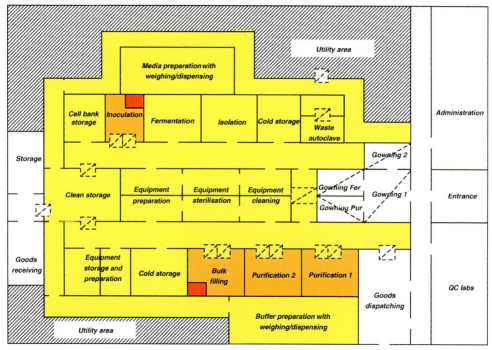

Figure 8.6 Room classifications in the production plant. White = controlled non-classified area; yellow = ISO 9; orange = ISO 8; red = ISO 5 in 8 (laminar-flow hood in class C) (Table 6.2).

The zones are connected by airlocks for personnel and material. These airlocks allow cleaning when transitioning between the zones and that personnel do not mix when transferring material from the up- to the downstream area.

Clean-room cascade

Usually the sequence of the clean-rooms follows a cascade (i.e. that one can transfer from one class to the next via an airlock). This is obligatory for sterile manufacturing according to European regulations (GMP Guideline Annex 1, cascade concept). However, in Figure 8.6, in inoculation and filling there is no class ISO 6 between the biological workbench (ISO 5) and the environment (ISO 8). To understand this, it should be noted that an API manufacturing plant is not comparable to a sterile production plant. Indeed, in API manufacturing one should do anything possible to render the product germ-free ('bioburden controlled'), yet absolute sterility is not required. Biotechnological APIs are terminally filtered; additional control is achieved by microbial testing.

The desired absence of microbial contamination in inoculation obviously aims at avoiding cultivation of the wrong organism. While inoculation has to be performed under consideration of aseptic working

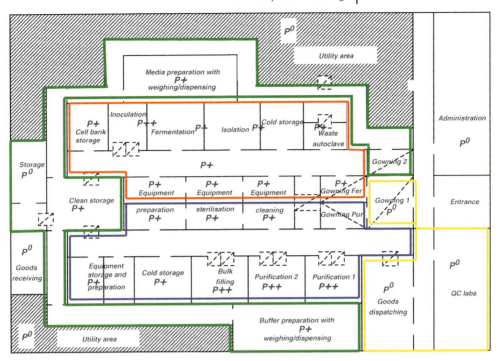

Figure 8.7 Air pressure steps and mapping of ventilation systems. Colored frame = one dedicated air conditioning system; P^0 = atmospheric pressure; $P+$ = overpressure over P^0; $P++$ = overpressure over P.

procedures, regulations for sterile manufacturing do not apply. Here, a good chance exists to discover potential microbial contaminations during fermentation or the subsequent purification process.

The cleanliness of the zones in Figure 8.6 is determined to a large extent by air quality. It is not sufficient to filter the air, but to create pressure gradients between the rooms forcing the air to flow from the cleaner to the less clean section. Moreover, it has become common to assign separate ventilation systems for the different areas of the production facility. In Figure 8.7 a typical pressure-gradient configuration is illustrated (P^0 means ambient pressure, P^+ and P^{++} higher pressure levels) and the assignment of ventilation systems to the rooms. The colored lines indicate an area which is supplied by one HVAC system. There is no air recirculation between these areas.

Until now the zoning concept has only been contemplated from the perspective of product protection. When working with biological material, sometimes substances are handled that have the potential to threaten the environment. Here, regulations of environmental and occupational health law come into play. Just as product protection

Room air

Pressure gradient

Environmental protection

requires the classification into cleanliness classes, ambient protection requires categorization into so-called Biosafety Levels. There are four levels: (S-1: no risk; S-2: low risk; S-3: medium risk; S-4: high risk) expressing the potential risk of the genetically modified organisms or protein to harm the environment. The classification goes along with staggered protective measures for environment and working personnel. Biotechnological production of proteins from recombinant organisms mostly qualifies for a low level S-1 or S-2, since there is no direct risk of infection for humans, neither are the modified organisms or proteins stable in ambient conditions, nor does the genome contain a sequence dangerous to humans or environment. Despite of the often recognized harmlessness, it is established practice to inactivate genetically modified material by chemical or thermal processes (Section 8.4.3) prior to dumping it into municipal or industrial waste treatment systems.

Biological safety class

The presented example has helped to introduce principal GMP design criteria. The following sections will offer more basic concepts that enable us to balance the requirements of facility purpose, flexibility and cost pressure.

8.3
Basic Concepts for Production Plants

In the preceding section it was described how to include GMP aspects into plant layout. While GMP is always a 'must', there are other design concepts that decisively determine the plant layout, but which leave more degrees of freedom than GMP pre-settings.

Each production plant is unique – there is no such thing as 'the biotech plant'. There is no prototype and very few chances for touching up afterwards. Therefore, conceptual planning is essential when trying to come up with a design that accommodates the company's current and future needs.

There are three principal variables for manufacturing plants that are in continuous competition: investments, operating costs and desired flexibility.

These variables face each other with pairwise conflicts of interests – decreased operating costs often are a consequence of investments in optimization projects; flexibility often is not tied to the immediate purpose of the plant, upfront investments in flexible solutions are difficult to justify. Likewise behaves the conflict between flexibility and operating costs. Adding flexibility usually leads to increased unit costs, since the measures guaranteeing flexibility increase operating expenses (e.g. in the form of idle costs). Figure 8.8 shows the area of conflict for investment decisions in the framework of these parameters.

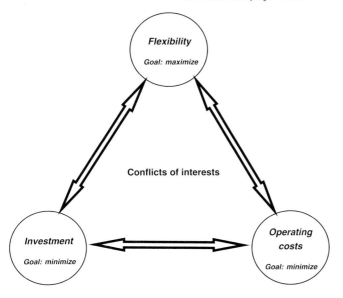

> Investment: equipment, ground and building, qualification/validation, training
> Manufacturing costs: personnel, material, building operation
> Flexibility: scale-up and expansion possibility, multi purpose

Figure 8.8 Conflicts of interest when optimizing a production plant for the criteria of investments, operating costs, complexity and flexibility.

A further optimization factor is technical complexity. Complexity should be adequately minimized since it comes with failure and time risks for facility commissioning and regulatory acceptance. Whether a technically complicated solution makes sense depends on the maturity of the application. If there is experience with the technology and not too much flexibility is required, a high degree of automation does have significant advantages for process consistency and labor costs. In this case, operating costs are kept low by labor savings and quality improvements. On the other hand, high complexity usually limits flexibility since the systems cannot simply be switched over to different processes. If there is less experience with the technology and high flexibility is required, the disadvantages of technical complexity prevail. The terms of technical complexity and flexibility are further described in Table 8.1, and analyzed as to their risks and chances for process quality and costs.

Technical complexity

The following sections introduce some concepts that have a decisive impact on the relation between operating costs, investments and

Design concepts

Table 8.1 Chances and risks for realization of high technical complexity and flexibility.

	… is high if:	Chances of high implementation	Risks of high implementation
Technical complexity …	– the degree of automation is high – cleaning is made by stationary CIP installations – all vessels are hard-piped – manufacturing documentation is electronic (electronic batch record) – functional areas are integrated in one building	– consistent processes – low labor costs – low risk of human error	– high validation effort – high capital investment – low flexibility – longer times for construction and commissioning activities due to limited ability to parallelize – lack of experience with new technology
Flexibility …	– the building is expandable – technology can be adapted easily to other products – the support functions allow to expand production – the facility can make several products in parallel or one after the other	– use for different processes/products – expandability (grow with product)	– upfront investment in flexible concept – suboptimal costs for individual product

Example: If there is high technical complexity, the chance of having consistent processes is high; at the same time the validation effort poses a risk.

flexibility, and are therefore discussed every time a concept is worked out. Background will be provided for the design concepts:

- Single-product plant versus multi-product plant.
- Integrated design versus fractal design.
- Flexible piping versus hard piping.
- Disposable equipment versus stainless steel.

The design concepts overlap – a multi-purpose facility can be built in fractal style, with hard piping and disposable equipment. The optimal combination of the basic concepts always depends on the individual case.

The following sections will discuss how these concepts impact the optimization factors described in Figure 8.8. Table 8.2 summarizes this discussion and gives an overview of the consequences.

8.3.1
Single- and Multi-Product Plants

Many facility engineering projects start with the question whether the plant should be built for one or more products. The decision mainly depends on the required capacity and the probability that the forecasted product quantities materialize. A marketed product with a high sales volume is a clear indicator for a monoplant; the facility can be tailor-

8.3 Basic Concepts for Production Plants

Table 8.2 Impact of basic design concepts (rows) on optimization parameters (columns).

	Flexibility	Investment	Operating costs[a]	Complexity	Remark
Multi-product plant	↑↑	↑	↑	↑	
Fractal design	↑	↑	→	→	property, construction and piping expensive
Flexible piping	↑	↓	↑	↓	
Disposable equipment	↑ (↓)	↓	↑↑	↓	limited scalability

Arrows: ↑↑ = very high; ↑ = high; → = medium; ↓ = low.
[a]Comparison only possible for 100% usage.

made for the product, yielding optimized unit costs. The answer is also relatively clear if several development projects are coming up with the same technology platform requiring smaller product volumes; in this case a multi-purpose plant is the obvious solution. The situation is more difficult if the pipeline contains several projects with different technologies, each of which promises future high-volume business. Due to the risk of project failure it remains unclear which process should form the basis of plant design.

Multi-purpose and dedicated plants are different in many aspects. The design of the monoplant aims at achieving production capacity for the product as quickly and economically as possible, and adapts to the product lifecycle. It constitutes a local optimum for the purpose.

In multi-product plants, the flexibility has to be built into the floor plan. The equipment should be exchangeable or it should be possible to connect it in different configurations. The design has to fit to the highest possible capacity at each single process step, requiring definition of model processes for design purposes. Primary goals for multi-purpose plants are the avoidance of cross-contamination and reduction of associated product change-over times. Cleaning procedures and installations play a major role in these facilities. Like any other capital investment, multi-purpose plants should support production as long as possible. However, if at a later timepoint new products are launched, they are subject to regulatory inspection comparing the plant to the most recent standards since a new product is considered to be manufactured. That means that they have to be kept at the state-of-the-art with follow-up investments, whereas monoplants enjoy a certain right of continuance.

Design with model processes

Right of continuance for old plants

The multi-product concept can be realized in two ways: by campaigning or parallel processing (Section 6.1.2.2). In the campaigning mode, the main design differences compared to the monoplant are in the storage areas. Not only different raw materials and products have to be stored, also equipment that is dedicated to the different products has to be stored outside the currently used manufacturing space. The cleanability of the whole plant has to be guaranteed by design.

The parallel mode generates synergies mainly in the utility generation. Basically, two independent monoplants have to be placed side by side to avoid cross-contamination issues.

8.3.2
Fractal and Integrated Configuration

One design principle that promises maximum flexibility as to expandability is the so-called 'fractal concept'. Here the functional units are lined up to a central corridor, also called a spine or backbone. Figure 8.9

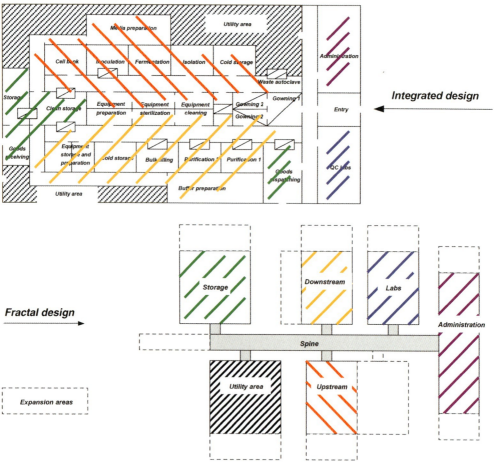

Figure 8.9 Integrated and fractal construction. Expansion areas are dashed.

compares the fractal to the integrated concept. The spine contains the supply pipes and enables personnel and material traffic between the units. The units can be expanded independently from each other and even space for additional units can be planned. It becomes obvious that this concept needs more space than the integrated one. Moreover, if the integrated concept is realized in vertical instead of horizontal integration, the additional space and facade area, hence building costs, needed by the fractal concept is significant. A further disadvantage is that the fractal concept has longer routes between the functional sections, resulting amongst others in higher piping costs.

Apart from being easily expandable, the fractal concept offers the advantage that several construction sections can be built in parallel, which can lead to a shorter construction time. Moreover, changes that come late during construction have only a minor impact on other areas and can therefore be implemented more easily. Obviously the integrated concept does not have this flexibility. A role model for the fractal design is the LIP facility (Large Insulin Plant) in Frankfurt-Hoechst. *Parallelization of construction stages*

Often the terms 'modular' and 'fractal' are used synonymously by error. 'Modular' does not relate to the layout of the plant, but to the construction mode. In modular construction, the building modules are prefabricated, transferred as complete packages to the construction site and integrated into the building. Fractal as well as integrated facilities can be built in the modular mode. In the most advanced concept of modularization, complete facilities are delivered as prefabricated and factory-tested units. The advantage of the modular route lies in the shorter building time that is achieved by parallel work on different sections. *Modular versus fractal construction*

8.3.3
Flexible and Fixed Piping

Flexibility inside a plant can be realized by keeping the connections between different process steps flexible, which is achievable with both fixed piping as well as flexible hose connections. The main disadvantages of fixed piping are:

- Each potential combination of the equipment should be anticipated and connected upfront.
- The cleaning of the closed equipment means significant additional piping for cleaning liquids and steam supply; moreover, cleaning validation can be laborious.
- Cross-contamination has to be avoided by complex valve configurations.

These significant drawbacks have resulted in the fact that many biotechnological facilities are not hard-piped, but use flexible connec-

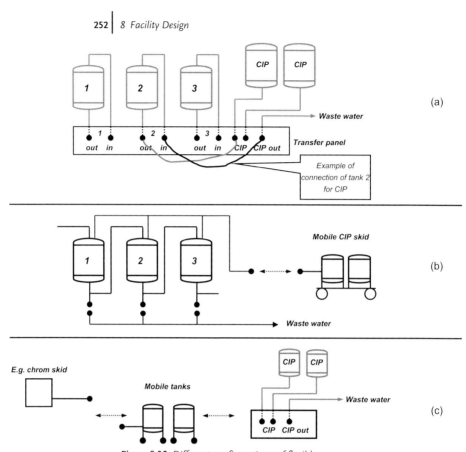

Figure 8.10 Different configurations of flexible piping. (a) Transfer panel with fixed tanks and CIP skid. (b) Fixed tanks with mobile CIP skid. (c) Mobile tanks with central fixed CIP station.

tions between the apparatus. There are different possibilities to realize these connections which are illustrated in Figure 8.10.

CIP (Section 8.5)
Transfer panels

- The parts of the plant are connected by so-called transfer panels. For connecting two pieces of equipment, the pipes have to be attached to the flanges of the panel. Tanks and pipes are cleaned via CIP connections in the panel.
- The equipment is connected with fixed pipes but cleaning is performed with a mobile CIP skid.
- If the tanks are small (less than 250 l) they can be rolled to a fixed CIP installation. Process piping can be held flexible between the tanks.

The decision for one or the other option largely depends on the size of the tanks to be connected. Large tanks cannot be moved around and usually have to be equipped with fixed CIP installations, whereas smaller tanks offer the option to be moved around.

For flexible hoses or short pipe-connections used with transfer panels, cleaning success can be verified visually; however, cleaning of the pieces in a shared cleaning area is prone to cross-contamination, which has to be avoided. Manual tank connection is a further source of error, since wrong tanks could potentially be joined. This risk is minimized by installing contact switches at the transfer panels. *Risk of contamination in the cleaning area*

Flexible piping opens the process to the environment. Only fixed piping can guarantee a closed system at any time, which is a great advantage for the assessment of contamination risks. The closed process can be performed in rooms with lower cleanliness class than the partly open counterparts. If organic solvents are deployed, practically always closed installations have to be used to ensure explosion protection. The same applies to other substances which are harmful to humans and environment; here, the fixed and closed arrangement provides the best protection. *Fixed piping guarantees closed processing* *Organic solvents*

8.3.4
Steel Tanks and Disposable Equipment

A further step towards flexibility is the use of disposable process equipment. For filters or plastic tubes this is established practice, but there are even storage bags up to 3000 l and fermentation bags up to the 500 l scale made of plastic.

The obvious advantage of disposable bags is that they do not have to be cleaned. Up to a certain scale, the bags provide maximum flexibility at low investment. The bags do not need to be maintained and the contamination risk is very low.

On the other hand scalability of the bags is limited; they can only be used once and it has to be verified that the bag material is compatible with the product. The manual effort is pretty high since all connections have to be made by hand, which again lends itself to contamination. A particular concern is whether the material leaches substances into the product solution or interacts with the protein, potentially leading to precipitation or other degradation phenomena. A very significant argument against disposable technology is the dependence on the supplier, which can fail due to bankruptcy or quality problems. *Drawbacks of disposables*

Performing processes in plastic containers is unique to biotechnology since the conditions are basically 'life friendly' and do not have extreme pH values or temperatures. Moreover, there is a certain tradition of plastic bags from the blood industry that in many ways has been a model for the original biotechnology processes.

Typically, disposable technology is deployed for productions with quickly changing products and processes. The smaller the scale of the process, the more disposable components can be used.

8.4
Clean and Plant Utilities

The utilities necessary to run a facility consist of the clean utilities, which come into contact with the product or the manufacturing equipment, and the plant utilities, which do not come close to the product. Clean utilities include water, sterilization steam, air, clean gases and cleaning chemicals. Plant utilities comprise heating and cooling loops, pressurized air for process control, deep-freeze and waste systems as well as electrical supply.

This section begins with a description of the clean utilities (Section 8.4.1). Due to its important role, pharmaceutical water is described in detail. Shorter sections are dedicated to the supply of clean steam and technical gases. Subsequently, plant utilities are described (Section 8.4.2).

8.4.1
Clean Utilities

8.4.1.1 Water

Water is one of the most important raw materials in biopharmacy. It is used in all process stages from cell banking through fermentation and purification down to sterile filling from where it enters the patient's blood system. Water is needed in multiple roles and in large quantities, and has to fulfill high-quality requirements. The primary source of pharmaceutical water is municipal suppliers providing drinking water of very different qualities.

Water is the preferred habitat for many microorganisms. Therefore, water generation, distribution and control are of essence for biopharmaceutical quality. In many cases water systems are the focus of regulatory inspections. If only small quantities of high-quality water are needed, the water can be bought packed in bags ready for use; if larger quantities are required, self-generation, as described in this section, should be considered.

Water users

Water Users in a Production Plant By far the biggest consumers of water in a manufacturing facility are cleaning processes. The second largest users in an API facility are the media and buffer preparation for fermentation and purification, respectively. In a sterile manufacturing facility, buffer preparation for formulation and glass washing equipment add to that. Further smaller users are the cleaning areas for small parts. Other than for cleaning applications and ultra/diafiltration steps, the actual production units like fermentation or purification usually do not need water.

Not every process or cleaning step requires the same high quality; hence, since the generation of high quality costs money, the qualities are categorized in classes each of which has its own application range.

Water Qualities and Fields of Application Different water qualities differ in their purity with regard to particles, hardness minerals, heavy metals, organic substances, microorganisms and other additives.

Water qualities

Usually the plant receives drinking water that can vary in contaminant burden depending on the season and regional characteristics. A well-known contaminant is the different chlorine contents that are added by municipal suppliers for water disinfection. On the consumer side, quality is driven by patient safety requirements; a parenteral drug has to be free from bacteria and endotoxins. Therefore, a rule of thumb applies: the closer the process step is to the final product and the fewer steps follow that can eliminate relevant contaminants, the purer the water has to be.

Manufacturing process consistency also drives water quality. A sensitive mammalian cell culture can react differently to different water qualities and varying ion contents influence chromatography. Particle overfreight can lead to filter blocking.

Provisions for guaranteeing patient safety are stipulated in pharmaceutical bodies of legislation like the pharmacopoeia. The water qualities defined here ('compendial waters') have to be used for the manufacture of drugs. While this rule very clearly ordinates the use of WFI for pharmaceutical manufacturing (final formulation), the choice of water for API manufacturing is left to the manufacturer – which of course needs regulatory approval for its suggestion. Due to this freedom of choice very different water qualities can be found in API manufacturing.

Water for injection
Compendial water qualities

Table 8.3 gives an overview over the used water qualities and their typical areas of use. The qualities WFI and aqua purificata (AP) as well as the recently added aqua valde purificata (AVP) are defined in the pharmacopoeia.

The non-compendial qualities can vary due to regional characteristics since they depend on the quality of the incoming water and do not follow a standard specification. They can be used for cleaning purposes or fermentation of insensitive organisms.

Although increasing water quality is associated with increasing generation costs, it can be advantageous to use the same water quality for all areas in the facility. The reason is that each water type needs a separate distribution system that can be more expensive than just using one quality. Ultimately, the investment costs have to be balanced against the operating costs. For this reason the use of AP water for almost all applications except for the ones requiring WFI is very common. WFI is, for example, mandatory for the final rinsing in equipment cleaning

Different water qualities require separate distribution nets

Table 8.3 Water qualities and their areas of use.

Nomenclature	Properties	Typical area of use
Classified according to pharmacopoeia (compendial water)		
WFI (water for injection)	like AP but higher specifications regarding absence of microbial contaminations and endotoxins	final formulation for parenterals; final rinse in tank and equipment cleaning
AVP (highly purified water)	like AP but higher specifications regarding absence of microbial contaminations and endotoxins	final formulation for non-parenterals; purification of parenteral forms, CIP
AP (purified water)	specifications for conductivity, total organic carbon, microorganisms (colony-forming units), endotoxins, pH value, diverse chemical parameters	final formulation for non-parenterals; purification of parenteral forms, CIP
Not classified according to pharmacopoeia (non-compendial water)		
Non-compendial suitable water – usually named after the last treatment step (e.g. deionized water or water from reverse osmosis)	drinking water quality and reduced ion content, disinfectants, microorganisms, particles, etc.	CIP, fermentation
Drinking water	Specifications for particles, hardness minerals, heavy metals, organic substances, other additives, microorganisms	CIP, sometimes fermentation

(Section 8.5) or final formulation. The use of untreated drinking water can no longer be observed in modern facilities.

Water Generation: Pre-Treatment and Final Treatment On its way from the feed to the use point, the water passes different treatment steps.

Usually, the plant receives water from the municipal source with drinking water quality. Its origin determines what treatment the water needs to reach pharmaceutical quality. The final treatment step determines quality, yet its ability to reduce contaminants is limited and it even has to be protected from these contaminants. Consequently, water pre-treatment has to ensure sufficient reduction of contaminants. The final step and water source drive the pre-treatment of the water; typical steps are:

- Chlorination or ozonation for reducing microbial burden and subsequent dechlorination/deozonation.
- Particle filtration.
- Softening.
- pH adjustment.
- Desalting.

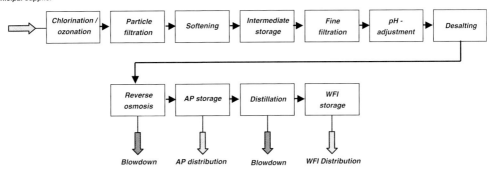

Figure 8.11 Typical plant for generating pharma-grade water from drinking water.

After these pre-treatment steps have been performed, the last treatment step can generate the desired quality. For AP, no special final step is obligatory, yet reverse osmosis or deionization by membrane or ion exchanger is very common.

In Europe, the final step for WFI generation has to be distillation. In the US, WFI can be produced by reverse osmosis; however, due to the limited usability outside the US this is not seen very often. The feed-water quality for final treatment is not regulated; mostly AP quality is used in order to ensure a high safety margin for depletion of contaminants.

Final step for WFI generation is regulated

Figure 8.11 schematically illustrates the typical stages of a water treatment plant. Starting with the intake of water from the municipal water supplier, the water passes several pre-treatment steps before being fed to the final treatment, and ultimately into the storage and distribution systems for AP and WFI. Typically, intermediate storage tanks can be found between the pre-treatment steps since the capacities of the equipment cannot always be adjusted exactly to fit each other. Furthermore, the tanks facilitate switching between equipment that is run in parallel, like filters, in order to perform cleaning or maintenance work.

Storage and Distribution Systems Storage and distribution systems for water have to be kept germ-free under all circumstances. Once a system is contaminated, such a source of impurity is extremely difficult to localize and eliminate. Since all manufacturing depends on the water system, water is the Achilles heel of production. Water purity is ensured by intensive monitoring and design specifications, which aim at both minimizing the contamination risk and ensuring that the system can be cleaned after having been contaminated. These criteria are:

Self-sanitizing hot storage	• The water should be stored under conditions that aggravate microbial growth (self-sanitizing hot storage at over 80 °C). Therefore, the system has to be heat-insulated and local cold spots in the tank as well as in the piping system have to be avoided.
Recirculating storage	• The water has to be constantly recirculated and may not stand in the ductwork. The pipes should be designed free of dead volume (6D rule, i.e. the branch pipe may not be longer than 6 diameters of the main pipe). Water under non-self-sanitizing conditions (i.e. no hot storage) should not remain standing in the tank for over 1 h, but be recirculated.
Self draining	• The piping system should be self-draining, that means having a slope to the draining spots.
	• The system should be steamable.
Subloops	• If the system has many points of use, several separate subloops should be established in order to keep the damage of potential contaminations limited. Moreover, subloops offer better hydrodynamic control.
Points of use	• The points of use are the spots that are open to the environment. Here, the danger of introducing contaminations is imminent. Points of use should be flushed before withdrawal and sampling.
Water cooling	The requirement to store the water under self-sanitizing conditions (i.e. to store it hot) stands in contrast to the commonly used temperature: the water for buffer and media preparation as well as many cleaning steps is needed at ambient temperature. Only certain CIP steps really need hot water. The necessary cooling can be performed at different sites: the whole loop as well as the storage tank can be cooled during withdrawal; point-of-use heat exchangers for local flow-through refrigeration are very common.

Despite of these constraints there are many different ways to design a water system. Determining design factors include:

- Regulatory requirements for the pharmaceutical form.
- Application (equipment cleaning or product manufacturing).
- Overall demand, concurrence of withdrawal and required temperature levels.
- Storage capacity, generation capacity, municipal water capacity and quality.
- Maintenance periods potentially resulting in the need for redundant installations.

Figure 8.12 provides an example for the essential parts of a supply installation of AP and WFI. The schematic also gives information about the design of subloops and use points with local heat exchangers. In addition to the water loops, the principal heat flows realizing the temperature levels are indicated.

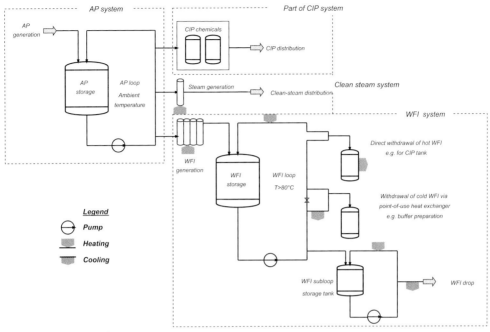

Figure 8.12 Elements of a water system, AP system, CIP system, WFI system and clean steam system. Use points are drawn schematically only. WFI is generated by distillation. AP is stored at ambient temperature. WFI storage at 80 °C.

AP water can indeed be stored at ambient temperature. It also becomes apparent that in regions with continuous cold withdrawal the temperature can stay below the desired level of 80 °C for longer periods. There are WFI loops that are run constantly at ambient temperature (ambient WFI, between 10 and 65 °C) which is closest to the use temperature, but provides the best living conditions for microorganisms. Cold loops (cold WFI) which are run between 4 and 10 °C can also be found. Again bacterial growth is aggravated under these conditions. The desired temperature can be achieved by mixing the hot and cold WFI. If the loop runs below 80 °C, it has to be frequently sanitized, which means heating the water over 80 °C. Cold WFI needs to be sanitized several hours weekly, while ambient WFI requires daily sanitization of several hours.

Storage at ambient temperature
Sanitisation

Automation   Water systems can be automated to a large extent. With the exception of sampling for microbial testing, no manual operation is necessary. Online analytics are available for conductivity, total organic carbon, pH value, ozone concentration, flow, temperature, and pressure and tank level.

Online analytics

Table 8.4 Phases of the qualification of a WFI system.

Phase	Target	Duration	Sampling effort	Production possible?
Operational qualification	identify system capacity develop SOPs for cleaning, operation and maintenance demonstrate that quality-compliant water can be generated and supplied	14–28 days	each use point daily	no
Performance qualification – phase 1	demonstrate that quantities can be supplied according to SOP specifications demonstrate that quality can be supplied according to SOP specifications	14–28 days	each use point daily	yes
Performance qualification – phase 2	demonstrate long-term stability demonstrate control of seasonal parameters	1 year	each use point weekly	yes

Use management

Therefore, the level of automation is usually high. Automation challenges occur if combinations of package units of different vendors are connected. Likewise, the control of withdrawal management is difficult, ensuring that there is always sufficient water for withdrawal and maintaining the required flow in the piping system.

Water systems are directly quality relevant

Qualification Water systems are 'direct impact' systems – a water-quality defect has a great potential to affect product quality adversely (Section 9.1.3). Therefore, water systems need to be qualified. Table 8.4 provides an overview over the effort for qualification of a water system. After having completed design and installation qualification (Section 6.1.3), the operational qualification is executed over a timeframe of 14–28 days. Note the high sampling and testing effort. It is essential for qualification that the sampling procedures are well trained, otherwise analytical artifacts can jeopardize successful completion. Production can start after operational qualification in parallel with performance qualification which extends over 1 year to account for seasonal fluctuations. Typical municipal water quality changes with the season.

The successful qualification demonstrates that the water system is good for routine production; after that the sampling effort can be reduced significantly.

8.4.1.2 Clean Steam

Clean steam is used for sterilization purposes. The main users are autoclaves for small parts and cleaning of piping and tanks.

The specification of clean steam is that its condensate has to have WFI quality. In most cases the steam is generated from AP or WFI by one-step distillation and fed into the steam pipe network. The network is constantly purged with fresh steam in order to keep the temperature of the ductwork high, impeding microbial growth in cold spots. It is therefore necessary to design the system in such a way that the steam condensate that forms by cooling in the inner side of the pipes runs by itself to the deepest spots of the network and is purged through steam traps. The system has to be installed with a suitable slope.

8.4.1.3 Gases and Process Air

Gases and process air are mainly used to aerate fermentation. They are also used to pressurize tanks to push liquids out, circumventing the shear stress caused by mechanical pumps. One can also find liquid surfaces covered with gas blankets to support sterile processing.

Carbon dioxide, nitrogen, oxygen and sterile air are mainly used. While nitrogen and oxygen can be used to mix an air-like atmosphere, carbon dioxide is used to adjust the pH value during fermentation. Sterile air can be used for aeration and to operate control instruments.

Gases can be purchased in sufficient quality from specialized suppliers. On site they often have to be dehumidified and sterilized. Pressurized air is mostly produced using oil-free compressors.

8.4.2 Plant Utilities

The main plant utilities are cooling and heating media and electricity.

Heating and refrigeration of process equipment and HVAC units usually is supported by heat-transfer systems containing thermo oil or a water–glycol mixture, depending on the temperature level. The temperatures of these systems are maintained by primary energy-media like heating steam, cooling-tower water or chiller loops. The additional step of the heat-transfer systems helps to buffer the transferred energy separating the 'dirty' energy generation from the 'clean' process environment.

Tables 8.5 and 8.6 give an overview of the principal users of heating and cooling media as well as their generation from the energy sources (oil/gas and electricity). Liquid nitrogen for deep-freezing applications is usually sourced from specialized companies.

Figure 8.13 illustrates the heating and cooling loops. Due to economical reasons often more than one temperature level for refrigeration purposes can be found.

The electrical supply of a plant should always be installed with high reliability. Considering the list of users, it becomes obvious that electricity plays a key role for maintaining the GMP status, process control and work safety:

Electrical supply

Table 8.5 Utility users and suppliers.

User	Use point	Utility
Air conditioning	cooling register	cooling loop (<10 °C)
	heating register	heating loop (>60 °C)
AP/WFI water	loop heating	steam
	loop cooling	cooling loop (>30 °C)
	distillation/condensation	heating steam, cooling loop (>30 °C)
	point-of-use heat exchanger	cooling loop (>30 °C), cooling loop (<10 °C)
Clean steam generator	still	heating steam
	steam trap	cooling loop (>30 °C)
Waste water treatment	thermal inactivation	heating steam, cooling loop (>30 °C)
Process	equipment cooling/heating	heating loop (>60 °C), cooling loop (>30 °C), cooling loop (<10 °C)
Cell bank storage	storage tanks	liquid nitrogen

- HVAC units.
- Pumps and compressors for water and energy loops and their generation.
- Process pumps.
- Illumination.
- Control systems for process, media and premises.

Reliability

The chain of electrical supply has to be checked for failure risks, which should be minimized as far as possible. Since a failure of the main supply can never be ruled out, a risk analysis should identify the critical systems depending on electricity; for those, 100% redundant emergency supply by batteries or an independent power aggregate should be installed.

Table 8.6 Generation of utilities.

Utility	Generated from	Primary energy source
Heating steam	steam generator (oil/gas fired)	oil/gas
Heating loop (>60 °C)	heat transfer from steam to heating loop	oil/gas
Cooling loop (>30 °C)	electrical cooling tower cools cooling-tower water, transfer to cooling loop	electricity
Cooling loop (<10 °C)	electrical chiller cools refrigerant, transfer to cooling loop	electricity

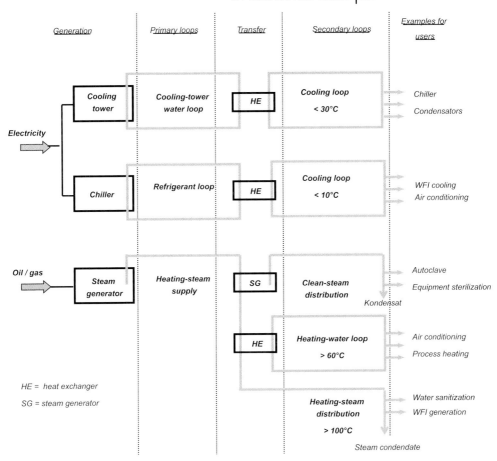

Figure 8.13 Heating and cooling loops in a production plant.

8.4.3
Waste Management

Biological production generates a lot of waste water. The main focus lies on the biological substances in these water streams. Water from fermentation usually contains genetically modified organisms as well as DNA and RNA and traces of therapeutic proteins.

A risk assessment should identify to what extent these substances are hazardous for health or the environment. Depending on this, the waste is classified into a biological safety class that determines the waste treatment.

Most proteins degrade outside of their controlled process environment and do not jeopardize health when coming into contact with the skin or when being inhaled or swallowed. Also potentially released modified genes are not able to infect other natural living cells. Mammalian cells are not able to survive outside the bioreactor. Bacteria and yeasts theoretically could propagate; however, it is unlikely that they would survive the conditions of the waste water sewer, the waste treatment plant and the discharge into free water. In comparison to traditional industrial waste waters, the hazardous potential of biological processes is low.

Inactivation of residuals

Despite the relative harmlessness, the discharge of genetically modified organism-containing waste streams is limited. Usually, waste streams are biologically inactivated; this can be limited to the waste streams from fermentation and isolation since it is only here that genetically modified organisms are present. In the purification area they are already separated. Water is inactivated by collecting it in tanks and either heating (thermal inactivation) or exposing it to extreme pH shifts by alternately adding acid and base causing cell lysis (chemical inactivation, pH swing).

In most cases this treatment is sufficient to purge the waste water into the municipal sewer system since the concentration of hazardous substances is very low. If larger amounts of organic solvents are used (e.g. for HPLC chromatography), separate collection of these streams may become necessary. These waste streams have to be put through the usual industrial treatment routes like recycling or incineration (Table 8.7).

With respect to genetically modified organism contaminations, for solid waste the same holds true as for liquid waste. The contaminated solid disposables are inactivated in an autoclave prior to leaving the facility and then enter the common treatment routes.

Exhaust air

Exhaust air from fermentations can become a real olfactory nuisance, but usually is not hazardous to health or environment. Depending on the location of the facility and municipal regulations, it may be necessary to deodorize the air by exhaust air scrubbers.

Table 8.7 Overview over types of waste and usual disposal routes.

Type	Genetically modified organisms contact?	Example	Usual treatment
Solid waste	yes	filter from fermentation	autoclave before waste leaves the building
	no	disposables from purification	normal disposal routes (incineration)
Waste water	yes	supernatant, cell debris, CIP waste from fermentation	thermal or chemical inactivation in building, afterwards further industrial treatment or direct draining to municipal sewer
	no	buffer residues and CIP waste water from purification	no inactivation necessary; depending on freight of substances hazardous to water: industrial treatment or direct draining to municipal sewer
Exhaust air	not applicable	fermentation exhaust air	exhaust air scrubber for odor reduction
		exhaust air containing organic solvents	exhaust air scrubber for reduction of solvent emissions

8.5 Equipment Cleaning

Installations for equipment cleaning are amongst the most elaborate in a biotechnological manufacturing plant. The piping of a fixed tank largely consists of cleaning installations and many utilities are used predominantly for cleaning purposes. Generally, all product-contacting parts have to be cleaned, including tanks, pipes and process equipment (e.g. chromatography columns or filtration units).

The equipment is cleaned after each production run. If it has not been used for a longer time, it can become necessary to clean before use. The cleaning process after use typically starts by flushing with cold water to deplete rough impurities. The water has to be cold to avoid proteins denaturing and sticking to the vessel walls. Often, as a second step, hot sodium hydroxide (caustic) is used, which is flushed with hot water afterwards. The caustic is even more efficiently removed if the water step is again followed by an acidic step. The last step (final rinse) is made with WFI. The success of the cleaning process is demonstrated by analyzing this final rinse water after it has left the tank. The complete cleaning protocol is: cold water – caustic – hot water – phosphoric acid – hot water (final rinse with WFI). Prior-to-use cleaning of previously cleaned tanks is usually performed with suitable water like AP or WFI. In order to guarantee sterile conditions – especially in fermentation – pipes and tanks are steamed with clean steam after cleaning. *Cleaning process*

Final Rinse

The cleaning process is successful if all spots in the tank are free of residues. This is achieved by distributing the cleaning solution through sprayballs all over the vessel interior. The solution can either run once through the tanks, or be recirculated multiple times through the sprayballs. Cleaning is the largest water consumer in a plant. It is a very important measure to assure quality and has to be validated.

There are three widely used abbreviations for different types of cleaning:

- *CIP (cleaning in place)*. General term for automated cleaning, yet 'CIP' is also used in a narrower sense to describe the cleaning of fixed tanks by fixed cleaning installations. *CIP*

- *COP (cleaning out of place)*. Denominates the cleaning of mobile tanks which are transferred and connected to a shared CIP station with fixed cleaning installations. *COP*

- *SIP*. There are different interpretations of 'SIP'. Most often it is meant to mean 'steaming in place', which – in analogy to CIP – describes the steaming of hard-piped equipment and ductwork with clean steam. One can also find 'sterilization in place' or 'sanitization in place' which is essentially the same. *SIP*

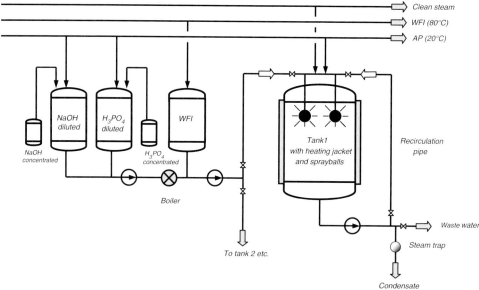

Figure 8.14 Typical configuration of a CIP/SIP system. NaOH = sodium hydroxide; H_3PO_4 = phosphoric acid.

Figure 8.14 shows a typical equipment configuration for a CIP/SIP installation. The tank to be cleaned has sprayballs and recirculation lines. It can be directly fed with purified water at ambient temperature for the first flush. The following steps are served by the installation (tanks for WFI, caustic, phosphoric acid) shown on the left side of the figure (CIP skid). The chemicals are supplied in concentrated form and diluted on site with AP. Before going into tank 1, the cleaning media are heated in the flow heater. The final rinse is made with WFI, which – for reasons of contamination safety and capacity – is not fed directly from the loop but via a break tank. The cleaning installation can be used for other tanks, for which the feed has to be shifted to tank 2. The tank is steamed directly from the steam pipe. This configuration is only one of many different possibilities to design a CIP installation.

8.6
Clean-Rooms

Biotechnological production processes are mostly performed in cleanrooms. Section 6.1.2 has highlighted the special significance of such premises. Cleanliness of ambient air is a safeguard for the absence of contaminations in the product, be it particles or microorganisms. If the processes are closed, the clean-room air forms another barrier against the access of contaminations into the core process; if processes are open, the product has direct contact with the room air. Design and finishing of clean-rooms aims at good cleanability, avoiding hidden corners for enhancing cleanability and selecting materials of construction that release as few particles as possible.

Based on Sections 8.6 and 8.2, in which the basics of the GMP-compliant clean-room design have already been described, the following sections briefly describe the technical installations of clean-rooms and HVAC installations.

8.6.1
Separation of Zones by Clean-Room Design

The separation of the clean-room area into zones with different cleanliness should guarantee a stepwise – therefore safe – approach from the normal to the clean environment. The zones can be separated by different measures:

- Displacement flow is used in the direct vicinity of surfaces to be protected. By suitable technical design, the air flux is directed in such a way that particles potentially emerging from other areas cannot reach the area to be protected since they are pushed away by the air blowing against them. Typical examples are the realization of laminar flows through ceiling-mounted air outlets, the use of biological workbenches for sterile open processes or the use of mobile air showers for connecting flexible hoses. In a wider sense, the realization of a pressure difference between two rooms in which the room with higher cleanliness class is operated at overpressure relative to the rooms classified lower also generates a displacement flow (Figure 8.7). *Displacement flow*

- Different zones can simply be separated by physical barriers like walls. It is important that these walls are air-tight, easy to clean and allow access to the working area. Isolator technology is an example of a physical barrier, but also transparent curtains of an air shower for separating class A from class B in sterile manufacturing. *Physical barriers*

- The most sensitive spots between two zones lie where material or persons cross the zone limit. Goods that are transferred into the cleaner environment have to be cleaned without contaminating the cleaner area, which is achieved by airlock systems. Airlocks maintain the pressure difference between the rooms; at the same time they are a physical barrier and provide a defined space for depletion of potential contaminations. *Airlock systems*

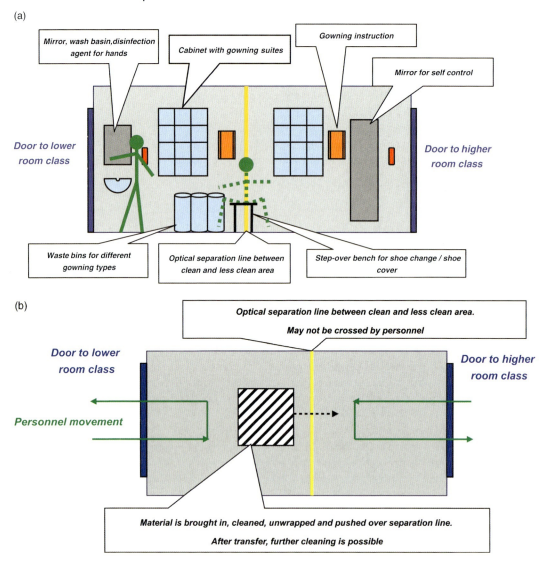

Figure 8.15 (a) Typical personnel airlock (side view). (b) Procedure for bringing in material through the material airlock (top view).

Figure 8.15 illustrates typical airlock systems for personnel and material transfer. The airlock doors are interlocked in such a way that only one can be opened at the time. Figure 8.15(a) shows a personnel airlock. The person entering the clean-room has to gown-up into clean-room garments (Table 6.2) and disinfect their hands in the less clean area. After that, they transfer over the step-over bench putting on another pair of overshoes. The bench marks the border between the less and the more clean area of the airlock. Prior to opening the clean-room door there is a possibility to check the outfit by looking in a mirror. In the material airlock (Figure 8.15b) the material is brought into the airlock from the less clean side and wiped down and/or disinfected. The person leaves the airlock to the less clean side. After a certain time period, which should enable the air flow to reduce air particle freight, a person from the cleaner area enters the airlock and brings the material into the clean-room. Also here, cleaning can precede the transfer. Both operators do not leave their areas which are separated by the yellow line on the airlock floor.

8.6.2
Finishing of Floors, Walls and Ceilings

Floors, walls and ceilings have to be air-tight and easy to clean. They may not release particles and, moreover, should be suitable to accommodate installations like control panels, piping penetrations, windows or air in/outlets. Floors struggle with additional requirements regarding non-slip and electrical conductivity in areas with ex-classification. All in all, walls and floors should be plain, without corners and without cracks or splices.

The walls separate the clean-rooms from surrounding technical installations; they are pressurized above ambient pressure. To maintain this pressure they have to be as air-tight as possible. Behind the walls, technical areas are located which should be accessed from the unclassified area only. In older plants it can sometimes be observed that technical installations are maintained from the clean-room area. In modern plants this is avoided due to the contamination risk and the need for additional training of technical staff.

Ceiling-mounted installations like air-inlet filters and illumination are relatively high maintenance. For maintenance and later building modifications, the ceiling should be walkable. *Ceiling installations*

Maximal flexibility in clean-room design can be achieved by building the clean-rooms with removable light walls in an outer building shell providing the architectural structure. With this concept the clean-rooms can be redesigned later without affecting the building structure. Different suppliers offer system walls for this so-called 'room-in-room concept' (Figure 8.16). *Room-in-room concept*

8.6.3
HVAC Installations

HVAC stands for heating, ventilation and air conditioning, and denotes the entity of technical installations for aeration, humidity *HVAC*

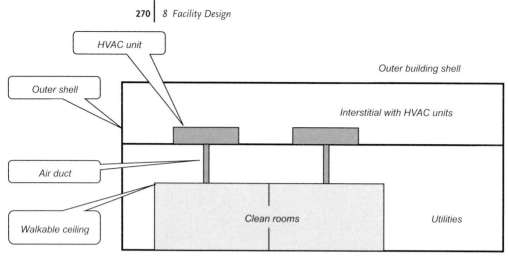

Figure 8.16 Example of a room-in-room concept.

control, temperature and pressure control, and air-flow control in clean-rooms.

In the clean-room, specially designed air in/outlets take care for the desired airflow pattern. Depending on the cleanliness class, air inlets are designed for turbulent or laminar flow (ceiling diffusers, perforated ceiling cassettes). Before flowing into the room, the air is purified using HEPA filters (high-efficiency particulate air). If the supply stream is filtered it is denoted as 'HEPA-in'; in certain cases also the outgoing air is filtered, which is called 'HEPA-out'. The air ducts connect these filters that are close to the clean-rooms with the HVAC units. The latter recirculate, filter, cool and heat the air, and control its humidity. In order to separate different building areas it is mandatory that each section has its own HVAC unit. Due to its large volume, the air ductwork occupies a significant part of the building volume.

HEPA filter

Room air has to fulfill certain requirements regarding temperature, humidity, pressure and particle load. This is achieved by exchanging the room air continuously; the number of room air changes per hour is one important parameter for the achievable room air quality. Recirculating air is brought back into the room through the HVAC unit; fresh outside air is added here as needed. See Figure 8.17.

Room air exchange

8.6.4
Qualification

Clean-room conditions have to be qualified. That means it has to be demonstrated that clean-rooms comply with their specifications. In the

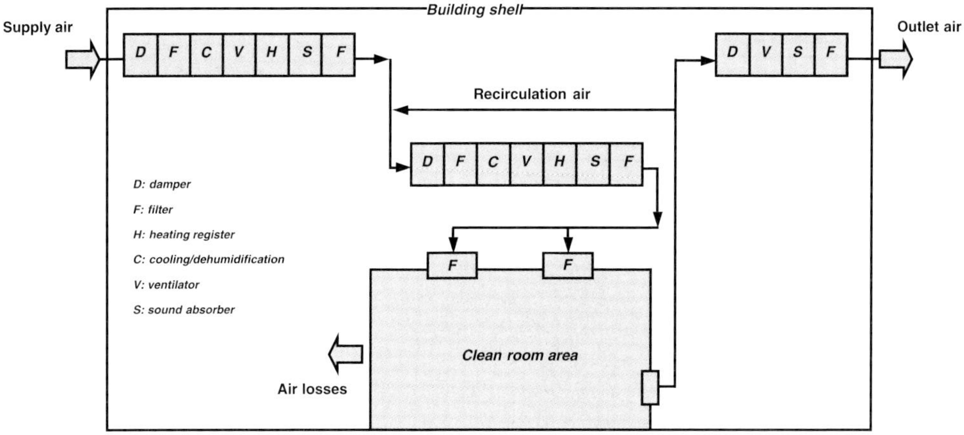

Figure 8.17 Typical configuration of a HVAC system for clean-rooms.

Supply air — D F C V H S F — Building shell — Outlet air — D V S F

Recirculation air

D F C V H S F

Clean room area

Air losses

D: damper
F: filter
H: heating register
C: cooling/dehumidification
V: ventilator
S: sound absorber

operational quantification phase, flow patterns are visualized with the help of smoke studies. Particle counts are performed as well as investigations concerning microbial contaminations. Maintenance of the specified conditions has to be further demonstrated during routine operation of the rooms. Deviations from specifications have to be handled by a deviation management process. *Flow visualization*

8.7
Automation

Automation means to replace human interventions by technical installations and controls. Automation extends over different areas:

- Replacement of manual work by machines or equipment.
- Control of processes by programmable control systems instead of human interventions.
- Surveillance of processes by on-line analytics instead of laboratory-based testing.
- Electronic instead of paper-based process documentation.

Generally, automation results in higher process consistency and avoidance of the human 'risk factor'. In turn it is based on a deep process understanding that recognizes correlations between the measurable parameters, the manipulable control parameters and their impact on product quality. An important aspect of automation is the availability of suitable analytical methods that allow assessing the process in adequate time and sufficient quality. *Automation capability*

The degree of automation of biotechnological processes is very diverse. Completely avoiding manual labor – also for cleaning – means that the facility must be totally hard-piped, requiring high investments (Section 8.2). Typical examples for hardly avoidable manual interventions are crystallization processes after which the crystalline phase has to be processed manually due to its consistency. Another example is the collection of chromatography fractions, their testing in the laboratory and the subsequent processing. In this case a suitable analytical method for efficient and fully automated testing is missing. The common discontinuous batch processing and the high sensitivity of the processes as to variations in their parameters make automation even more challenging.

Despite these obstacles, there are fully automated facilities. Especially in the fermentation area, in which large liquid quantities have to be processed under sterile conditions, automation is well established. Here, the usual configuration of a centralized process control system which controls, monitors and documents all steps of a process can be found.

Partial automation

In contrast to fully automated plants with an integrated process control system there are partly automated systems in which subprocesses are controlled by their own local systems. Examples for this configuration are chromatography or fermentation skids purchased from suppliers that include control systems (package units). The control panels can be accessed locally. In this case documentation can be paper based; however, if the systems are compatible with centralized documentation software, at least documentation can be electronic. If fast product changes are desired it is often easier to operate the facility completely decentralized, avoiding the effort of programming and computer validation. On the other hand, one has to accept higher labor costs.

Package units

Electronic batch record

The full electronic process documentation (electronic batch recording) prevails only slowly due to the relative inflexibility and validation challenges (Section 6.1.5).

A significant advantage of automated process monitoring is the seamless recording of parameter and specification excursions. Automatic alarms and statistical process analysis enhance failure analysis and increase process safety.

In addition to process automation there are other important electronic systems:

Building automation system

- The building automation system controls facility installations including HVAC, particle counters, and water and utility systems. With the exception of microbial monitoring and WFI analytics, this can be done fully automatically.

Laboratory information management system

- The laboratory information management system serves to archive analytical data from IPCs and release-relevant data from quality control laboratories.

There is a clear trend towards more automation in biotechnology, since the advantages of process consistency and safety prevail especially for large productions. However, the path to more automated solutions requires a stronger standardization of processes and technical installations.

8.8
Quality Control Laboratories

The responsibilities of quality control include testing incoming material, IPCs, product-release analytics, stability testing and environmental monitoring. Investments in premises, equipment and staff to manage all these tasks can become significant. Special consideration should be given to the increased sampling schedule during qualification, and cleaning and process validation, since analytical capacities can become limiting in these phases.

The design of laboratories will not be further detailed here. Obviously, demand planning is essential, providing a good projection of number and types of analysis needed. Apart from the special requirements for handling of biological substances, quality control laboratories in biotechnological production do not differ from other quality control laboratories.

8.9
Location Factors

The question where the plant shall be located is addressed at the beginning of an investment decision for a production plant. This section discusses some important location factors for biotechnological manufacturing plants.

Cost

- Prices for fixed assets. The prices for fixed assets can differ regionally. *Fixed assets* While special equipment is mostly available at world-market prices, fluctuations occur due to differences in costs for labor, construction material and installations. Usually the impact of these differences is not very high and therefore not decisive.

- Tax incentives and subsidies granted for attracting investors can *Subsidies* make up a big part of the investment, and can therefore be a decisive factor in the choice of the location.

Labor costs
- Labor costs are essential if processes rely extensively on manual labor. In the pharmaceutical industry this can be the case in the final manufacturing, where 100% checks of vials or packaging processes with relatively low qualified personnel are performed.

Currency risk
- Independence of currency risk. Currency risks can be mitigated if production is where the market is. An example may be provided for a drug which manufacture in Europe costs €15. The market price for the American consumer is US$50 and the exchange rate US$1.2 for €1. The difference between manufacturing costs and sales price is €26.7 (US$32). If the rate changes to US$1.4 per €1 the remaining revenue is only about €20.7 (US$29). If the drug would have been manufactured in the US, the profit for the company would have been basically unchanged.

In recent years, financial aspects have motivated relocation of chemical, API and pharmaceutical manufacturing to the Asian region. This is not yet the case for biotechnology, which could be attributable to the high sales prices of biotech drugs, the demanding technologies and the legal questions around protection of intellectual property. With biotechnology evolving more and more into a mature industry, the cost aspect will also play a larger role for biotechs in the future.

Personnel

The question for sufficiently qualified personnel plays a major role for processes with a high share of manual labor. The argument about the labor force can be used in two directions: on the one hand, it is advantageous to settle in an area with highly specialized labor potential; on the other hand, this potential has been created in most cases by industries competing in the same field. The competition in these regions can result in a higher salary level and increased staff fluctuation.

Permitting

Legal foundations and practical handling of environmental and building permitting processes can have a great influence on investment decisions. If the responsibility for the decision is not with a reliable authority, the investment can be linked to hardly calculable risks for the company.

Synergies with Existing Facilities or Units

In many cases the realization of synergies with existing facilities or units like quality control laboratories or process development is a decisive argument for site selection. The scientific and technical knowhow established at one site, as well as the existing (and positive) experience with regulatory authorities, minimizes the risk and the volume of a capital investment significantly.

Synergies can also be identified with foreign companies that are close and that offer a service needed to run the plant. In the biotech sector, local clusters of cooperating highly specialized companies show their strengths.   *Biotechnology cluster*

Logistics

Since the freight volumes of pharmaceutical products are usually very small, logistical aspects do not play a role in determining the manufacturing location. Only for products with an exceptionally short shelf-life (several hours to days) like radio-pharmaceuticals or cellular therapeutics can vicinity of international airports be important.

Knowhow and Intellectual Property Protection

The safety of company knowhow is an important factor for site selection. The risk that knowhow – be it patented or not – gets copied and used for commercial purposes has in the past hindered high-tech industry settling in some countries outside the western economic hemisphere.

Other Risks

Risks such as the probability of natural catastrophes (earthquakes, hurricanes, flooding) or political instability of a country have to be considered.

Market Access

Due to strategic reasons, like getting market access, it can be important to invest in or close to the target market.

Language and Culture

Obviously, language is one of the most dominant location factors. A company would maybe easily import goods from a foreign country; however, investing into fixed assets without being anchored in the culture and the language remains a challenge. Independent of the location, English documentation or authorized translations are a prerequisite for obtaining a import license by the US FDA, which controls the largest single market worldwide.

9
Planning, Construction and Commissioning of a Manufacturing Plant

Planning, construction and commissioning of a manufacturing plant constitute one of the most demanding and interesting areas of biotechnological production.

Each manufacturing plant is custom-made and designed individually to accommodate the projected demands of the company. The challenge is to find the compromise between functionality and flexibility, on the one side, and the financial constraints as to capital investment and operating costs, on the other. The planning process has to anticipate future trends in GMP requirements, process technology and capacity demand, and create the basis for a long operating time and optimized usage of the asset.

The present chapter introduces the principles of an investment project. It starts in Section 9.1 with an explanation of the phases of the planning, construction and commissioning process. Section 9.2 discusses typical time schedules of engineering projects, Section 9.3 elaborates on cost estimates and Section 9.4 is dedicated to the organization of such a project. Since engineering projects are usually executed with specialized external engineering companies, one focus of this chapter lies on how to build and maintain contractual relationships.

Section 9.5 discusses success factors for the execution of an engineering project. Section 9.6 deals with the legal framework for the construction of an industrial plant.

9.1
Steps of the Engineering Project

The pathway to plant construction is paved with technical and organizational challenges. The principal steps are:

- Identification of user requirements. This includes the collection of data as to the desired capacity up to qualification requirements. *User requirements*

- Specification of building and equipment. The realization of user requirements results in the specification of the technical and construction components. The interplay of these components forms the functional plant. Understanding and anticipating this technical complexity is a prerequisite for completing the project within schedule and budget, and with the desired functionality. *(Technical specifications)*

- Cost planning and control. In the planning phase building costs are estimated that have to be met in the later realization phase. *(Cost planning)*

- Implementation and management of an organization for planning and execution of the project. This ranges from small teams for concept engineering to the coordination of associated subcontractors. *(Organization and execution)*

These targets are achieved by a methodology that balances the speed of progression between the slow and safe sequential route and the quicker and riskier parallel route.

The method splits the engineering project in three distinct phases: planning, construction, and commissioning and qualification. The subsequent validation usually falls under the responsibility of the production or quality unit and is not considered as being a part of engineering. However, it constitutes the actual endpoint of a successful construction project, since only in validation it is demonstrated that the investment is capable of delivering the desired product quality and quantity. The target and main workflows of these three phases will be explained in the following sections.

9.1.1
Planning

The planning of a facility ranges from creating user requirements up to the final purchasing documents. User requirements could look like this:

- Five kilograms of frozen drug substance shall be provided according to process A from timepoint X on.
- The plant shall be capable of extending the capacity of process A by 100% and accommodating process B.
- The facility has to comply with pharmaceutical regulatory guidelines of the US, the EU and Japan, and with applicable environmental, safety and occupational health laws as well as building regulations.
- Ideally the facility should be built on an existing site.
- Storage and laboratory capacities are to be established.
- Maximum cost of goods: €400 per gram of product from process A.
- Maximum budget for capital investment: €70 million.

Based on these rough wishes, more defined requirements as well as detailed equipment and construction drawings emerge in the planning phase.

In the course of the planning process, the cost estimate for the plant becomes more and more accurate, and serves as the basis for budgeting before construction activities begin.

Planning is structured in three phases: concept engineering, basic engineering and detail engineering. The accuracy of both technical documents and the cost estimates increase with each of these phases. Table 9.1 provides an overview over the phases and their principal targets. *Concept, basic and detail engineering*

The most momentous decisions of the design process are taken in the concept phase. If a wrong decision taken in this phase propagates through to plant operation, the follow-up costs can be significant. In comparison, a planning mistake can be corrected relatively easy in the detail specification. The earlier a mistake is detected, the smaller its consequences. Therefore, the concept should be developed with utmost diligence; mostly small teams of experienced personnel are deployed here. The effort increases when entering the basic engineering, but the financial risk is still fairly low. For detail engineering, large and costly teams are formed. Figure 9.1 schematically illustrates this cost–risk relation.

The quality of the design process is assured by processes like change management and quality reviews that are performed throughout the entire engineering phase. At the end of the detail engineering stands the final cost estimate, the budget approval for binding orders and the starting signal for construction.

9.1.2
Construction

The construction of a plant starts with earthworks. After erecting the structure, equipment and larger machines are brought in. They are followed by ductwork and electrical installations, and finally the interior work. Contractor firms are selected prior to construction; during construction execution, experience and organizing ability are important to ensure that timelines and cost stay in the projected framework.

Meshing activities, in particular, cause overall delays if they do not work as planned. One option to reduce interdependencies is modular construction (Section 8.3.2). The phase of construction lasts 12–18 months and concludes with mechanical completion; it is followed by the functional check of the technical systems. *Mechanical completion*

9.1.3
Commissioning, Qualification, Validation

Functionality of the facility is checked in three steps:

Table 9.1 Phases of facility engineering.

Phase	Target	Duration	Main documents (deliverables)
Feasibility study	get first idea of technical feasibility, rough cost estimate ±50%	NA	study report, target definition[a] (user requirement specification[b]) process: process description organization: project milestones
Concept engineering	cost estimate ±30%	3–6 months	basis of design[c] technology: process flow diagram[d], list of main equipment process: material and energy balances, utility requirements building: rough layout, site selection organization: engineering team, contractor pre-selection, estimate of planning costs, schedule of activities
Basic engineering	cost estimate ±15%	4–6 months	functional description and design specification technology: piping and instrumentation diagram[e], equipment lists, rough specifications[f], functional descriptions, automation system, room book[g] process: material characterization relevant for regulatory permission (toxicology, biological assessment, emission assessment) building: floor plans, equipment layout qualification: validation master plan[h], GMP flows organization: detailed execution plan, documentation for regulatory permits
Detail engineering	cost estimate ±5%, construction plans, most accurate cost estimate before approval of financial funding	6–8 months	complete plant design, 'ready to order' technology: purchasing documentation, isometric drawings (piping) building: construction drawings qualification: qualification and validation protocols, design qualification

[a] Planning document that describes: what is needed?
[b] Planning document that describes: how should it function?
[c] Planning document that describes: how should it be designed?
[d] Graphic schematic of the process showing important steps as well as details concerning material and energy flows between the steps.
[e] More detailed diagram than the process flow diagram. Contains all apparatus, valves, pipes and required pipe slopes, material specifications, measuring instruments, information on process controls, and safety-relevant alarm spots.
[f] Description of apparatus, valves and plant elements regarding dimensions, materials of construction, surface finishing, internals and attachments
[g] Description of rooms with requirements regarding air quality and cleanliness class. Additionally, it contains aspects relevant for occupational health and energy loads inside the building
[h] High-level description of required validation work

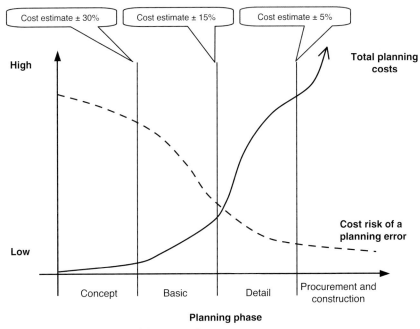

Figure 9.1 Planning costs and failure cost risk in the planning phases.

- The technical functionality of the system is checked in commissioning. For example, a tank is filled with water and a test is performed whether pumps, stirrer, valves, heating installations and control systems work according to the desired function. This check follows previously agreed acceptance programs. Commissioning is controlled by the technical team of the ordering party.

Commissioning

- Qualification is an extended commissioning process for facility parts that are considered to be critical for product quality. The difference from commissioning consists of the extended documentation of testing procedures and failure corrections. Documented evidence is created that the plant has been constructed as specified. The quality department of the company is involved and carries a part of the responsibility for qualification.

Qualification

- Validation demonstrates that the plant can indeed generate the desired results. Hence, validation is performed with 'real' product. The responsibility for validation lies in the hands of the production unit since only there is the necessary knowhow available.

Validation

Sections 6.1.3 and 6.1.4 elaborate on the differences between qualification and validation; Section 8.4.1.1 lists the steps for the qualification of a water system.

Since qualifying means more work than commissioning, it is worthwhile reducing the number of quality-critical systems as far as possible. The identification of these systems starts very early during process development. The process description should contain critical process parameters and their acceptance criteria as well as alarm thresholds. These limits, which are based on laboratory-scale data, have to be aligned with the technical equipment of large-scale production.

Critical parameter

Critical process parameters point to the critical technical systems in the plant: the latter are directly responsible for the control of critical parameters. Apart from these systems there are others with a direct influence on product quality. For example, the water used is not a critical process parameter, yet an error in the water supply has a direct impact on product quality. The systems relevant for product quality are denoted as 'direct impact' systems (ISPE, 2001). Other technical systems are 'indirect impact' systems. These include the primary heating and cooling loops that do not come into contact with product tanks. The 'direct impact' systems have to be qualified, the 'indirect impact' systems not; the latter have to commissioned.

Critical technical systems

Direct impact system

Whether a system belongs to the one or the other category is evaluated by assessing its projected quality impact during routine production, the ability to control, the efficiency of the alarm systems and the potential consequences of a system failure. This procedure of identifying quality-critical spots early on is called the 'enhanced design

---

Box 9.1

Example for the identification of critical parameters and qualification

Process development finds out that the protein solution has to be maintained in a temperature window in order to avoid that a thermally unstable protein degrading at higher temperature or agglomerating at lower temperatures. The target temperature is 30 °C and it may only deviate by 3 °C up- and downwards. This implies that the allowable surface temperature of the storage tanks and pipes is limited. The critical parameter 'temperature' has been defined with the acceptance limits: minimum 27 °C, maximum = 33 °C. Since these values may not be exceeded, an alarm has to be triggered before reaching these limits in order to initiate adequate counter measures.

Here, temperature measurement and the jacket heating of the tank are the critical systems. The heating medium may not exceed a certain temperature to avoid that the surface temperature harms the protein. The suitability of these systems is checked in design qualification; during qualification the systems are checked with water.

review'. It starts with design qualification that allows reviewing the technical documents for quality-critical systems and continues with the usual qualification cascade (installational, operational and performance; Section 6.1.3). *Enhanced design review*

In many cases, plant commissioning is the contractual acceptance of the delivered service. In order to save time, the technical system often undergoes a first check at the assembly site ('factory acceptance test') and after the installation on site an additional check ('site acceptance test'). *Factory acceptance test/ Site acceptance test*

Since most technical systems are functionally interlinked, commissioning and qualification have to follow a certain sequence. Qualification usually starts with building systems and clean media. This is followed by equipment and computer systems. Validation starts with analytical methods followed by computer validation, and ends with cleaning and process validation.

Qualification and validation are GMP-relevant steps, which means that documented evidence for their execution has to be created. In detail, these documents are the validation master plan, qualification and validation protocols and associated reports, as well as a failure documentation listing and commenting on the observed deviations from the expected course.

The personnel effort during qualification and validation is fairly high. In validation, a huge analytical effort is added to that. Due to their general technical character, qualification activities can be supported by an external workforce; however, validation requires certain specialized knowhow that cannot be easily provided by external staff. Also, the actual process knowhow lies here, which often should not be disclosed to individuals foreign to the company. *Personnel effort*

Validation starts with first technical runs (also called 'engineering runs') to gain initial experience with the product. After that the actual validation starts. A process is considered to be validated if three consecutive runs result in the desired product without going through major deviations. These three conformance runs (or 'consistency batches') are a prerequisite for the submission of the documents to obtain regulatory approval for manufacturing. *Engineering runs*

Validation runs

9.2 Project Schedules

Time and execution schedules are the most important organizational tools of plant design and construction. They illustrate:

- When and which resources have to be provided.
- Which activity depends on other activities.
- When the plant is going to be ready.

Critical path

The essential benefit of the plan is that it illustrates the functional relationships and dependencies of the individual disciplines.

In the interest of pushing the timepoint of investments as close as possible towards the generation of sales (Section 11.1.1), it would be beneficial to work in parallel on the activities in order to shorten the engineering and construction time. However, due to technical reasons and in order to mitigate financial risks, many activities are performed sequentially. The chain of these unconditionally sequential activities is called the 'critical path'. Activities along this path are rate limiting for the entire project.

Lead times

An important rate-limiting step is the purchasing of special equipment; equipment can have lead times of between 6 and 12 months (long-lead items). Therefore, these parts are often ordered upfront and the project plan is hurried ahead in order to achieve the overall goal.

Figure 9.2 shows a typical project schedule for the engineering, construction and licensing of a production plant. The major activities are indicated as well as the important milestones. The entire timeframe

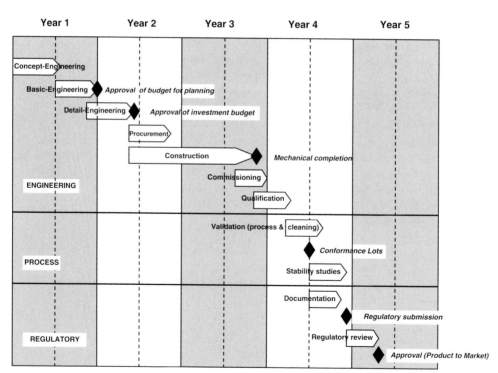

Figure 9.2 Typical project schedule for the planning and construction of a production plant.

spans 4.5 years. The activities can be arranged into technical, process-driven and regulatory activities. Stability tests demonstrate that the product from the validation runs indeed has the expected shelf-life. Usually comparisons with older research data and 3–6 months of real-time data are sufficient at the timepoint of submission. Longer-lasting studies can be filed afterwards during the regulatory review period.

9.3
Cost Estimates

One of the most important tasks of engineering is estimating the expected capital investment costs.

The cost estimate becomes more and more accurate with planning progressing. Table 9.1 shows that only rough ideas for the plant exist in the feasibility study; the accuracy of the cost estimate corresponds to this fuzzy picture, which in this phase is primarily based on experience and benchmark figures.

In concept engineering, technical documents are added. The equipment list and the rough layout of the building are essential foundations of the next step of the estimate. In this phase, requests for non-binding proposals can be obtained from component suppliers or architecture firms. Together with assumptions about missing details, the proposals allow putting together a rough cost structure. *Rough Cost Structure* Disciplines like automation and piping are considered by supplemental allowance factors. In basic engineering, specifications are ***Allowance factor***: pretty accurate and allow estimating within a ±15% range. This is *the individual sections are* mainly due to the more comprehensive documentation of construc- *estimated per their* tion parts that had been considered with factors previously. Also, non- *fraction of the total work* binding proposals can now be obtained for these parts.

It is only the accuracy of detail engineering that makes it possible to obtain fixed and negotiated offers based on complete equipment lists and specifications as well as construction drawings. An accuracy of ±5% is expected for this phase. This last estimate is the basis for the internal company release of the budget.

Regardless of the accuracy, the early cost estimates especially contain a 'contingency' position, which usually is considered within 10% of the *Contingency* overall sum.

The structure of the cost estimate comprises the 'hard and software' of the building process:

- Building (earthwork, skeleton, facade, finishing).
- Building services (light, water, fire alarm and protection, air conditioning/HVAC, heating).
- Infrastructure civil engineering (streets, parking lots, water, waste water).

- Utility and infrastructure connections (water systems, steam and heating, waste treatment, refrigeration, piping).
- Clean utilities (pharma-grade water, clean steam, gases, air, ductwork).
- Process equipment (main equipment, valves, piping, small parts).
- Automation/IT (process control system, building automation system, laboratory information management system, IT).
- Electrical (electrical energy supply, telephone, surveillance and communication systems).
- Engineering services (concept, basic, detail).
- Construction services (management, labor).
- Commissioning and qualification (external and in-house services).
- Contingency.

Value engineering

The term 'value engineering' denotes a process in which the planning process is scrutinized under different aspects: (i) are all elements that are planned really needed and (ii) do these elements fulfill the intended purpose or are they planned far beyond that? Experience shows that each planer builds in some safety margins which – summed up over the whole project – result in extensive spending and offer the chance to save costs when reduced appropriately. Moreover, only in the planning phase is the understanding developed about which systems and functionalities are necessary to achieve the goal. This acquired understanding helps in reviewing and modifying the status of the engineered technical solutions.

9.4
Organization of an Engineering Project

The organization of a planning and construction project comprises internal organization as well as coordination of the contractor firms involved. Due to the intermittent demand for specialized personnel, almost every project is guided by an internal core team involving external companies on demand. The principal disciplines are represented in the core team; they shall be described briefly in Section Section 9.4.1. After being selected (Section 9.4.2), the contractors are bound to the project by legal agreements, which define the relationship between client and contractor; Section 9.4.3 highlights the options for such agreements.

9.4.1
Expert Groups Involved

A well-functioning project team unifies the expert groups and is headed by a project manager responsible for costs, quality and timelines. A typical team includes the following functions:

- Project manager (plus controller and/or assistant).
- Production.
- Process development.
- Process engineering.
- Utilities and infrastructure.
- Architecture.
- Technical building systems.
- Quality assurance.
- Quality control.
- Process control.
- Electrical installations.
- Health, environment and safety.
- Regulatory affairs.

Typically, the team reports in regular intervals to a higher level team consisting of company management and often also contractor management (steering committee).

9.4.2
Role and Selection of Contractors

Engineering projects are usually executed with the help of specialized contractors. One can find engineering companies that focus on parts *Contractor* of the organization and execution, while others like to work on GMP-relevant activities (e.g. GMP design and qualification support). In the construction phase, the different functional sections are assembled by different companies; additionally component suppliers for equipment are involved. Consequently, the project requires diligent selection and coordination of contractors as well as careful contract design. Coordination can be handled by another external company, which may act as a general contractor and is responsible for the performance *General contractor* of the associated subcontractors.

The companies are selected in the bidding process, which should *Bidding Process* follow these steps:

- Selection of potential candidates (long list).
- Initial contact to explore whether there is interest in the project.
- Personal presentation of the contractors, their references and capabilities.
- Pre-selection of a smaller number of candidates (e.g. three), which continue in the bidding process (short list). *Short list*
- Closing of a confidentiality agreement.
- Detailed presentation of the project to the contractor. Request for non-binding proposal.

- Evaluation of the received proposals and qualitative impressions. Here, it is important to assure that the proposals are comparable (e.g. that the contractors have the same understanding of the task and that the cost structure of the proposals are comparable).

- Further reduction of the candidate list if necessary.

- Start of negotiation with at least two companies.

Term sheet
- Drafting of a term sheet laying down the technical and financial principles of the intended contract. After this term sheet ('Letter of Intent'), the actual work can start if time is of the essence; however, this rush can weaken the negotiating position of the client.

- Drafting and negotiation of the agreement.

The agreement as well as the practice of project management shape the collaboration of the partners (Section 9.5).

9.4.3
Contracts and Scope Changes

An agreement defines a balance of interests between at least two parties, i.e. the client and the contractor. It consists of a mutual commitment to service and consideration.

The chances for the companies lie in the attendance to the same interest. The chances that stand in relation to certain risks that can be addressed and partly minimized in the legal agreement. These risks include:

Technical risk
- *Technical risks*: malfunction of the warranted performance. A way to mitigate this risk is to select the contractor diligently and monitor their performance constantly.

Commercial risk
- *Commercial risks*: insolvency of the client. The contractors should assure themselves of the solvency of the client.

Cost risk
- *Cost risks*: can arise when necessary and maybe unexpected supplementary work has to be done. This can result from insufficient resource planning or a too optimistic labor estimate at the timepoint of agreement closing. They affect the client and contractor equally; however, the burden can be put more on the client side (fee-for-service contract) or the contractor side (lump-sum contract) depending on the type of agreement.

Time risk
- *Timeline risk*: the services can be delayed either by inadequate resource allocation or due to lack of cooperation on the part of the client. Despite the fact that contractual agreements may exist for this case, the best way to project success is the early recognition of such root causes.

- *Quality risks*: defaults in the provided services. This risk is addressed by legal clauses for guarantee and liability. *Quality risk*

- *Change risks*: frequently during execution of a project, changes to the technical specifications or the organizational flow will occur that have to be handled by the parties. The way to manage such scope changes should be agreed upon in the contract, otherwise they can lead to severe disturbance of the relationship. *Change risk*

- *Execution risks*: personnel fluctuation in the team as well as inadequate execution structures and tools. This risk can be mitigated by suitable legal agreements. *Execution risk*

The contract usually consists of a legally formulated part, and the appendices defining the technical scope and financial considerations. Figure 9.3 illustrates the logical structure of the legal part (Braganz, 2001). The left side shows contractor obligations; the right side, client obligations. First, the service of the agreement should be defined; it is usually much more detailed in the technical appendix. The main

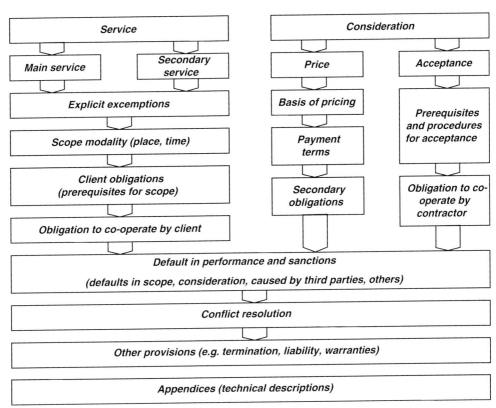

Figure 9.3 Recommended agreement structure.
Modified from Braganz (2001).

Main service

Secondary service

service denotes the principal subject matter of the agreement (e.g. 'construction of a manufacturing plant'); the secondary services have to be accomplished to achieve the main service (e.g. 'coordination of subcontractors'). For the sake of clarity, non-obvious exemptions should be explicitly mentioned, such as 'qualification of the clean rooms is not part of the service'. The scope modalities describe when and where the service should be provided. The 'client obligations' define the prerequisites for the service that have to be provided by the client, such as the final process description or the building ground with infrastructure feeds. The 'obligations to cooperate by client' define the services that the client has to provide during service. This can include the provision of personnel for review and approval of construction drawings.

Obligation to cooperate

The right upper side of Figure 9.3 starts with the consideration usually being monetary. One branch of this structure defines the price, the basis of accounting (e.g. hour sheets, material lists or expense reports) and conditions of payment (e.g. invoice dates and date of payment). The right branch defines the procedures for service acceptance and the duties of the client to perform the acceptance.

The broad band in Figure 9.3 contains provisions for disturbances of the cooperation, meaning deviations from agreed services or compensation. In almost all complex projects this happens one or more times. It is wise to agree upon conflict resolution mechanisms for such cases. Typical provisions stipulate a controlled escalation of the conflict up to management level. Finally, the place of jurisdiction is set forth here; however, the path to court often is ineffective and therefore rarely used.

Conflict resolution

The remaining sections contain rules for warranties and liability, term and termination, and confidentiality. These are often associated with the highest risk for both client and contractor. It is the mutual dependence that makes the termination rights and consequences one of the most disputed issues when negotiating a contract.

Termination of the contract

The contract regulates the services; if these primary obligations are not rendered, it leads to secondary obligations. Malperformance can have different legal forms:

- Impossibility of service.
- Delay of service provision.
- Bad performance.
- Change in services.
- Additional services.

The disturbance of the agreed service and consideration relationship leads to compensation claims to the originator, be it the client or the contractor.

Claims management

Claims management is an important part when making or operating the contract. Due to the complexity of the tasks, scope changes are unavoidable, yet have the potential to lead to conflict. If they compromise the basis of the cooperation, the project – or parts of it – is jeopardized. A contractual agreement about how to detect and treat such deviations is as important as their timely and forceful registration. First, the service should be defined as precisely as possible as to type, scope, location and time in order to be able to detect and quantify deviations.

Before signing the contract, the fairness principle should prevail since contracts at the sole expenses of one party lead to conflicts and project threats in the mid term. Upon contract execution, the work scope and conflict resolution mechanisms should be clear and well defined. During service provision, evidence for claims should be collected in a timely manner and discussed promptly with the other party. *Fairness principle*

There are different basic forms of contracts regulating the distribution of risks between the parties. In general there are two types of contracts. *Type of contract*

In one case, the work hours provided by the contractor are reimbursed by the client. The risk for the contractor is low. Since there is no real incentive to complete the work efficiently, this type of contract requires intensive control on the part of the client. Control tools can be regular meetings, the delegation of staff to work at the offices of the contractor and frequent checks through monthly worksheets. It is also common to agree on an upper limit for payments ('cap'). This type of contract is made if the scope cannot be completely defined beforehand; typically, this is the case in the concept and basic engineering phase as well as for support services for qualification and validation. *Fee for service*

In contrast to that is the concept of the delivered work, being independent of effort. With this type of contract the contractor absorbs a part of the risk, which is reflected in certain calculation surcharges of the proposal. The risk assumption does not only relate to the labor, but also to the quality and the timely completion of the work. In this case, the client has less control over the result and the path towards it. This result-oriented scheme is beneficial if the scope of work can be clearly defined upfront; typically, this is the case for detail engineering and construction services. Of course the risk to the contractor is limited. The warranties are only valid if the scope does not change during the service provision. In the case of scope changes, the change management process has to ensure that implications for the project are acknowledged. The most distinct form among these contracts is the delivery contract for turn-key facilities at a fixed price (lump-sum turn-key). Hereby, the contractor accepts full responsibility for a fixed price. This risk distribution can be found in chemical plant engineering, especially if the contractor owns the process knowhow. In comparison, in the biotech area there is too little experience and the processes too poorly standardized, so that the lump-sum turn-key concept has not yet been established. *Lump-sum turn-key*

9.5
Successful Execution of an Engineering Project

There are essential factors for the successful execution of an engineering project.

Process definition — An indispensable prerequisite for efficient plant engineering is that the manufacturing process is finally defined. Due to the tight timelines of product development, this is not always the case. Clear ranges for the specification limits of equipment and system design should be provided.

Scope definition — Every planning should build on a clear scope definition. Unclear tasks and fuzzy concepts do not justify the effort of basic engineering. If it becomes clear that the target is not fixed, the conceptual phase should be extended with adequately loose contractual commitments and strong in-house expertise and staffing.

Continuous monitoring of the provided service in terms of quantity and quality is an important prerequisite to discover and discuss issues early on in order to avoid conflicts building up and jeopardizing project progression. One part of this is that the content and spirit of the agreement is clear to the working team.

Acceptance of services — Speedy acceptance of services enhances the relationship with the contractor and maintains the awareness of the client's team of their own responsibility to reach the goals.

Analytical laboratory — An important prerequisite for the successful execution of the validation phase is a well-functioning and adequately equipped analytical laboratory.

Team cooperation — Maybe the most important success factor, however, is well-functioning team cooperation. Especially in teams with international composition, different national as well as company cultures and professional attitudes meet, each of which can have very different communication behaviors. Understanding the expectations and characters on the other side, enhanced in a team-building workshop, for example, can significantly reduce friction losses due to inefficient communication.

9.6
Legal Aspects of Facility Engineering

Apart from drug jurisdiction (see Chapter 7), there are other legal areas that affect biotechnological production. These are the health, safety, environment and building regulations that need to be considered when building or operating a manufacturing plant.

The laws and regulations of each jurisdiction cannot be listed in this book; therefore, this section only provides a brief overview of the regulated areas for which rules usually exist in each country.

A remarkable basic difference in the legal approach exists between the US and Europe. While technical measures are widely described in rules and standards in Europe, there is more room for flexible solutions in the US. An example can be seen in terms of fire protection measures that can be discussed with insurance companies in order to reach an optimum between the insurance premium and the investments into constructive fire-protection measures. This approach goes hand in hand with product liability, which is usually much higher in the US than in Europe. It allows more freedom in designing the product; however, if damages occur due to malfunction, these are compensated much higher than elsewhere.

9.6.1
Health, Safety and Environmental Law

Environmental protection and occupational health are the central issues for the licensing and operation of an industrial plant.

There is some overlap between these areas since the working place is the direct environment of the plant. Exposure of staff is not desirable just as exposure of the environment. However, the person working close to the equipment obviously is exposed to a higher risk and sometimes can even be considered a risk factor. Therefore, the regulations of occupational health are shaped very clearly to the immediate protection of individuals, while environmental law also takes the long-term impacts on nature and human beings into account.

Environmental law should ensure that no substances hazardous to health and nature can enter the environment. In general, it regulates the protection of the three compartments – soil, water and air – as to the types of emissions: *Environmental law*

- Solid waste.
- Liquid waste.
- Gaseous waste.

These types of wastes can generate any kind of chemical, toxic or biological effect (poisonous, acid, mutagenic, carcinogen, infectious) in the compartments. In soil, the issue of relics (i.e. the conservation of environmental damage over decades) has to be considered. Moreover, environmental law takes care of questions around noise protection. Questions of optical disturbance are mostly handled in building law. *Past pollution*

Industrial plants that generate emissions need regulatory permits everywhere in the world; the regulatory procedures are set forth in environmental law. A further field of this law is the treatment of unexpected operating conditions or accidents generating unexpected emissions. *Regulatory obligation to obtain a permit*

In the area between environmental law and occupational health, regulations are located for:

Device safety
- *Safety of technical devices*: like pressure vessels, electrical installations or hoisting devices.

Handling of hazardous material
- *Handling of hazardous substances and chemicals*: like gases, explosive materials and flammable, acid or toxic substances.

Biological substances
- *Safety when handling biological substances*: this includes rules for handling and waste management of genetically modified substances and toxic or infectious materials.

Transport safety
- *Transport safety*: this includes the design and construction of tanks, and marking and permit for transports over public traffic ways.

Occupational health rules cover the items:

General occupational health and safety
- *General occupational health and safety*. This includes all measures to protect the working staff from damage to their health in the workplace. Examples may be:
  - Protection against contact (hot surfaces or liquids; deep freezes like liquid nitrogen; rotating and moving mechanical parts; acid or toxic substances, infectious material, etc.).
  - Breathing protection (dust; organic solvents; toxic, acid or carcinogenic gases; gases like nitrogen or oxygen; infectious material, etc.).
  - Protection from unintended ingestion of hazardous material.
  - Noise protection.
  - Protection from extreme temperature conditions.
  - Protection from consequences of breakdowns or accidents (emergency exits; personal protective equipment; training; availability of support personnel).
  - Handling of loads.
  - Protection from burdens like deprivation of daylight and other requirements concerning the workplace.

Social occupational safety
- *Social occupational safety*.
  - Working time.
  - Maternity protection.
  - Juvenile labor.

- *Law of accident insurance*.

9.6.2
Building Law

Building law contains regulations for visual appearance, structure, safety and other aspects relating to construction.

Building permits usually have to be obtained separately from the environmental permit. They mostly depend on municipal development schemes and therefore the responsibility lies with local authorities. Further areas of regulation are:

- General building provisions (distance and traffic spaces, building height, visual appearance).
- Fire protection (alarm systems, extinguishing installations, emergency exits, building materials, ignition sources).
- Structural analysis including special rules (e.g. earthquake, wind and snow loads).
- Heat insulation.

There are other rules that have an impact on construction. For example, the rules for flammable liquids require that tanks are located outside the building at a distance that depends on their volume and hazardous potential.

Part Seven
Economy

One essential goal of production is to project and realize the profitability of an investment project or a manufacturing process. The challenge is to keep the costs of product supply as low as possible and at the same time minimize the risks for quality and availability.

Production costs can make up a big part of the overall costs of a pharmaceutical company. However, investing in equipment and premises also produces financial challenges. A mid-size production plant for a biotechnological drug can cost up to €100 million and takes more than 4 years from investment decision to manufacture of marketable goods. Therefore, the decision about a production concept requires a good understanding of the relevant economic relations.

The following chapters provide a strongly simplified and biotech-focused view of these topics.

Chapter 10 elaborates on the basics for calculating manufacturing costs of biotechnological pharmaceuticals. The chapter starts by contemplating the lifecycle model for drugs, which allows capturing the principal mechanisms of the market. The subsequent Chapter 11 describes how capital investments should be approached and introduces the basics of investment appraisal. Chapter 12 is dedicated to production concepts. Here, the focus is on make-or-buy decisions – when is it beneficial to out-source production and when should an in-house solution be preferred? The basics of decision making are explained by means of a comparison between timelines of external and internal manufacturing, cost aspects, and other strategically relevant topics. The chapter closes with a look at the possibilities to optimize the costs after market launch by process and supply-chain optimization.

10
Product Sales and Manufacturing Costs

The profit of a company is calculated as the difference between the sales and the effort required to achieve those sales. The sales depend on the quantity of the product sold and the sales price achieved. Manufacturing costs are one part of the relevant effort, in addition to sales and marketing, administration, and research and development. The cost–sales ratio is often taken as an easily accessible index to get a first impression of the profitability of a product.

Section 10.1 describes the basics of sales generation and pricing of drugs. The subsequent sections look at cost calculations and methods of profitability assessment of products.

For the sake of clarity it should be noted that this section intends to illustrate the essential perspectives when costs of goods for biotechnological product are calculated. On purpose it simplifies financial terms (e.g. when using the term 'profit' it does not specify whether this is meant before or after taxes). For more detailed accounting literature, the reader should consult dedicated monographs.

10.1
Lifecycle of a Drug

The pharmaceutical market is based on the same principle as other markets. Marketing, innovation and competition for market share determine the scene; however, the pharmaceutical market has some special features: *Characteristics of the pharmaceutical market*

- Prices are often regulated by official bodies.

- Drug development is subject to official control, lasts relatively long and is expensive. Regulatory bodies perform the risk–benefit assessment that is a prerequisite for market approval. This assessment can only be done after extensive development work has been completed.

- Production and administration are subject to official surveillance.

- Clients (doctors, pharmacists), users (patients), buyers (wholesaler) and payers (health insurance) are not identical, and require specific marketing strategies. Moreover, direct promotion is partly forbidden due to ethical reasons.

- Bargaining drugs often touches ethical principles (basic right for medical care, quality of life, 'two-class medicines', etc.). The ethical responsibility requires special aptitude when dealing with these questions.

Lifecycle curve

Phases of the lifecycle

Even though these specialties shape the market environment, drugs – like other products – go through certain phases in their lifetime which can be described by the lifecycle curve (Figure 10.1). The scheme connects timewise to Figure 4.6 and starts with product launch. The dashed line symbolizes the sales development; the solid line symbolizes the cumulated balance between in- and out-payments. The sales curve increases slightly in the introduction phase and reaches a steep slope in the growth phase. During introduction the market has to become familiar with the product; the growth phase is characterized by a broad acceptance and stable prices. In the maturation phase, the product is well established and serves a large market; however, due to its relatively

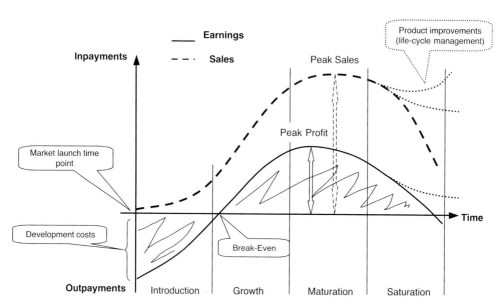

Figure 10.1 Lifecycle curve of a drug. Dashed line = sales; solid line = earnings (sales – efforts); hatched area = cumulative profit.

old age it cannot gain significant growth rates, since the market is already saturated or competitive products have taken market share. In the saturation phase, these influences finally prevail and the sales volume decreases.

The earnings curve primarily follows the sales curve. However, due to the high development costs the product starts with a negative balance that is not compensated for by positive sales until the break-even point. *Break-even point* Only here does the product start to make profit for the company. The profit reaches its maximum in the maturation phase and declines in the saturation phase due to the typically rising marketing effort.

Obviously, not all products follow this simple curve, but it helps to illustrate the basic market characteristics. The total profit that the company makes from this particular product is equal to the hatched area under the earnings curve; companies aim at maximizing this area.

A determining factor for the beginning of the saturation phase is the patent protection of the product. In principle, after patent expiry, *Patent protection* generic drugs have the chance to enter the market. Due to the limited effort for research and development these products are usually much less expensive than the original preparations. It is currently subject to intensive discussion whether complex biological generics can be brought to the market without clinical trials (Section 3.6).

Box 10.1

Patents

There are two categories of patents: process patents and product patents. Process patents can relate to new manufacturing processes (e.g. protein purification) or working methods (e.g. methods for protein analysis). Application patents also belong to the group of process patents; they can be granted for new indications of known drugs. Process patents do not protect the produced substance, but only its manufacture or application. In contrast, product patents protect the object itself; this can be an appliance (machine, work equipment) or a material (e.g. a pharmaceutical ingredient or the formulation of an API).

Patent protection lasts 20 years; for products requiring particularly intensive efforts for research and bearing high cost risks, this time-frame can be prolonged by a supplementary protection certificate for another 5 years. Usually a patent application is filed in the early research phase. Until marketing approval, another 10–12 years expire; that means that 10–15 years remain to obtain a return on the investment *Supplementary protection* that helps to finance other ongoing research projects. *certificate*

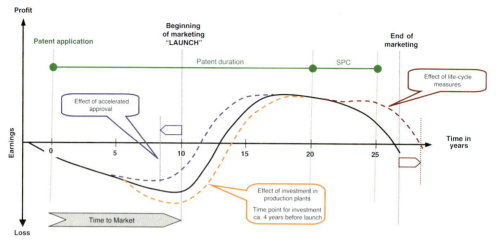

Figure 10.2 Overview over the profit/loss situation of a drug during development and commercialization. SPC = supplementary protection certificate.

Time to market

Figure 10.2 illustrates this relation. Time to market is of essence in drug development since positive payments that compensate for the spending of development only start with market launch. A further incentive for a short time to market is a potential competitive advantage if the own product reaches the market before others do.

Development effort

The development effort can be so high that drug development for complicated and rare diseases does not pay itself back in the market place. In order to provide patients with rare diseases the access to sophisticated treatment options, regulatory authorities can, upon request by the company, grant a drug the 'status for rare diseases' (orphan drug status, Section 7.5). This status goes along with facilitations in the approval process and marketing. Time to market can be shortened and spending for clinical trials can be reduced. Moreover, the status guarantees market exclusivity for a couple of years after launch. A comparable goal is aspired by the 'fast-track status' offered

Fast track status

by the FDA for drugs with urgent unmet medical needs. Here, the authority offers intensive collaboration for the design of the clinical trial, giving the manufacturer a higher security for their investment, and the chance to pass the regulatory review faster and reach the market place earlier.

While these measures shift the break-even point to the left, necessary investments shift it to the right. They come into play about 4 years before launch when the question about how the supply of the commercial market should be secured is discussed (Figure 12.1).

The profit curve can be prolonged if the product is kept attractive by lifecycle management activities. Common lifecycle management measures increase the comfort of physicians or patients, like:

Lifecycle management

- Simpler administration (pulmonary instead of intravenous, pre-filled syringes instead of vials).
- Longer administration intervals (every 3 days instead of every 2 days).
- Improved storage conditions (ambient temperature instead of refrigeration).
- Other dosages that justify price adjustment.

As for other products, pricing is geared to the market situation. The mostly used competitive therapy (gold standard) is the principal benchmark against which the benefit, but also the price of the development, has to be measured.

Pricing

The quantity sold in the first place goes by the available market. Since a drug can be approved for one or several indications, the total market size is determined by the patient group suffering from these indications. This total market has to be developed by appropriate marketing measures since doctors and patients need to be informed about the availability and capability of the new drug. Apart from one's own development other competitive therapies can enter this market. It is important to differentiate one's own product sufficiently in one or all aspects of safety, efficacy or comfort of use (target profile) from the competitive therapy in order to obtain a reasonable market share.

Target profile

Physicians also have the possibility to use approved drugs in other than the approved indications. This practice, which is denoted as 'off-label use', is pretty common in cancer therapy in which therapy options are limited and the clinical aspects are closely related. Off-label use can significantly increase the marketable product quantity.

Off-label use

10.2
Position of the Manufacturing Costs in the Overall Cost Framework

The position of the manufacturing costs in a company becomes apparent when contemplating these costs together with other cost positions and also the revenue side. Figure 10.3 provides one possible perspective on qualitative coherences between sales, manufacturing costs and investments. The company sells goods on the market at a certain price and generates its revenues. On the other hand, the company puts effort into development, production and marketing of the products. The difference between the effort and sales is the profit that enables the company to develop new markets.

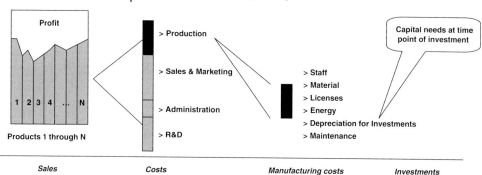

Figure 10.3 Relations between investment, manufacturing costs, sales and profit.

The sales depend on the market situation and the direct promotional effort for sales and marketing. The aim of the company is to maximize sales and to minimize effort; one important contribution is optimizing manufacturing costs.

Overhead costs

In a company that has more than one product on the market, the costs are composed of those attributable to each individual product unambiguously and those that are not. The former are, for example, manufacturing costs, sales and marketing costs, and development costs, which clearly relate to one product. Research and development costs that are spent on planned products cannot yet be allocated to marketed products. They accrue just as the general administration costs, as overhead costs of the company and have to borne by the disposal of the marketed products.

Consequently, for a profitability assessment of a product, the whole company has to be taken into consideration. Only the complete picture allows setting profitability goals for the individual product since its sales have to carry the overhead costs.

Manufacturing costs/ capital costs

The individual product costs encompass manufacturing costs, which on their part can be distributed into personnel, material, energy, maintenance, other dues and capital costs. The capital costs consider the costs of investments. These enter the calculation as depreciation, expressing the loss of value of equipment and premises, and capital costs (interest), expressing the costs of deployed capital. Figure 10.4 shows a schematic of the accruing costs. Sales costs are generated by distribution, marketing and sales. Royalty payments may also be allocated here; however, if they are manufacturing royalties, they appear in the manufacturing costs. Ultimately, costs also accrue for administrative activities like accounting, regulatory affairs, legal and other activities. The manufacturing, sales and administration costs add up to the net costs. Thus, the net costs are a measure for the

10.2 Position of the Manufacturing Costs in the Overall Cost Framework | 305

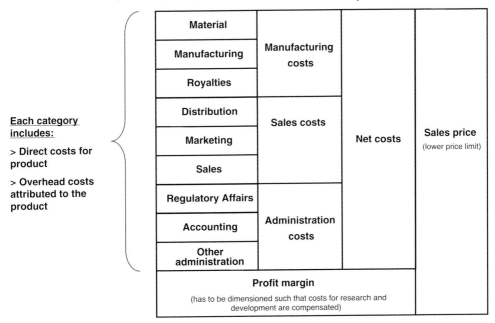

Figure 10.4 Basic scheme of costs (overhead calculation).

assigned product costs. They are one possibility to identify the lower price limit.

10.2.1
Basic Principles of Cost Calculation

Cost calculation is one topic of managerial cost accounting. The appraisal and allocation of costs is a principal instrument for planning and controlling. Depending on the goal of the cost calculation, there are different types of cost appraisal and assignment in order to analyze payment flows.

Cost accounting can encompass the whole factory, a part of it or a single object (e.g. the product). The focus of this work will be the calculation of product manufacturing costs; calculation methods for factory or area costs can be found in the literature.

The following basic principles of cost calculation should be known and differentiated from each other.

10.2.1.1 Nominal Accounting – Actual Accounting
Actual accounting is made by retrospectively analyzing the costs of production. The projected costs have to be estimated for nominal accounting – this is common practice for tender preparation. What

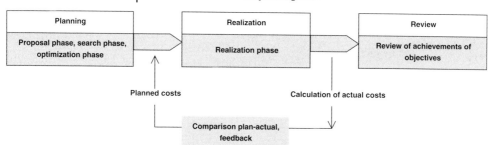

Figure 10.5 Steps of the decision-making process (Plinke and Rese, 2002).

type of calculation is made depends on the stage of the decision process as illustrated in Figure 10.5.

Nominal calculations are needed in the planning phase, while the review phase is supported by unit-cost calculations (Section 10.3.3).

10.2.1.2 Cost Accounting – Profit and Loss Accounting

Cost accounting takes into account the accrued or projected costs only. In profit and loss accounting, work results and costs are compared, and the success identified. Hence, profit and loss accounting always includes cost accounting.

10.2.1.3 Direct Costs – Overhead Costs

Direct costs are costs generated by one output unit (e.g. product batch) and are assignable directly to this unit. This can encompass raw materials, packaging materials and labor time of operator personnel. Allocating costs to a single unit becomes somewhat more complicated if it has been manufactured in a multi-purpose facility. Services that benefit the whole area should also be found in the product unit costs. Examples for such services are energy consumption for maintaining air quality, installation maintenance, environmental monitoring, facility rent and so on. These costs accrue for all products and are therefore overhead costs. It is not possible to assign them directly to the individual products. Hence, allocation keys have to be identified (e.g. the different use times of the plant). As far as possible such an allocation should be made according to the costs-by-cause principle, yet the calculation effort should not lead to such detail that accounting gets uneconomical.

Cost allocation key

Costs-by-cause principle

These terms should be clearly differentiated from the fixed and variable costs.

10.2.1.4 Fixed Costs – Variable Costs

Plant utilization

Variable costs are those that change depending on a selected reference value. This reference in most cases is the usage of the facility, which can be measured as the percentage of actual batches versus nominal batch capacity. Variable costs rise with increasing usage since, among others, more material is consumed.

Fixed costs can also vary, but not depending on usage. Often the fixed costs are those spent to maintain operability of the plant. These can include staff, machine and energy costs. Changing fixed costs is subject to an active management decision and can include shutting down parts of the facility. It can be observed that fixed costs increase quickly, yet their decrease in the case of disuse progresses much slower – an effect known as cost inertia.

Cost inertia

In the following the term 'costs' stands for the overall efforts of the plant (expressed in Euros) and the term 'unit costs' for the manufacturing costs of the individual product unit (expressed in Euros/unit).

When identifying costs of a product, the costs of providing an operational plant have to be considered. The following calculation example shows the significance of the fixed costs in the case of low facility usage:

Box 10.2

Calculation example of the influence of low facility usage

A plant is contemplated as having a nominal capacity of 20 kg protein, but only produces 4 kg. The fixed costs are €10 million. Variable costs are €1 million/kg.

If the plant would be run with 20 kg, total costs would amount to €30 million. The costs for 1 kg of protein would be €1.5 million.

Since the plant only generates 4 kg, the total costs amount to €14 million. The costs for 1 kg of protein increase to €3.5 million.

The example shows that the 'dilution of fixed costs' rises with increasing facility utilization (fixed-cost degression). In this example, management should try to place other products in this plant or relocate the production and to reduce capacities. In biotechnological production any relocation comes at a high cost, and can easily last 2–3 years if a transfer of analytical methods is necessary and a comparability study is required (Section 12.3.3). The freedom to operate is fairly limited here.

Figure 10.6 qualitatively illustrates the relation between variable and fixed costs. The left ordinate shows the fixed and variable costs as a function of facility utilization. In this case the variable unit costs are constant so that the variable costs increase linearly with utilization (proportional progression). Often the curve is flattened to the right since economy-of-scale effects allow the variable costs to decrease (degressive progression).

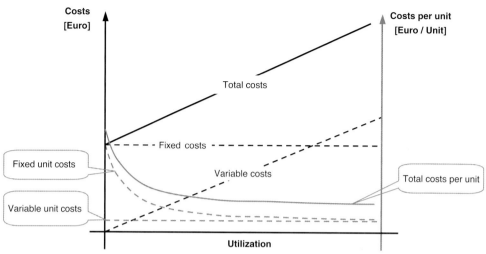

Figure 10.6 Influence of plant utilization on fixed, variable and unit costs.

The right ordinate shows the unit costs against utilization. If the full fixed costs are charged to the product, the resulting unit costs are very high at low utilization. With rising utilization, the relation of fixed to variable costs decreases and the unit costs approach a lower limit above the variable unit costs.

10.2.1.5 Relevant and Irrelevant Costs

Cost information is often used to decide between different options. In decision situations, the costs are denoted as being relevant if they are directly affected by the decision; hence, relevant costs express the difference between decision options. Costs are denoted as irrelevant if they are independent of this decision. Typical so-called irrelevant costs are those invested in advance performance like development work ('sunk costs'). In each decision situation it should be analyzed whether costs are relevant or irrelevant.

10.2.1.6 Cost Type, Cost Center and Cost Unit

Cost accounting has some principal goals in the company:

- Determination of success on the level of product, division and company.
- Provide calculations for the assessment of the optimal option.
- Control of profitability.
- Pricing, and calculation of lower and upper price limits. Determination of internal transfer prices.

Registration of costs according to cost type	Allocation of costs to cost centers	Allocation of costs to cost units
COST-TYPE ACCOUNTING	COST-CENTER ACCOUNTING	COST-UNIT ACCOUNTING
Which costs accrue?	Where do costs accrue?	Who carries the costs?
Examples: Testing material Manufacturing salaries External services	Examples: Fermentation area Purification area QC-department	Examples: Development department Product Y Product X

Figure 10.7 The areas of cost accounting.

In order to fulfill these tasks, data are collected for the following basic questions:

- What costs have accrued?
- Where have the costs accrued?
- Who bears the costs?

The first question is answered by cost-type accounting, the second by cost-center accounting and the third one by unit-cost accounting. Figure 10.7 provides an overview over the questions and gives examples.

The following sections will elaborate on cost-type accounting. The determination of manufacturing costs will be traced back to the deployed goods and labor, and typical cost structures will be discussed.

10.3
Manufacturing Costs

By detailing the cost types appearing in manufacturing cost accounting the section 10.3.1 bridges the gap between managerial cost accounting and the technologically dominated environment of the production floor. Section 10.3.2 provides details on cost types in biotechnological processes. Finally, Section 10.3.3 outlines how nominal accounting can rely on known cost structures to provide estimates of product costs and to evaluate whether a production option is profitable.

Salaries		Personnel costs		
Social benefits				
Manufacturing material	Essential constituent of products	Material costs	Operating costs	Manufacturing costs
Manufacturing aids	Minor constituent of products			
Operating aids	No constituent of product but consumption			
Reparation		Maintenance costs		
Inspection	Control of condition			
Maintenance	Preventive "reparation"			
External services	Analytical services, cleaning services etc.			
Other costs	Office material, communication etc.			
Premises	Rents, maintenance, depreciation etc.			
Energy	Heating, cooling, electricity, water, clean media, gases, waste etc.			
Tools	Tools prone to abrasion			
Interest	Use of invested capital		Capital costs	
Depreciation	Consumption of investment goods			

Figure 10.8 Types of manufacturing costs.

10.3.1
Cost Types

Cost types

Cost of goods are composed of operating costs and capital costs. Figure 10.8 shows an example for cost types; usually they can be selected freely based on the target of the information and the dataset that needs to be analyzed. While most of the positions of operating costs are self-explanatory, allocating concrete components mainly in the material section is not always that unambiguous. This is OK, since the types should only provide a framework for cost appraisal. For example the finished and packed product can be seen as the unit for which cost of goods should be calculated. In this case, the main components are the API, the vial and the packaging material. If one looks at the 'API' component, the operating aids could be chromatography gels and fermentation media. Chemicals for buffer preparation could be considered as manufacturing aids and cell banks or formulation buffer as manufacturing material. The classification has an assortative character only and no influence on the actual calculation of costs. If needed, this classification can be amended by additional positions.

In biotechnology, apart from actual manufacturing, there are significant costs for quality control and quality assurance. Quality assurance costs are mainly personnel costs; the quality control costs can also include high material or investment costs. Capital investments are expressed by cost positions for interest and depreciation.

10.3.1.1 Depreciation

Depreciations express the consumption of the usability of capital assets. They are related to limited-life items of the operating fixed assets (e.g. buildings or technical equipment).

The common reasons for decreasing usability can be technical abrasion or obsolescence. It is also possible that usability is reduced by non-technical events like expiry of an exclusively used patent.

The primary goal of considering depreciations is to include the expenses for assets into the company's balance over their useful lifetime. The value of periodic depreciation – the depreciation per time period (e.g. per year) – is determined by:

- The depreciation amount is the value that has to be depreciated. *Depreciation amount* It is calculated from the depreciation basis less the liquidation value of the asset at the end of the useful life time. The liquidation value is the *Liquidation value* assumed achievable sales price of the asset at the end of the useful life time. In most cases, in which the installation is not sold long before its operating life expires, this can be set to zero. Consequently, the acquisition value can be taken as the depreciation amount.

- The useful economic lifetime of the asset determines the period or *Useful economic lifetime* operating time over which the depreciation amount is depreciated. Short-lived investments like IT equipment typically are depreciated over 3–5 years, technical equipment over 10 years and buildings over 20–30 years, depending on their type of structure.

- The depreciation method is the algorithm applied to distribute the *Depreciation method* depreciation amount over the depreciation time. The simplest method is linear depreciation; here the periodical values are assigned to the periods according to the formula: $K_A^t = A/N$, where K_A^t is the depreciation value over the period, A is the depreciation amount and N is the useful economic lifetime. Other algorithms are geometrical or arithmetical degressive depreciation.

10.3.1.2 Interest

One way to look at interest is that it expresses the consumption of the usability of invested capital. Bound capital can be regarded as an asset whose usability is reduced by having invested it into fixed assets instead of being liquid and able to generate profit elsewhere. The interest evaluates this loss of flexibility. In this evaluation, the actual interest rate at which the capital was raised – in the case of credit financing – is only one part; other parts may include additional risk factors.

Box 10.3

Calculation example for calculation of depreciation and interest

As an example, a production plant is contemplated that was built in 2001 for an investment of €70 million. The linear depreciation is €7 million per year. The value of inventory in raw materials, semi-finished and finished goods is €2 million. The interest rate is assumed to be 3.5%. It becomes clear that depreciation and interest significantly contribute to the cost of goods.

All numbers in million Euro	2007	2008	2009	2010	2011	2012	2013
Operating fixed assets	70	63	56	49	42	35	28
Operating circulating assets	2	2	2	2	2	2	2
Operating assets	72	65	58	51	44	37	30
Interest (3.5% annually)	2.5	2.3	2.0	1.8	1.5	1.3	1.1
Depreciation	7.0	7.0	7.0	7.0	7.0	7.0	7.0
Costs per year	9.5	9.3	9.0	8.8	8.5	8.3	8.1

Interest rate for calculation

Operating assets

Principally the interest rate is geared to the best alternative financial investment, usually the average interest rate for long-term financial investments given by the prime rates of central banks.

The interest rate is applied to the total operating assets of the contemplated period. This is composed of the operating fixed assets (book value of investment for the production plant) and the operating circulating assets (e.g. inventories). The calculation basis is the book value – the acquisition value of the asset diminished by the accumulated depreciation values up to the considered period.

10.3.2
Typical Costs of Biotechnological Manufacturing Processes

Cost structure in biotech production

According to Figure 2.2, typical biotechnological processes can be split into fermentation, purification, formulation and packaging. A rough rule of thumb says that about 20–40% of costs are generated in fermentation, 30–60% in purification, and 10–20% in formulation and packaging. These ranges contain quality control costs. Administrative tasks like regulatory affairs and quality assurance typically only contribute below 5%. Assuming further that approximately 30% of overall costs are indeed costs of goods, it becomes apparent that, for example, an improvement in purification costs by 20% results in an overall cost impact of 3%. On the other hand, this optimization means technological and regulatory effort that needs to be justified by the cost benefit. Indeed, manufacturing costs are not the only key parameter for a patented product. Quality consistency, supply safety and short time to market are equally, if not more, important; hence, there is not always room for optimized process development.

Table 10.1 lists typical cost positions that are common in the manufacturing areas of fermentation, purification and filling/packaging. They provide a checklist that can be used to model costs. For a complete cost picture, more information has to be collected:

- Quantity structures for the processes like number and size of fermentations, purification runs, and cleaning cycles.
- Process times.
- Lifetimes of membranes and chromatography gels.
- Consumption of chemicals.
- Consumption of energy.
- Analytical program.
- Degree of automation for cleaning, column packing, filtration, media and buffer preparation, and documentation.
- Cost basis for determining the imputed costs.
- Costs of personnel, energy and used raw materials.

It is conceivable that costs can be calculated retrospectively with this detailed information; however, projective cost modeling must obviously be based on comprehensive process knowhow. The next section shows how the prospective estimate of manufacturing costs can be facilitated.

10.3.3
Methods of Calculation

One of the principal tasks of production is to evaluate the economical feasibility of a process design on the basis of technical/scientific aspects. The manufacturer is faced with this task when putting together a proposal; the client needs to do this when checking the proposal and the production department when deciding between different options for manufacturing.

From the point of view of production, the most important instruments of profitability assessment are calculations of the costs of a single product unit or the individual order. Table 10.2 provides an overview of different types of unit calculations. Unit calculation can be made as the actual or nominal account. The actual account is denoted as product unit costing. It deals with collecting data concerning direct costs and

Calculation

Table 10.1a Cost positions of fermentation.

Process step	Cell bank	Media preparation	Fermentation	Cleaning	Quality control	Storage
Activities	storage generation	receiving raw materials storage dispensing preparation sterilization feed	inoculation fermentation transfer to purification	media tanks, fermenter	sample testing sample storage calibration for raw materials, environment, clean utilities, intermediate	media powder intermediate
Personnel	– laboratory	– logistics, clean-room	– clean-room	– clean-room	– laboratory	– logistics, clean-room
Process raw materials	– master cell bank	– media powder – chemicals	– working cell bank – medium	–	–	–
Processing aids	– cell bank medium – pharmaceutical water	– pharmaceutical water	– chemicals – sterile gases (O_2, N_2, CO_2, air)	– chemicals (acid and base) – pharmaceutical water hot and cold – clean steam	– laboratory reagents – analytical standards	–
Other auxiliary materials	– clean-room garment	– clean-room garment	– clean-room garment	– clean-room garments – disinfection and cleaning agents for all rooms	–	–
Maintenance	– laboratory	– clean-room, tanks, analytical devices	– clean-room, tanks, analytical devices	– clean-room, tanks, analytical devices	– laboratory, analytical devices	– storage
Rooms	– laboratory, clean-room	– storage, dispensing, media preparation	– fermentation suite	– storage rooms for dirty, cleaned and sterilized equipment – cleaning room	– laboratory, storage room	– storage room, cold storage, refrigerators

10.3 Manufacturing Costs | 315

Process step	Cell bank	Media preparation	Fermentation	Cleaning	Quality control	Storage
Energy	– electrical power – liquid nitrogen	– water – heating steam – electrical power	– heating steam – electrical power – gases – cooling water – sterile compressed air	– heating steam – electrical power – cooling media	– only in laboratory scale	– refrigeration – electrical power
Tools/others	–	– filter	– filter	–	– measuring devices, external services	–

Table 10.1b Cost positions of purification.

Process step	Cell separation/ isolation	Buffer preparation	Purification	Cleaning	Quality control	Storage
Activities	centrifugation filtration	receiving of buffer powder storage weighing/ dispensing preparation feed	chromatography (column packing) tangential flow filtration crystallization other processes	buffer tanks product tanks	sample testing sample storage calibration for raw materials, environment, clean utilities, intermediate	buffer chemicals intermediates fill and freeze API
Personnel	– clean-room	– logistics, clean-room	– clean-room	– clean-room	– laboratory	– logistics, clean-room
Process raw materials	– harvest from fermentation	– buffer ingredients – chemicals	– buffer	–	–	–
Processing aids	– pharmaceutical water	– pharmaceutical water	– chemicals	– chemicals (acid and base) – pharmaceutical water hot and cold – clean steam	– laboratory reagents – analytical standards	–
Other auxiliary materials	– clean-room garment	– clean-room garment	– clean-room garment	– clean-room garment – disinfection and cleaning agents for all rooms	–	–

(*Continued*)

Table 10.1b (Continued)

Process step	Cell separation/ isolation	Buffer preparation	Purification	Cleaning	Quality control	Storage
Activities	centrifugation filtration	receiving of buffer powder storage weighing/ dispensing preparation feed	chromatography (column packing) tangential flow filtration crystallization other processes	buffer tanks product tanks	sample testing sample storage calibration for raw materials, environment, clean utilities, intermediate	buffer chemicals intermediates fill and freeze API
Maintenance	– clean-room, equipment and machines	– clean-room, tanks, analytical devices	– clean-room, tanks, analytical devices	– clean-room, tanks, analytical devices	– laboratory, analytical devices	– storage
Rooms	– clean-room	– storage, weighing, buffer preparation	– purification suite	– staging rooms for dirty, clean and sterilized equipment – cleaning room for equipment	– laboratory, storage	– storage, cold storage, refrigerators
Energy	– electrical power – waste-water treatment (heating steam, chemicals) – heating media	– water – heating media – electrical power	– heating media – cooling media – electrical power – sterile air	– heating steam – electrical power – cooling media	– only in laboratory scale	– refrigeration – electrical power
Tools/others	–	– filter	– chromatography gels – filter membranes	–	– measuring aids, external services	–

10.3 Manufacturing Costs | 317

Table 10.1c Cost positions of filling and packaging.

Process step	Formulation	Buffer preparation	Padeaging	Cleaning	Quality control	Storage
Activities	thawing, buffer exchange, sterile filtration, filling, lyophilization, capping	receiving buffer powder, storage, weighing/dispensing, preparation, feed	labeling, patient information, folding box	filling equipment, product tanks	sample testing, sample storage, calibration for raw materials, environment, clean utilities, intermediate, final product	buffer chemicals, API, packaging materials, final product
Personnel	– clean-room	– logistics, clean-room	– logistics	– clean-room	– laboratory	– logistics, clean-room
Process raw materials	– active ingredient – formulation buffer	– buffer ingredients – chemicals	– label, folding box, patient information	–	–	–
Processing aids	– pharmaceutical water	– pharmaceutical water	–	– chemicals (acid and base) – pharmaceutical water hot and cold – clean steam	– laboratory reagents – analytical standards	–
Other auxiliary materials	– clean-room garment	– clean-room garment	–	– clean-room garment – disinfection and cleaning agents for all rooms	–	–
Maintenance	– clean-room, equipment and machines	– clean-room, tanks, analytical devices	– packing machines	– clean-room, tanks, analytical devices	– laboratory, analytical devices	– storage
Rooms	– clean-room	– storage, weighing, buffer preparation	– packaging room	– staging rooms for dirty, clean and sterilized equipment – cleaning room	– laboratory – storage room	– storage, cold storage, refrigerators

(Continued)

Table 10.1c (Continued)

Process step	Formulation	Buffer preparation	Packaging	Cleaning	Quality control	Storage
Activities	thawing buffer exchange sterile filtration filling lyophilization capping	receiving buffer powder storage weighing/dispensing preparation feed	labeling patient information folding box	filling equipment product tanks	sample testing sample storage calibration for raw materials, environment, clean utilities, intermediate, final product	buffer chemicals API packaging materials final product
Energy	– electrical power – waste water management (heating steam, chemicals) – heating media – sterile nitrogen or air for lyophilization	– water – heating media – electrical power	– cooling media – electrical power	– heating steam – electrical power – cooling media	– only in laboratory scale	– refrigeration – electrical power
Tools/others	– form sets for machines	– filter	– form sets for machines	–	– measuring aids, external services	–

Table 10.2 Types of calculations.

Type		Cost account (Section 10.3.3.1)	Profit and loss account (Section 10.3.3.2)
		considers costs only; does not evaluate the success of the object	in addition to cost account also considers success; hence, information on revenue is required
Full account, actual account	considers all costs, unit and overhead; retrospective status report	product unit costing	unit profit and loss account
Full account, actual and nominal account	considers all costs, unit and overhead; makes forward looking statements regarding expected costs and profits	standard costing	profitability calculation
Partial account, actual and nominal account	considers relevant direct costs; makes statements regarding expected costs and success	marginal costing, direct costing	unit contribution margin account

allocating overhead costs. Forward-looking nominal accounts serve to determine order prices and can be performed as:

- *Standard costing*: consideration of all costs.
- *Marginal costing*: consideration of relevant variable unit costs.

Both methods are cost accounts and do not allow evaluating the success of the unit or the order, respectively.

If the success of the object is to be assessed, revenues have to be considered in addition to costs. If all costs are considered, this is done in the profitability calculation; in the case of marginal costing, this is done in the unit contribution margin account.

The present section elaborates on the basics of each of these methods. It starts with a view of the cost accounts, and continues with the profit and loss accounts. In both cases the actual accounts lead over to the nominal account methods.

10.3.3.1 Cost Calculations

Product unit costing determines the share of the overall operating costs which have to be borne by the single product unit, also denoted as net costs. In a one-product plant this task is easy: all accruing costs are borne by one product. The challenge of product unit costing is to distribute the costs in a fair way, according to the input involved, to the cost units in a multi-product plant. Ultimately this is the only way to assess whether a single product is profitable for the company. At the same time the detection of costs itself should be performed with appropriate effort.

Overhead costs

Costs are always composed of direct costs and overhead costs (Section 10.2.2). Direct costs can be attributed unequivocally to the product and can, for example, be raw materials that have been bought for this product only. This computation should be easy with well-organized data collection. The net costs of a product are composed of the direct costs and the proportionate overhead cost. A very simplistic approach to allocate the overhead costs, regardless of the area where they accrue, is to distribute costs according to the respective output quantity, yet it is common practice that each area uses its own key for allocating such costs. This method is called differentiating overhead calculation.

Allocation key

The allocation keys can, for example, be production quantities, processing times, energy and material consumption, but also the share of external services or the achieved sales (for sales and administration). The real effort cannot always be reflected (e.g. if one product is associated with a higher legal or accounting effort than another, this effort cannot be enumerated by its sales). However, it is questionable whether the effort of registering product-specific work hours of individuals is an economically meaningful solution to that problem. The most differentiated registration of direct costs is the machine-hour rate calculation; here each machine constitutes a cost center in which the direct costs are identified and key values are set based on that for the individual products. The so-called overhead calculation works with the principle of cost allocation, which stipulates that all costs in the company have to be allocated to the cost units. Figure 10.9 schematically shows how cost allocation works in this case. Costs accrue in the areas of the company (here production, administration and sales), which can be split up into direct costs for the three products P1, P2 and P3, and overhead costs. The areas in this example are cost centers. Usually, the areas are further differentiated in complex cost center structures. A prerequisite for this differentiated form of cost analysis is an accurate cost center control with so-called cost-distribution sheets in which costs are captured and structured.

Overhead calculation

Combined production

Production processes in which several products are generated from one manufacturing run (combined production) deserve special consideration. A typical biotechnological example for combined production is the fractionation of blood in which several products are made in one batch. Other than that, combined production does not have any significance for pharmaceutical proteins and will not be considered further.

While actual calculations contemplate the status quo, nominal calculations provide forward-looking statements about costs. Accurate prediction of real costs is a fundamental requirement for economic optimization. Typically, projections are provided for:

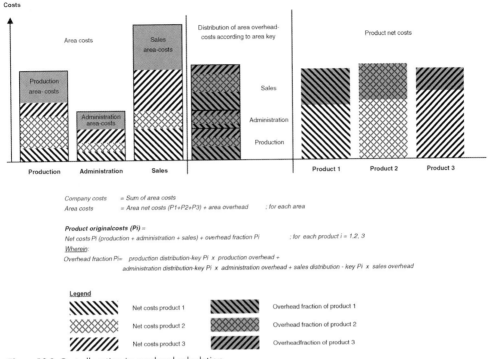

Figure 10.9 Cost allocation in overhead calculation.

- Profitability assessment of a product during development.
- Compilation (contractor) or examination (client) of a proposal.

Several factors render this projection challenging:

- Neither the product nor the manufacturing process is finally defined.
- Due to time pressure and the necessity to streamline the proposal compilation itself, short-cut methods have to be used, introducing the risk of inaccuracies.
- The quantity of orders is difficult to project; therefore, the overhead can be distributed over an insecure quantity of units.
- Long-term orders come with changes to prices for raw materials during the term of delivery.

Uncertainties in standard costing

These insecurities are accounted for in different ways:

- Proposals are issued with staggered bindingness (orientation offer, non-binding offer, binding offer) depending on the information depth of the request for proposal.
- Calculation methods with staggered accuracy can be used.

- Planning risks are deferred to the client. This can be achieved by listing the basic assumptions of the proposal, and by explicitly stating the risks and adjustment mechanisms.

Simple methods of calculation correlate the costs with the produced quantity, which is considered to be the only parameter influencing the cost. Methods of this type are not suitable for biotech products since each manufacturing process is different.

Cost functions

For pharmaceutical proteins, more accurate methods like the calculation with cost functions should be used or the detailed calculation, which is more accurate but much more labor intensive. The calculation with cost functions relies on correlation functions derived from empirical data capable of predicting costs for a certain process step. This method can, for example, be used for typical standardized manufacturing processes of monoclonal antibodies. Figure 10.10 shows such a standard process and the possible impact factors. A basis for this method is the comprehensive evaluation of data from former production runs with similar products. The maturation of the biotechnological industry will give rise to platform technologies and standardized processes enabling the calculation with cost functions to become a powerful prediction method.

Cost-structure analysis

Currently, however, a large variety of manufacturing processes can be found in the innovative biotechnological environment that require cost projection on the basis of detailed process modeling. Thus, all aspects listed in Table 10.1 have to be captured and predicted as well as possible. This very labor-intensive procedure can also be used for cost-structure analysis when developing a product. The analysis helps to identify

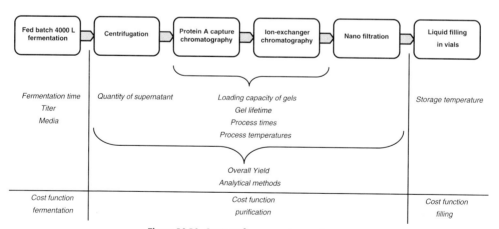

Figure 10.10 Options for parameters and calculation segments for the calculation with cost functions in biotechnological manufacturing.

saving potentials early on and to direct development efforts to the cost-sensitive aspects.

10.3.3.2 Profit and Loss Calculation

Up to now, cost calculations were considered in which the costs of an object (unit or order) are detected. However, the ultimate economic interest is to identify the profit of the company which is composed of the profits of the individual pieces. In addition to the cost data, profit-and-loss calculation needs information from the revenue side.

The unit profit is calculated as the difference between the revenues and costs according to the following definition:

$$\text{Unit profit} = \text{benefit} - \text{costs}$$

Therefore, unit profit depends on capacity usage. In this definition, the benefit is, for example, the achieved price; the costs are obtained from cost unit accounting, where 'costs' represents the total variable and fixed costs (fixed costs depend on capacity usage).

As illustrated in Figure 10.6, fixed unit costs depend on utilization and contribute significantly to the unit costs. Due to this correlation the unit profit also depends on the utilization. Consequently, it is only valid for the contemplated utilization case; if one order more or less is performed, the profit of a unit changes.

This correlation makes it difficult to interpret the unit profit. It constitutes an average value and has limited ability to assess the real success of a unit. A further disadvantage is that the performance is solely assessed on the basis of the achieved revenues; strategic impacts of a successfully executed order like follow-up deals can be easily overlooked. These strategic considerations should play a role when deciding on production options even if they do not appear in the unit profit account.

An improved statement can be made with the unit contribution margin account. This method circumvents both problems of full-cost accounting – the allocation of overhead costs and the proportionalization of fixed costs. Figure 10.11 illustrates this difference between the full-cost calculation and the contribution margin calculation. In contrast to the full-cost calculation, the method starts from the revenue side and calculates the relevant cost of the order. The contribution margin results from the difference of revenues and relevant order costs. Hence, it expresses to what extent costs that are not affected by the decision are covered by the contribution margin sum. The variable or direct costs are usually taken as the relevant costs. The contribution margin is not a net profit since the fixed or overhead costs still have to be subtracted; however, a unit or order is only good for the company if it can deliver a positive contribution margin.

Contribution margin

In each case it should be carefully assessed whether the variable or direct costs really are the relevant costs. In cases where fixed costs would

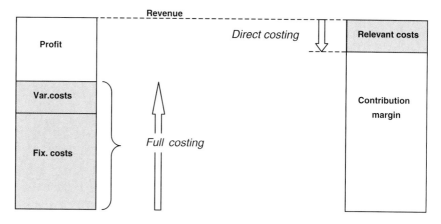

Relevant costs = variable costs OR direct costs

Figure 10.11 Left = full costing; right = direct costing for the calculation of the contribution margin.

rise (e.g. if a production plant is constructed or maintained exclusively for this purpose), the associated fixed costs are indeed relevant. As a quick check, when compiling the costs one can ask which costs cease to exist if the object of calculation was not there. The answer to this question provides the relevant costs.

Direct costing helps in answering the often raised question as to the allowed costs of a production and as to the price decision for an order. The calculation is suited to fix a lower price limit; the actual price is determined by the market place. Different cases of price decision can be distinguished based on the degree of utilization:

- In situations of under-utilization the relevant costs are the lower price limit. If the price is equal to the relevant costs, the contribution margin for the company is zero; this means that the situation of the company does not change whether it accepts the order or not.

- In situations of full capacity usage of the production plant, the lower price limit for order A depends on the alternatives that the plant has to generate a positive contribution margin. In this case the lower price limit is made up by the relevant costs of order A plus the best alternative that has been displaced by order A.

- In situations where the contribution margins are pre-set by company goals, the lower price limit is composed of the relevant costs plus the nominal contribution margin.

These considerations of company success conclude the overview of the costing methods. The next chapter will elaborate on investment decisions that enter into the cost calculations via interest and depreciation.

11
Investments

The focus of the preceding chapter was put on manufacturing costs – it was shown that the capital invested in production facilities finds its expression in the product costs in interest and depreciation.

Apart from having this impact on production costs, investments bind large capital values that the company consequently cannot invest in other projects. Mostly, this capital commitment is long lasting since *Capital commitment* procured goods can only be sold accepting losses. Moreover, investing means advanced payment, expecting return only in the future. The capital demand for the necessary advance financing has to be satisfied in a way that the company does not miss its profitability and liquidity goals. The check for the ability to fund projects is made by financial planning, *Capital demand* which goes hand in hand with investment planning in order to secure the solvency of the company.

Investment decisions are decisions for the future of the company; *Investment planning* they are based on more or less good projections of the development to come. In summary, one can say that investments:

- Constitute a long-term capital commitment that can be relatively high and difficult to change.
- Require advanced funding for which the return comes much later.
- Require decisions on the basis of insecure, forward-looking projections and therefore are fraught with risks.

This chapter mainly deals with investments in assets relevant for production. Section 11.1, apart from providing an overview of the goals of investment, introduces a broader perspective on investments. Investments are appraised by means of qualitative and quantitative criteria. The former are captured in value–benefit analysis, as outlined in Section 11.2. Section 11.3 describes the most important methods for financial evaluation of investments.

11.1
Basic Principles

This section clarifies the features and the process of how investments are prepared. The structured view facilitates putting together evaluation criteria and the formulation of the investment problem in economic terms.

11.1.1
Investment Targets

Investments are made with the aim to provide the production factors essential for operation. General targets of investments are:

Profitability
Liquidity
Security

- *High profitability*: return from the deployed capital.
- *Sufficient liquidity*: solvency of the company at any time.
- *Security*: risk associated with the investment should not lead to a threat to the company.

Independence

- *Independence*: the company should retain the ability to act and decide independently.

Lifecycle of capital investments

The investment is supposed to generate profit for the company along its lifecycle. Figure 11.1 illustrates the typical lifecycle of a capital investment in production facilities. At first the investment itself causes high costs (out-payments) at the timepoint of investment. The production of goods generates costs that lead to further out-payments. After the plant has started production, return flows back into the company. In the biotechnological field, the time between investment and return can easily be 3–4 years for larger investment projects. This period has to be bridged by interim financing.

The solid line shows the course of the total of cumulated in- and out-payments. Depending on the slope, this curve crosses the x-axis at the timepoint at which the cumulated in-payments exceed the cumulated out-payments. The time that the investment needs to reach this point and the balance at the end of the lifecycle are important evaluation criteria for the investment (Section 11.3). At the end of the lifecycle, the plant is decommissioned or sold. The latter results in a one-time in-payment reflecting the liquidation or residual value of the plant.

11.1.2
Types of Investments

Investments appear in many different forms. A structured approach looks at investments from different perspectives and classifies them accordingly. These perspectives could be:

- The object of investment.
- The effect of the investment.

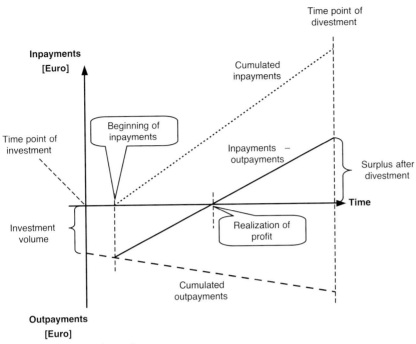

Figure 11.1 Lifecycle of capital investment.

- Other features like volume and interdependencies with other fields.

The distinctive criteria provided in the following tables can help to identify the essential features of an investment and develop criteria for its evaluation. Usually an investment will exhibit more than one of these features.

11.1.2.1 Classification According to the Object of Investment

Table 11.1 shows three objects according to which investments can be differentiated. Apart from the fixed and finance investments, from the perspective of production the non-fixed investments are of interest; these include one-time costs for the implementation of suitable staff for operating the installed fixed assets.

Salaries do not count as investments. The detection of expenditures for the investment usually is very simple; however, identification and allocation of revenues related to expenditures is more or less challenging depending on the type of asset. Non-fixed investments usually are difficult to evaluate quantitatively; for fixed assets this depends on the degree of involvement in the production process. Obviously, the return of large plants for single products can be easier to define than invest-

Table 11.1 Classification of investments according to the investment object.

Type of object	Examples	Detection and allocation of returns
Fixed investment	production facilities, premises, machines (capital assets) intermediates, raw materials (working assets)	often difficult for small components which are part of a multi-step process easier for large investments for one product
Finance investment	participations, shares	often fairly easily possible
Other non-fixed investments	personnel (headcount, training) research and development (project funding) marketing (advertisements, image improvement)	very difficult

ments in smaller parts that are integrated in a complex manufacturing workflow. However, since the projection of this benefit usually is the core of a budget request, detecting and assigning of earnings plays a central role for investment appraisal.

Box 11.1

Examples for detection and assignability

(i) A functioning analytical device that is used for the measurement of several products should be replaced by another one. Reasons can be that sample preparation shall be facilitated, official regulations require modernization or that method robustness will be improved. In any case the revenues of the modernization are difficult to identify.

(ii) A single-product facility should be constructed. The revenues can be clearly correlated with product sales.

11.1.2.2 Classification According to the Effect of Investment

Table 11.2 lists criteria according to which investments can be classified as to their effect.

New and expansion investments are characterized by their singular nature, they are also called net investments. Asset maintenance, rationalization and change-over investments typically are follow-up investments to reconstitute the assets. Together with the net investments, they are called gross investments.

11.1.2.3 Classification According to Other Criteria

The other criteria of classification encompass rather finance-oriented factors like allocation of revenues, duration and volume of capital commitment, as well as frequency, dependency or turnover rate, as explained in Table 11.3.

Table 11.2 Effect of investments.

Type of effect	Description
New and expansion investments	single investments: construction and expansion of plants
Asset maintenance investments	maintenance of performance: replacement of defect equipment with similar new one
Rationalization investments	increase of performance: replacement of old equipment with more modern one (e.g. automation)
Change-over investments	change-over of facilities from one product to the other (e.g. refitting of a fermentation suite)
Disinvestments (structural measures)	shutdown of facilities or discontinuation of business areas

Table 11.3 Other features of investments.

Feature	Description
Allocation of returns	expenditure projects: returns are not directly assignable like a waste water plant surplus projects: returns are directly assignable (e.g. a chromatography column for the purification of a product)
Duration and volume of capital commitment	strategic investments: long term capital commitment >5–10 years (e.g. construction of a production facility) tactical investments: mid-term capital commitment <5 years (e.g. procurement of a chromatography skid) operative investments are made on facility level to maintain operability and are short term ('routine investments')
Frequency	single investment versus series of investments
Dependency	isolated investments without effect on other areas versus interdependent investments with effect on other areas
Turnover rate	fast turnover (e.g. raw and auxiliary materials) medium turnover (e.g. equipment and machines) slow turnover (e.g. buildings, real estate)

11.1.3
Decision Processes

The capital commitment of investments requires a profound basis of decision to minimize the risk for the company. It begins with the recognition of the necessity to invest, through option screening and evaluation, and finally decisions, as well as execution, and control after the investment has been made.

Investment decisions start with the investment proposal; this can be illustrated by the example of a manufacturing plant (Figure 11.2). The proposal can be motivated by the lack of capacity in existing plants. In

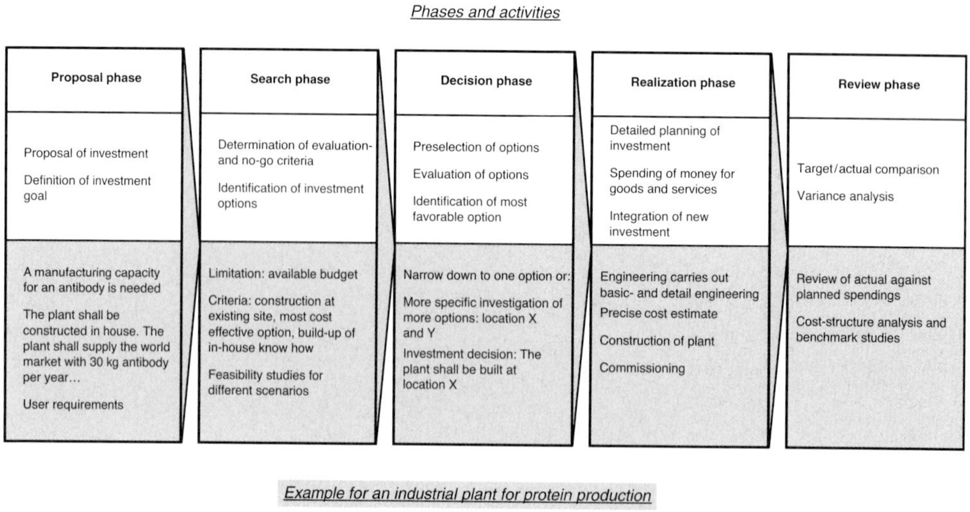

Phases and activities

Proposal phase	Search phase	Decision phase	Realization phase	Review phase
Proposal of investment Definition of investment goal	Determination of evaluation- and no-go criteria Identification of investment options	Preselection of options Evaluation of options Identification of most favorable option	Detailed planning of investment Spending of money for goods and services Integration of new investment	Target/actual comparison Variance analysis
A manufacturing capacity for an antibody is needed The plant shall be constructed in house. The plant shall supply the world market with 30 kg antibody per year… User requirements	Limitation: available budget Criteria: construction at existing site, most cost effective option, build-up of in-house know how Feasibility studies for different scenarios	Narrow down to one option or: More specific investigation of more options: location X and Y Investment decision: The plant shall be built at location X	Engineering carries out basic- and detail engineering Precise cost estimate Construction of plant Commissioning	Review of actual against planned spendings Cost-structure analysis and benchmark studies

Example for an industrial plant for protein production

Figure 11.2 Decision process for investments with example for a facility construction project.

Process of investment decision

this proposal phase the requirements for goals of the investment are formulated, while the path to achieve these goals is still open. At this stage even switching to external sourcing cannot be excluded. The latter also means an investment for the company since a certain amount of money has to be calculated for capacity reservation regardless of whether goods will be purchased afterwards or not (Section 12.3). In practice, external manufacturing means to invest in use rights.

Investments in rights to use

The proposal phase is followed by the search phase in which the investment options are defined, typically by means of short studies and feasibility analysis. Here, it is important to define evaluation and no-go criteria ('showstoppers'). Evaluation criteria can come from very different areas, such as strategic orientation, finance, synergy effects and qualitative experiences from former projects (Section 11.2). No-go criteria are absolute limits that may not be violated by the option in question. Typical no-go criteria are maximum budget or time limits of project execution. No-go criteria help to pre-select investment options. For this purpose, essential characteristics of the investment as to time schedules and cost have to be roughly estimated. Typically, one or two alternatives are retained for further investigation at this stage.

No-go criteria

Compared to the preceding phase, the realization phase is characterized by highly increasing cash demand. Detailed planning of the investment is the basis for construction and purchasing activities (Section 9.1), in which long-term capital commitments are made. The realization phase ends with integrating the new plant into the production environment. In the biotechnological area, in particular, facility qualification and process validation mean high effort. In the case of investments in external capacities this stage would include costs for technology transfer. Finally (and ideally), target and actual costs are compared in the review phase in order to evaluate the decision and obtain performance feedback.

The terms of the phases are very different. Significant costs are generated for the first time in the realization phase for engineering services and equipment. One therefore strives to pushing these measures as far out as possible and to shorten this phase as much as possible. On one hand, a later timepoint comes with higher security in the basis of planning (availability of clinical data, mature manufacturing process) and the timepoint of capital spending is moved closer to that of expected revenues, resulting in a smaller gap (compare Figure 11.1). On the other hand, the assets have to be capable of delivering the desired target (e.g. production of a protein). Hence, the realization phase is shut in from the front and back end. For biotechnological production plants it can last between 3 and 5 years.

Duration of decision phases

The terms of the search and decision phases are determined by the beginning of the proposal phase of the investment. If the need for investment is recognized early enough these phases can be worked through relatively calmly as long as the company provides the necessary resources. In the narrowest case, these phases last between 6 and 12 months depending on the investment volume.

The review should be made shortly after the investment has been made and does not have a clear endpoint. The *ex post* contemplation is as important as the *a priori* since lessons can be learned for future projects. The main difference between the two perspectives is not in the method, but in the security of available data.

Decision situations

In decision situations of investments, typical questions often occur that can be differentiated from each other fairly well:

- *Profitability of a single investment.* It should be decided whether an investment is economically meaningful. This question is answered by a full investment appraisal (Section 11.3).

  Profitability of single investments

- *Selection problem.* It should be decided which of the different possible investment options is the best. This question can be answered by taking an isolated look at the part of the investment that is relevant for the decision (i.e. the share of costs in which the options are different). However, this calculation does not prove that the best option is profitable for the company. It is recommended to always look at all cost positions to avoid missing a relevant one. Following this recommendation, the selection problem becomes a comparison between two single investments.

  Selection problem

- *Replacement problem.* It should be decided whether an apparatus or another asset should be replaced by another. The answer to this question requires comparing the costs and performance of the old

  Replacement problem

versus the new option. The question of residual value arises for the replacement problem. If the residual value is not zero, it has to be included in the calculation.

Questions whether to invest in in-house solutions or whether to rely on already existing external resources are special cases of decision situations:

Make or buy
- *Make or buy.* Here it should be decided whether the product will be bought or made in-house. This decision can be associated with investments either in one's own assets or into use rights of external capacities. In the case of under-utilization of one's own facilities, no investments have to be made. However, if due to a strategic decision a foreign investment is preferred despite the under-utilization of one's own suitable facilities, this own plant has to be divested. In general, the make-or-buy decision will come with investments on one's own side that have to compete with the external offer. Chapter 12 will further elaborate on make-or-buy situations.

Build or buy
- *Build or buy.* A closely related question comes up if the options are to build one's own production facilities or buy existing ones. In the latter case, the opportunity should be diligently reviewed by experts and will usually result in change-over investments.

The questions of investment decisions are answered taking into account qualitative and quantitative criteria, which are summarized in a report or presentation and used for decision finding. Qualitative criteria are captured in a value–benefit analysis (Section 11.2); the quantitative criteria are captured in the traditional investment appraisal (Section 11.3).

11.2
Value–Benefit Analysis

Value–benefit analysis is made to evaluate the qualitative features of an investment. The criteria are expressed on a suitable scale (e.g. school marks) and summarized in an assessment matrix. In most cases the analysis is performed early on in decision processes in order to identify potential no-go criteria.

Analyzing criteria is an important activity on the way to the investment decision since it generates a deep understanding of the decision-relevant factors. The following questions should always be asked for capital investments:

Criteria of value–benefit analysis
- Does the investment increase the ability for goods distribution? (Timeliness, reliability, delivery time, flexibility, customer service.)

- Can the marketing strategy be supported by the investment? (Market access, efficiency for advertising.)
- Does the investment improve the financial situation? (Currency risk, interest rate risk.)
- Is there enough qualified personnel for the investment?
- Can the asset rely on already existing or still to be constructed facilities? Are there synergies with other areas or suppliers of the company?
- Does the investment back up a long-term production strategy? (In-house versus external manufacturing.)
- How does the investment affect technical capabilities? (Versatility, specialization, automation, accuracy, capacity reserve, expandability, robustness, energy consumption, speed of work.)
- How does the investment affect the areas of health, environment and safety? (Safety standards, dust emission, noise emission, waste water, exhaust air, operability.)
- Are there social criteria that are affected by the investment? (Work monotony, work stress, work satisfaction, work interest, work autonomy, maintenance of workplaces, retainment of qualification, environmental friendliness, aesthetics, intellectual and physical demands.)
- What infrastructure supports the investment? (Transport possibilities, energy supply, storage possibilities, waste management.)
- Are there legal areas that make the investment necessary or pose limits? (Occupational health, environmental protection, patents, licenses, anti-trust laws, building codes.)

In principle, the qualitative criteria should be easy to operationalize (i.e. easy to describe, measurable and be independent from each other). The quantifiable criteria like tax incentives should find access into the quantitative appraisal.

The evaluation matrix can be set up in three shapes:

Evaluation scales

- The nominal scale assigns either a 'good' (also: 'fulfilled' or '+') or a 'not good' (also: 'not fulfilled' or '−') to the criteria. This rough classification is used in the screening phase.
- The ordinal scale allows more detailed assessment in which the options are scaled as to each criterion. Hereby, all alternatives are rated relative to the best option.
- Finally, the cardinal scale assigns 'marks' expressing how well the features meet the requirements. Since school marks are different in

Nominal scale

	Opt. 1	Opt. 2	Opt. 3
Crit. 1	+	+	−
Crit. 2	−	+	+
Crit. 3	−	−	−
Total	2 (+--)	1 (++-)	2 (-+-)

Ordinal scale

	Opt. 1	Opt. 2	Opt. 3
Crit. 1	1	2	3
Crit. 2	3	1	2
Crit. 3	1	3	2
Total	1 (5)	2 (6)	3 (7)

Legend:

Opt.	(Investment-) option
Crit.	Evaluation criterion
Wei.	Weight of criterion
Total	Position (value from criteria)

Cardinal scale without weight factors

	Opt. 1	Opt. 2	Opt. 3
Crit. 1	2	1	−1
Crit. 2	−1	0	1
Crit. 3	0	−2	−1
Total	1 (1)	2 (-1)	2 (-1)

Cardinal scale with weight factors

	Opt. 1	Opt. 2	Opt. 3	Wei.
Crit. 1	2	1	−1	20%
Crit. 2	−1	0	1	50%
Crit. 3	0	−2	−1	30%
Total	2 (-0,1)	3 (-0,4)	1 (0)	

Figure 11.3 Different scaling methods in evaluation matrices of value–benefit analysis. The last line shows the respective rank and the point score in brackets.

every country it is recommended to work with percentage indications (feature fulfills requirement by $x\%$) or a classification of the type '−2, −1, 0, +1, +2'.

Figure 11.3 shows examples for the scaling methods. The example also shows a cardinal matrix with weighted criteria. The criterion with the highest significance gets the highest weight. Purposely, the example results in different rankings depending on the selected scaling method. The result obtained with the ordinal scale is different from the weighted cardinal scale, which is a typical outcome. In practice, the value–benefit analysis has high significance for the thought process in investment decisions and as a communication tool, yet the ranking of investment options obtained should be interpreted with utmost caution.

11.3
Investment Appraisal

Investment appraisal deals with the basic financial assumptions and consequences of capital investments. Figure 11.2 illustrates the process of investment planning. In the search and decision phases, the future development has to be projected as accurately as possible. Essential elements of this forecast include:

- The success or failure of the product in the clinical trial.
- The development of the product after market launch. This contemplation includes the competitive situation, one's own marketing efforts and the future price development.
- The deployed technology and potential improvements thereto.

Market prognosis is done for at least 10 years following product launch. The investment decision for a production plant usually has to be taken 3–4 years before launch (Section 9.2). These relatively long periods are due to the high development effort, the diligent process validation and the regulatory approval process, and are typical for biotechnological production. The projections that reach from the clinical success over market development up to the deployed production technologies have high uncertainties. In this section it will be shown how the planning of investments can account for these unavoidable uncertainties.

Uncertainties can be distinguished in risks and uncertainties. Risks can be quantified to a certain extent by analyzing historic data (objective probability). Obviously, a comparable situation from the past has to be available. In contrast, uncertainties lack a data basis for evaluation. Since they cannot be quantified, their impact has to be assessed on the basis of experience (subjective probability). *Risks and uncertainties*

Investment appraisal is dominated by five influencing factors. During planning the uncertainties in these factors have to be estimated as accurately as possible in order to optimize the quality of the forecast. The factors are: *Parameters of investment appraisal*

- *Purchase price of the asset.* Costs can usually be estimated pretty accurately; however, cost estimates for larger projects can cause spending in the million Euro range by itself. *Purchase price*

- *Liquidation value for the divestment of the asset.* This value is difficult to estimate since it mostly lies in the far future and the market for used capital assets is difficult to project. For real estates, it is much easier since prices are more stable. For facilities, one would typically assume that the costs for shutdown would compensate for the potential sales price; hence, the liquidation value is set to zero. *Liquidation value*

- *Profit generated during the usage.* This requires an estimate of the expected cost of goods and other costs as well as the revenues. Due to the effort for calculation and the market projection, this position is maybe the most insecure. It also includes the failure to obtain market approval. *Profit*

- *Useful economic life.* This factor describes the period over which the asset can be operated economically meaningful. This period only *Useful economic life* partly depends on technical abrasion; other important factors are technological and economical developments. The former can bring technologies to the market that can antiquate the implemented technology in terms of quality, costs or performance (e.g. computers). Another limit of the useful life comes into play when the sales of the product decrease to such an extent that the plant cannot be operated economically feasibly (fixed-cost problem).

Interest rate for calculation

- *The interest rate for calculation is the rate at which the used capital should return profit.* One can obtain an easy interpretation of this financial unit figuring that the capital would have been invested in the finance market. The interest rate paid there for financial investments would contribute to the profit of the company. If the company invests in fixed assets instead, it expects to gain higher profit. Obviously, the profit should be higher than that for an investment in safe public bonds. The interest rate for calculation depends on the company goals. The interest rate on the capital market, the rate common in the particular field or an internal rate achieved somewhere else in the company can give direction. In the case of debt financing, the interest rate for calculation must be above the interest rate of the borrowed money.

Business administration uses methods to estimate the consequences of uncertainties in these factors.

Correction method

In correction methods, the factors are modified by additions or deductions. For example, the purchase costs can be provided with a flat premium for contingencies. The liquidation value can be lowered compared to an assumed value as well as the expected profit and the useful lifetime. The interest rate for calculation can be adjusted to the risk assessment. In all cases a quantification of the risk by means of a risk analysis is advantageous (e.g. decision-tree method). The plausibility of the corrections is a special challenge for correction methods. Following a too conservative approach, a project can be 'calculated to death'; a too aggressive approach can result in unexpected extra payments.

Sensitivity analysis

A sensitivity analysis can be performed in order to identify the influence of the correction factors. In this analysis the influencing parameters are varied and their impact on the outcome of the calculation detected, revealing the impact of each parameter. The analysis is based on model correlations between the parameters and the evaluation criteria of the investment. Such a model of investment costs can also be used to identify critical values that have to be met in order to keep the investment profitable and can help to identify no-go criteria.

Calculation methods

The traditional investment appraisal knows eight calculation methods. The four methods of static appraisal are: cost comparison, profit comparison, profitability comparison and static payback time. Due to their mathematical simplicity, static methods allow a fast but inaccurate comparison of options. More accurate results can be achieved with the dynamic methods: net present value (NPV), internal rate of return, annuity and dynamic payback time.

While static methods look at one selected period (e.g. business year), the dynamic methods take into account the complete lifecycle of the investment. The higher calculation effort for dynamic methods lies in the compilation of a calculation model and the provision of input data for the whole term in question.

These methods will be explained in the following sections. For clarity, both the single investment and the replacement problem will be considered. The selection problem can be solved by comparing two single investments.

11.3.1
Static Methods

The static methods of investment appraisal allow getting a quick estimate of investment options. The methods look at the efforts and returns of a representative period (e.g. business year). Efforts include material, personnel, depreciation and others like taxes, rents and so on. Returns are generated revenues and royalties. Effects of capital interest over time are disregarded, which limits the power of these methods.

The methods will be further explained in the subsequent sections. Figure 11.4 shows a comparison between them. In the selected example, the different methods result in slightly different evaluations. While option 1 is superior in the cost and profit comparison, option 2 prevails

Average yearly relevant quantity	Unit	Option 1	Option 2	Calculation		
Depreciation	Euro/year	40,000	30,000	A =	10% of F	
Other expenses	Euro/year	120,000	145,000	B	from cost accounting	
Costs	Euro/year	160,000	175,000	C =	A + B	Cost comparison
Return	Euro/year	240,000	240,000	D	from sales appraisal	
Profit	Euro/year	80,000	65,000	E =	D - C	Profit comparison
Investment volume	Euro	400,000	300,000	F	from cost estimate	
Profitability	%	20	22	G =	E / F	Profitability comparison
Reflux of capital	Euro/year	120,000	95,000	H =	A + E	
Payback time	Years	3.3	3.2	I =	F / H	Payback time

Figure 11.4 Overview of static methods of investment appraisal. Comparison of two investment options.

in the profitability comparison. This is a typical situation for one-product plants differing in degree of automation. The high investment in automation (option 1) pays back in lower operational costs, therefore initially the costs are lower; however, this picture can change if the capital deployment is also taken into consideration.

11.3.1.1 Cost Comparison

In cost comparison, the manufacturing costs of the produced goods in different options are compared. The assessment assumes that the returns from both options are equal. This assumption is justified if the options are designed for the same nominal capacity. A typical cost comparison could be made for the selection of a certain production technology (e.g. multi-purpose versus single-use equipment).

For cost comparison it is recommended to make a complete calculation and not limit the data to the relevant costs, otherwise the influence of single cost positions can be easily overlooked. Cost comparison does not allow evaluating the success of the investment since the costs are not seen relative to the investment volume.

The calculation of costs has been described in the Section 10.3. Costs per year can be calculated with the formula:

$$K = B + \frac{A}{2}i + \frac{A}{n}$$

where K is the cost per year, B is the operating cost per year, A is the investment volume ($A/2$: mean investment volume), i is the interest rate for calculation (% per year) and n is the time (years; A/n: yearly depreciation). Division by the yearly quantity of units produced results in the unit costs.

11.3.1.2 Profit Comparison

Profit comparison considers the difference between costs and returns. The method allows stating how high the profit from the investment project will be – based on mean figures. Cost comparison is a prerequisite for profit comparison.

11.3.1.3 Profitability Comparison

This method gives information on the interest yield of an investment. The relation of the profit – calculated from costs and returns – to the investment volume is evaluated. Profitability is one of the highest performance goals of a company. To get a read on the average capital yield (return on investment), the average profit calculated in the profit comparison is divided by the investment volume. The percentage resulting from that gives a hint to whether the capital is well invested or whether another option would be more profitable.

11.3.1.4 Static Payback Time

The static payback time is the period after which the investment is amortized by capital reflux, whereas capital reflux is made up of returns and depreciations. Most companies have a clear goal for acceptable payback times (usually 3–6 years).

11.3.2
Dynamic Methods

The dynamic methods are different from the static ones in that they:

- Use in- and out-payments instead of efforts and returns as the basis of calculation.
- Take the whole lifetime of an investment and the effects of interest into consideration.

In-payments are the revenues from the company's products and – in the case of an investment in a production plant – the products of the respective plant. Out-payments are payments for construction and operation of the plant.

Basically, dynamic methods answer the same question: does the return on capital employed meet the expectations or not? This return is preset in the company's interest rate for calculation. The methods do not directly tell whether the investment is profitable, but they offer the information whether the target return rate has been achieved.

To run the algorithm of dynamic methods, the investment volume (purchase out-payment at timepoint 0) and the series of payments of in- and out-payments have to be determined. The cash flows of the future periods are then discounted with the interest rate of calculation to the present timepoint and summarized. This discount allows comparing cash flows from different years. The dynamic methods look at certain parameters at this one timepoint. If the timepoint of evaluation lies in the present, the discounted amounts are called 'present values'.

Present value

11.3.2.1 Capital Value

In the capital value method, the present values of the surplus of in- and out-payments (net cash flow) are contemplated. Therefore, the method is also known as the NPV method.

NPV

Figure 11.5 graphically and numerically shows the in- and out-payments for six periods following an investment in period 0 of €400 000. In the first period, the investment is not yet profitable; this changes in period 2. The example does not consider a liquidation value, which would add to the in-payments at the end of the overall period. The totals of each period are discounted to period 0 and summed up. This sum is the NPV.

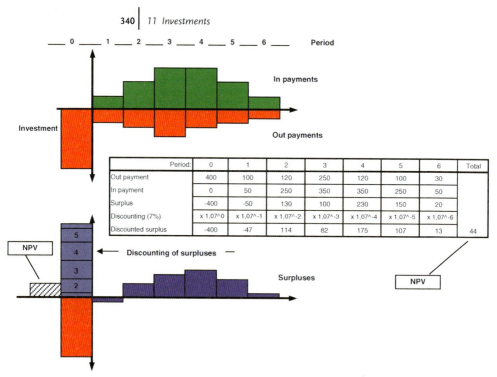

Figure 11.5 Calculation of the NPV from the series of payments of an investment (graph scale only roughly aligned with table).

A positive NPV means that the contemplated option outperforms the target interest yield (i.e. that the investment is advantageous). If the NPV is zero, the requirements are barely met; a negative NPV indicates that the requirements are not fulfilled.

11.3.2.2 Internal Rate of Return

The question of the NPV method is turned around to calculate the internal rate of return: it does not ask for the NPV at a given interest rate, but for the interest rate at which the NPV is exactly zero. This interest rate is denoted as 'internal rate of return'. If it is higher than the target rate, the interest yield of the employed capital is higher than the desired minimum and the investment is beneficial.

For a positive NPV, the internal rate of return is always higher than the interest rate for calculation. The internal rate allows a direct comparison with the profitability goal of the company.

11.3.2.3 Annuity

While the NPV method looks at the overall success of the investment, annuity looks at the success of the period. This is the decisive advantage of the annuity method, because the success of one period (e.g. one business year) can be more significant for financial planning than the overall number obtained by the NPV method.

To reach the annuity one has to calculate the NPV. Multiplication with the capital recovery factor results in a mean cash amount that could be drawn yearly out of the investment. How the capital-recovery factor is determined will not be further explained here. If the NPV is greater zero, the annuity is positive as well.

11.3.2.4 Dynamic Payback Time

The dynamic payback time indicates after what period of time an investment has amortized. It is based on the discounted values of the surplus payments. In the example given in Figure 11.5, by summing up the values in the line 'discounted surplus' one obtains a value of −19 for period 4 and of 88 for period 5. Hence, the investment has amortized after 5 years.

12
Production Concept

Production is responsible for supplying goods in the right quantity, at the right time, with the desired costs and quality. The basics for the quality and cost aspects have been covered in the previous chapters. The present chapter elaborates on the aspects of supply and timelines of the workflows relevant for production. We will look at the long-term strategic build-up as well as the operative organization of commercial market supply.

Capacity planning is based on product demand and process yield. How these parameters interact will be explained in Section 12.1. A production concept is always accompanied by the question whether an investment in one's own production facilities is favorable or whether an external source should be preferred. Section 12.2 starts with highlighting typical time schedules and questions of in-house manufacturing; in the following Section 12.3, principal aspects of external manufacturing are covered. In addition to contemplating project schedules and types of cooperation and contracts, challenges of technology transfer are also considered. Section 12.4 summarizes a comparison between internal and external manufacturing; here, typical make-or-buy decision situations are discussed with the aspects illustrated in the preceding chapters.

Capacity planning

Process development is mainly completed before the product is launched to the market; however, established processes also have optimization potentials worthy of being realized. Therefore, Section 12.5 points to the possibilities and consequences of optimization efforts. Since the product finds its way to the client through the supply chain, Section 12.6 shows basic principles of supply-chain management.

Optimization potentials for process and supply chain

12.1
Capacity Planning

In capacity planning, product demands and production capacities have to be aligned to create a secure and cost-optimized product supply. A meaningful demand forecast is the starting point for the

Manufacturing of Pharmaceutical Proteins: From Technology to Economy. Stefan Behme
Copyright © 2009 WILEY-VCH Verlag GmbH & Co. KGaA, Weinheim
ISBN: 978-3-527-32444-6

Fermentation capacity

implementation of a production concept. If the process parameters are known, such quantities can be recalculated into production capacities. In API manufacturing, in most cases, the fermentation capacity is targeted, since it can be assumed that downstream capacities have been adjusted to the upstream section. However, an additional check of the downstream capacities and technologies is indispensable. With rising productivity of the fermentation processes, the downstream area develops more and more into the bottleneck of protein manufacturing.

Table 12.1 shows that only a couple of key parameters (Section 2.2.3) are needed to obtain a good impression of the required production capacities. Examples are shown for typical process parameters of microbial batch fermentation and a fed-batch and perfusion cell culture.

Table 12.1 Calculation scheme and examples for processing times in fermentation.

	Unit	Fed-batch	Batch	Perfusion	Calculation
Quantity (pre-set)					
yearly demand for drug substance	g/year	4500	30000	30000	A
overfill	%	10	10	10	B
rejects/analytical demand	%	10	10	10	C
Process (pre-set)					
titer	g/l	0.3	1.2	0.2	D
overall yield	%	65	50	45	E
perfusion rate	volume/day	NA	NA	4	F
fermentation time (including turnaround time)	days/batch	14	6	40	H
Plant (pre-set)					
fermenter working volume	l	2000	10000	200	G
volume of supernatant	l	1600	8000	160	80% of working volume
operation days per year	days/year	240	240	240	I
Results					
maximum number of batches per year	batch/fermenter/year	17	40	6	J = I/H
drug substance to purification	kg/year	8.3	72.0	80.0	K = A * (1 + (B + C)/100)/E/10
supernatant to purification	kl/year	27.7	60.0	400.0	L = K/D
number of batches per year	batch/year	17	8	17	M = see below
capacity usage	%	101	19	289	N = M/J * 100
number of fermenters needed	–	2	1	3	O = ROUND UP (N/100)

[a] M = L * 1000/G for fed-batch and batch; M = L * 1000/(H * G * F * 0.9) for perfusion (factor 0.9 for consideration of turnaround time).
[b] Overfill = drug quantity in excess of the therapeutically required dose; turnaround time = time between two production runs needed for example, for cleaning; NA = not applicable.

Starting with the product amount to be delivered to the market, 10% is added to account for a possible overfill of vials. At this point one should calculate with the dose actually delivered and not the patient dose. A further 10% is added for rejects and analytical purposes. The resulting quantity is the starting point for the following calculations. Key parameters of process evaluation are the processing time, the overall yield and the fermentation titer. In line 'capacity usage', Table 12.1 shows a value that stands for the usage of a fermenter with given size (line G): 100% means that exactly one fermenter is used for 1 year; a number above 100% means that more than one fermenter is needed to supply the capacity. The 19% usage for the batch fermentation means that the fermenter is used only 19% of its time. When designing a facility, these numbers can be used to adjust the fermenter size. This would be done in the example for the fed-batch fermentation, since a capacity usage little over 100% would trigger the adjustment of the fermenter size instead of investing in a second fermenter that would remain idle for most of the time.

Capacity usage

The example illustrates that the manufacturing costs are indeed affected by process quality (i.e. fermentation titer and overall yield), but it is also important to consider the duration of the process. Yield and time should be optimized concurrently. Yield losses can be accepted if process time decreases in return. Examples of this could be the termination of fermentation prior to reaching the maximum titer or to run a chromatography column with higher flow, accepting less sharp separation. In both cases, processing time is spared at the cost of yield; however, this yield loss is overcompensated by the ability to perform more runs.

The dominant rate-limiting factor in fermentation is the cell growth rate. In chromatography, the loading capacity and the gel volume determine the number of runs, while the flow rate determines the duration of the single run.

Cleaning processes, which have to be performed between two batches, additionally occupy the equipment. Typical turnaround times are between 1 and 3 days depending on the process step.

The nominal capacity of a facility is limited by the maximum time that it is available for actual processing purposes. This time is reduced by other necessary activities like:

Nominal capacity

- Maintenance and repair work.
- Batch-to-batch cleaning for multiple batches.
- Product change-over cleaning in multi-product facilities.
- Re-validation activities (e.g. cleaning validation).

The 'net run time' of a production plant can be estimated to be between 220 and 280 days.

Decoupling and parallelization of process steps

This limited run time can be used optimally if process steps are decoupled and run in parallel. Concurrent operation of inoculation fermentation and main fermentation allows operating the main fermenter without interruption. Parallel processing can also be found in isolation if fast separation of proteases is required or in preparative HPLC chromatography if the capacity of one HPLC column is smaller than that of the other purification steps. Solutions from the individual runs can be pooled again for the subsequent process steps.

Decoupling process steps results in a higher flexibility as to production scheduling, which again enhances optimization of operations. Common holding points in the process are after isolation and purification (Figures 2.3 and 2.4). This decoupling allows using different locations for the steps, provided that the product is stable over the holding times. Refrigerated conditions are often necessary for storage of intermediates; sometimes they can be frozen and stored for several years.

Capacity limit/bottleneck analysis

The capacity limit of a plant is determined by the chain segment with the lowest capacity. This segment is identified by a bottleneck analysis in which the capacity of each single process step is carefully reviewed. It happens that storage capacities or supporting systems (e.g. water supply) show up as the bottlenecks of production. These gaps have to be recognized and corrected as early as possible in the planning process to achieve a balanced capacity usage throughout the process chain.

12.2
Dilemma of In-House Manufacturing

Figure 12.1 illustrates the typical development and launch schedule for a biotechnological product; it consists of four horizontally separated segments.

The upper segment shows the often rate-limiting clinical development. Below that the required product demand is drawn. One step further down, two possible options for the supply of clinical and commercial goods are shown.

The product should be launched as early as possible after gaining regulatory approval in order to maximize product sales. However, it is a regulatory requirement that the launch product – coming from the production plant – has been used in the clinical studies prior to dossier submission.

Facilities for commercial production are larger and more expensive than the smaller pilot plants for clinical manufacturing. The requirement for deploying the launch product in the clinical phase results in the question from which timepoint on one should produce in the 'final'

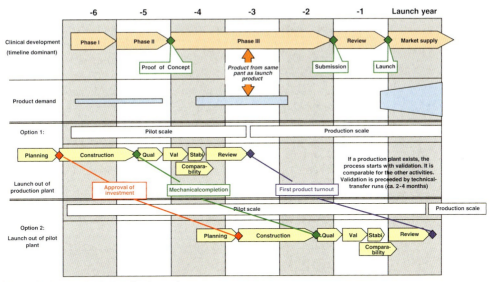

Figure 12.1 Typical schedule for the implementation of market supply for a biotechnological drug manufactured in-house. Qual = qualification; Val = validation; Stabi = stability investigation.

and more expensive facility. There are two approaches that are outlined in Figure 12.1. One is to launch the product out of the production plant, which means to invest early on in building up capacities or in blocking already existing high-volume capacities. In both cases more money has to be spent upfront than in the second option, but the risk of running short in the launch phase is circumvented. If process implementation in the large scale does not progress as expected, there is a risk of delaying the clinical trial, which would ultimately result in a delay of the launch timepoint.

Launch out of production facility

In the second option, the product is launched out of the pilot plant. Here, the timepoint of investment is shifted backwards to the period of highest certitude for decision. Obviously, the risk lies in a supply shortage of the market since the pilot plant may not be capable of delivering larger quantities. Since the return time of drugs is limited (Section 10.1), such a sales loss in the patent-protected period is painful. In this case additional production capacities are implemented after regulatory market approval.

In clinical development there are three decision points that are essential for the implementation of commercial market supply. The earliest timepoint is the 'proof of concept' or 'proof of principle' at the end of phase II of the clinical trials. At this timepoint information is available as to the expected dosing; the average probability that the product will make it to the market has increased to 75%. The second milestone is the timepoint of regulatory submission; here all the information regarding the product, production sites and processes has to be available. After that the period of agency review starts, which ends with market approval. Preparations for a quick launch should be completed now in order to realize sales without delay.

Going back to implementation of market supply, three phases can be distinguished. During planning, requirements for the plant are defined and the planning phases are performed, leading to the final design (Section 9.1). Make-or-buy decisions should be taken in this timeframe since delayed buy decisions can only be reverted by accepting the risk of project delays. If a 'make' decision has been taken, plant construction follows ending with qualification work. Here, a plant exists that has demonstrated that it functions technically, but has not yet proven its capability to produce the desired outcome consistently. The latter is provided by the subsequent validation. At this point also the process starts that has to be considered if a qualified plant already exists. When changing from one facility to the other, product comparability has to be demonstrated (Section 3.6). If the molecule is easy to characterize (e.g. no glycosylation, relatively small), an analytical study could suffice; for larger and more complicated molecules with a complex byproduct profile, a clinical bridging study may have to be taken into consideration.

Validation phase
Stability data

Additional time is needed to provide stability data of the product. Six months of stability data for the final formulation should be available at the timepoint of dossier submission. The material for these studies must be taken from the validation batches so that upon submission a consistent set of data is available. The regulatory approval lasts 6–18 months depending on the intensity of additional questions and document quality.

Approval of investment budget

The principal milestone of one's own manufacturing is the approval of the investment budget by the company.

Relations between the milestones
Dilemma of in-house manufacturing

There are some remarkable relations between clinical development and technical timelines for product supply. Due to the risk of losing the investment, an investment decision before proof of concept is usually hard to get. However, if the product is to be launched out of the plant to be erected, it would be necessary to start investing 2 years before proof of concept, which means that the risk of wasting the investment is around 70%. This dilemma can only be resolved if:

- Highly flexible multi-purpose plants are constructed that require only minor refitting.

- The processes are developed in standard technologies for which plants already exist. Such typical processes exist for fed-batch fermentation of defined expression systems with a standardized downstream sequence. These standards are also denoted as platform technologies.

Platform technology

- A broad portfolio of projects is available so that it is highly probable that at least one of the candidates fills the plant. In this case again it is necessary to develop the process to fit the technical plant.

Finally, the biotech-specific regulatory link between the launch product and the clinical trials forces one to focus process development on providing production capacities in short time. After launch it is still possible to optimize the process, while a delay of launch mostly goes along with severe, unacceptable sales losses. Therefore, production aspects should be included early on in process development.

12.3
Aspects of Manufacturing Out-Sourcing

Only a few companies have the ideal prerequisites for internal manufacturing – a broad product portfolio with an essentially standardized technology. Many companies circumvent the dilemma described above by outsourcing their manufacturing needs. At the contract manufacturer the number of client orders creates the broad portfolio allowing him to use the capacity of a multi-purpose facility.

Contract manufacturer

The advantages of external manufacturing are obvious:

- No need to invest in capital assets.
- No need to establish broad manufacturing expertise and staff.
- The quality of work at contract manufacturers can be predicted from reference projects. This is not always the case for newcomers in in-house manufacturing.

However, there are also drawbacks:

Drawbacks of external manufacturing

- There is no direct control of production and costs on the part of the client.
- Intensive interface management is often necessary. Different company, and sometimes national, cultures meet at these interfaces. Both companies may have essentially the same interest – a successful project – yet each company also has its particular interests. In this mixture of different perceptions and interests, contract negotiation and subsequent alliance management plays an essential role.
- A costly and mostly challenging technology transfer has to be made, which needs to be accompanied by the client with personnel of adequate expertise.

- Manufacturing costs can be higher than in in-house solutions, since the profit margin of the manufacturer has to be added. However, this comparison is not straightforward, since the high capacity usage and experience of an established contract manufacturer can balance out this difference.

- The early reservation of use rights of production capacity costs money. It is common practice that compensation payments are due to the contract manufacturer if projects are discontinued unexpectedly by the client. This corresponds to a misinvestment, yet it is mostly much smaller than investing in one's own assets.

This section highlights some special aspects of external manufacturing. It starts with the possible forms of cooperation, and then shifts to contractual agreements and aspects of technology transfer. Finally, time schedules for external manufacturing are contemplated.

12.3.1
Types of Cooperation

External manufacturing can be performed in the framework of different typical forms of cooperation, provided in Table 12.2. They reach from pure contract manufacturing to complete acquisition, which in a true sense cannot be understood as a cooperation any more, yet is listed here for the sake of completeness. In most cases real-world cooperation lies between these model types. Pure contract manufacturing is the

Table 12.2 Types and features of cooperation.

	Contract Manufacturing	Partnership	Alliance	Integration
Type of business	money for service	risk sharing and incentives through milestone payments	sharing of risk and profit	complete integration
Features	driven by price	joint development	joint independent company (joint venture)	acquisition of the manufacturer
Control of manufacturer	relatively poor, depends on contract and willingness to cooperate	medium, since the partner share interests	good, since partners share interests and personnel	complete
When to be applied?	if different contract manufacturers are suitable and comparable	if partners have complementary strengths	if manufacturer has key know-how	if manufacturer has key competencies in the own strategic field

predominant form. The client selects a contractor out of an offer of suitable options on the basis of the optimal price.

Pure contract manufacturing is beneficial if the market offers a choice of comparable offers. In this case the selection is driven by the price. The possibility to control the contractor depends on the shape of the contract and largely on the willingness to cooperate on the part of the manufacturer. Here, the control aspect principally relates to quality and price. Especially since in biotech the process determines the product, quality control of the delivered product alone is not sufficient to assure quality; it is mandatory that the client obtains a deep insight into quality assurance of the supplier. They can do that by performing quality audits or by reviewing the batch documentation. The control of the price is particularly important if the price base is 'cost plus profit', enabling the contractor to pass through manufacturing costs. In this case, the minimum cost control should be based on accounting sheets of labor hours and used materials. *Monitoring of contract manufacturers*

The next closer steps are the partnership and the alliance. In both, risks and benefits of the project are shared. The possibilities of control are enhanced since the partners have larger overlaps and more mutual insight. *Partnership and alliance*

12.3.2
Contractual Agreements

Cooperation between two partners starts with negotiating and closing of a contract. A prerequisite for cooperation is a common interest to the advantage of both parties. Since the parties also have diverging interests (e.g. everybody wants to get optimized profit out of the cooperation), mutual control is required as well as the formulation of provisions for the cooperation.

A cooperation agreement describes the common interest, the mutual services and considerations, and the provisions for special project situations like termination. The provisions should be shaped in a form that in cases of conflicts – which regularly occur when diverging interests clash – an independent institution (court) can decide upon them.

Legal agreements in the production fields typically encompass different fields of regulations:

- Product supply (Manufacturing and Supply Agreement).
- Quality assurance (Quality Assurance Agreement).
- Clinical safety (Pharmacovigilance Agreement).

If the project requires it, a Development Service Agreement can be closed. The contents of the agreements can be differentiated from each other:

12 Production Concept

Supply contract

- Provisions of the Manufacturing and Supply Agreement relate to services and considerations, forecasting and ordering processes, liability, confidentiality, term and termination, and basic quality measures. It should also describe the organization of the cooperation, the working teams and mechanisms for conflict resolution.

Quality assurance agreement

- Regulations of the Quality Assurance Agreement describe quality-assurance measures. These encompass audit rights, procedures for process changes and deviation handling, information obligations and rights, documentation requirements, release procedures, and applicable guidelines. The Quality Assurance Agreement has legal relevance as to drug law. Drug law requires that the marketing authorization holder implements quality-assurance measures; these can be partly delegated to the contractor by means of the Quality Assurance Agreement. Moreover, the supplied company is obliged to ensure that the supplying company fulfils the transfer duties. This can be accomplished by frequent audits.

Pharmcovigilance agreement

- Provisions of the Pharmacovigilance Agreement regulate procedures upon occurrence of unexpected clinical results in connection with the administration of the drug. Like the Quality Assurance Agreement, this agreement is drug-law relevant.

Depending on the urgency, complexity of the subject and contract volume, the negotiations can last 3–6 months (if not longer). In the negotiation sessions, situations that may arise during the term are anticipated as far as possible; however, it is impossible to regulate each situation. Therefore, it is very important that the parties share an agreed understanding of the contract spirit and that this spirit is lived in the day-to-day project work. Apart from this cooperative perspective, subjectively perceived contract violations should be articulated to save one's own position in case of a legal dispute (Section 9.4.3) and to give the other party the chance to react.

12.3.3
Technology Transfer

A rate-limiting and often underestimated step in outsourcing is the transfer of the manufacturing process from the client to the contractor. In most cases the client owns the technology; that means they are responsible for providing the process description.

Again, the biotech-specific feature (i.e. that the process defines the product) makes the transfer more difficult compared to small molecules. Together with the requirement to demonstrate comparability of the product before and after the transfer, it means that each transfer

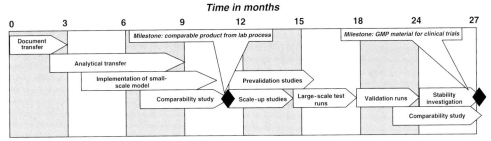

Figure 12.2 Typical timeline for technology transfer. Small-scale model = process in laboratory scale. Scale-up studies = check whether the lab process can be transferred to the production environment.

needs a comparability study. For this same reason, analytics play a major role in technology transfer.

Figure 12.2 shows an exemplary timeline for a transfer of a laboratory process with subsequent scale-up. Technology transfer begins with transferring documents of the process and the analytical methods. This is followed by a transfer of the analytical methods since they are crucial for assessing the success of the process transfer. Typical complications of analytical method transfer are:

- The methods may not be described precisely enough and their conduct may depend strongly on the experience of the laboratory staff.
- The results depend on the equipment, which may not be completely identical in the receiving lab.
- Acceptance criteria have not been defined accurately between the parties.

The laboratory process is transferred in parallel to the analytical methods. This so-called small-scale model serves to familiarize the relevant parties with the process and later for conducting pre-validation studies (e.g. identification of critical process parameters or virus validation). Also, an agreement on the acceptance criteria for equivalence with the original process is important here. *Small-scale model*

Analytical transfer

The completion of this 'small-scale transfer' marks the milestone at which product has been generated that is comparable between the newly established and the original process. This comparability study encompasses a parallel and complete characterization of the products from old and new process. *Comparability study*

Once the transfer to the laboratory scale has been completed successfully, process implementation at the large scale can be initiated. This path leads over first scale-up studies in which scalable technologies are tested for their specific process suitability. Typical examples for the transition from a laboratory process to the large scale are the *Scale-up studies* replacement of depth filters for cell separation and supernatant clarification by centrifuges and/or tangential-flow units. At these steps, the original process should only be changed if necessary. Deploying a different medium component can go along with changing cell metabolism and be critical for product comparability.

The studies are followed by first experimental runs on the large scale. If they are successful, process validation can start, which typically means that three consecutive runs under full GMP conditions are performed. Material from these runs has to be put on stability studies. Figure 12.2 shows a time of 3 months for the stability data. This is the minimum time if longer ranging stability data are present from earlier non-GMP runs and the comparability of these products has been demonstrated in respective studies. This product can be used for the next step of the clinical study.

In practice, process transfer should be closely monitored by experts on both sides and be the result of intensive cooperation. Frequent and regular meetings are mandatory to convey the relevant experience and knowhow.

12.3.4
Time Schedules

In comparison to Figure 12.1, Figure 12.3 shows the timelines for the transfer to a contract manufacturer.

The essential difference is that the 'planning', 'construction' and 'qualification' blocks from Figure 12.1 are replaced in Figure 12.3 by the '(manufacturer) screening and selection', 'contract negotiation' and 'technology transfer' blocks. The pivotal timepoint for approval of the investment budget (here for the contract with the manufacturer) in the outsourcing option shifts downwards only marginally. Also in Figure 12.3 the timepoint of decision for the manufacturer is shown as late as possible. It is beneficial to supply the complete phase III out of one manufacturing process, which may shift the timepoint 'Decision for CMO' 1.5 years further upwards and thus far before the clinical phase I.

Hence, the benefit of outsourcing manufacturing obviously cannot be found in the shift of the investment timepoint; indeed it is the lower investment risk that occurs at the early timepoint. One-time costs accrue for technology transfer and capacity reservation. Transfer costs can reach up to €20 million. Mostly reservation fees accrue, which are billed for securing capacity. If a project is discontinued, termination costs have to be paid, which are geared to the costs of the plant. Special and dedicated equipment for external manufacturing is usually paid by the client as well.

Taken together, these costs are still lower than one's own investments, making outsourcing attractive.

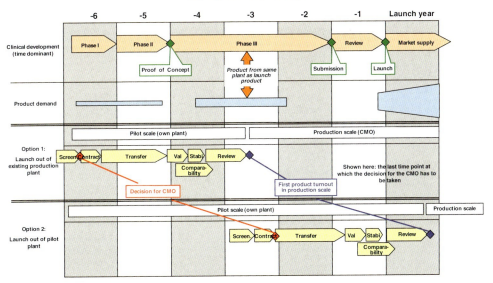

Figure 12.3 Typical schedule for the implementation of market supply for a biotechnological drug manufactured externally. Qual = qualification; Val = validation; Stabi = stability investigation; Screen = manufacturer screening and selection; Contract = contract negotiation; CMO = contract manufacturing organization.

12.4
Make-or-Buy Analysis

One central question of biotechnological research companies is whether a product should be produced in-house or by external specialized contractors, or even from other biotechnological companies that offer their capacity on the market. Obvious advantages of in-house production are:

- No payment of profit margin to the contract manufacturing organization.
- Retaining production knowhow can give competitive advantages.
- Better cost transparency and control of costs, resulting in enhanced possibilities for improvement.
- Better control of quality and organization.

Obvious drawbacks include:

- Necessity for one's own capital investments in facilities and personnel.
- Establishment of knowhow in a new field, which may remain inferior to the specialists knowhow in the same field and which may not be used in the future due to lack of follow-up projects.

Table 12.3 shows the essential aspects of make-or-buy decisions. It can be concluded that external manufacturing is beneficial if the market provides a wide range of suppliers. In this case, good prices can be negotiated, which can outweigh the drawbacks of contract manufacturing.

On the other hand, the ideal case for internal manufacturing is that the company has a broad product portfolio based on similar technology. Moreover, it is advantageous if structures for quality assurance, logistics and regulatory affairs already exist. Long-term advance financing often makes one's own manufacturing problematic for smaller companies. However, also for larger firms, the path to contract manufacturing

Table 12.3 Overview of in-house ('make') and external ('buy') manufacturing.

	Make	Buy
Manufacturing costs	difficult to calculate for new plants	mostly following the model: 'manufacturing costs plus profit'; if the client achieves the same manufacturing costs the CMO is more expensive due to profit margin
	for existing plants additional usage (dilution of fixed cost)	CMO can be cheaper
Investment costs	if no plant exists: high capital investments in equipment and staff at early timepoint	investment in transfer, capacity reservation and equipment at early timepoint
Time aspects	construction of a plant can be on critical path for product launch (time to market) investment timepoint before phase I	decision point in phase I
Transparency/control	full transparency and control	control of costs and quality depends on the contract and project management practice
Knowhow	proprietary knowhow can secure competitive advantages and be used for other projects well functioning and known team can minimize risks for investment decision (e.g. for technological realization or regulatory licensing)	depending on foreign knowhow is unfavorable
Effort	maybe construction of a plant built up and training of personnel	technology transfer contract negotiation and cooperation management
Typically applied if...	own production already existing *if no own production exists:*	competitive situation on the market place unique technology necessary which can only be used for this one project
	strategic expansion in this field of technology planned, if possible with platform technology	very unclear quantity demands; the higher flexibility of in-house production can make internal manufacturing more beneficial in this case
	broad portfolio for usage of plants sensitive knowhow needs to be protected	little risk acceptance for investments foreign costs smaller than own costs

CMO = contract manufacturing organization.

organizations can be financially favorable (e.g. if the demand projections are highly fluctuating or if the valuable resources are not to be invested in a technology that would remain unique in the company).

12.5 Process Optimization after Market Launch

The early fixing of the production process, which is triggered by regulatory requirements, poses the question as to what extent the process has to be optimized before launch and how much improvements can be postponed until after market entry.

As already mentioned, the main effort lies in showing comparability of the product before and after the process change. The risk is that indeed comparability cannot be shown. In the best case, this would show up in the analytical part of the study; in the worst case, safety deficiencies would show up only after implementation of the change, which could then lead to recall, marketing discontinuation or a limited indication, or at least to a weaker market profile.

It may also be the case that the market demand outstrips the forecasts and the comparability studies last longer than anticipated so that, in combination, the launch process does not suffice for supplying the market. Such a supply shortage goes along with sales losses.

Shifting process optimization too far to the back end means accepting this risk, yet retaining the freedom of implementing a process – suitable for market supply – later on, saving time and money in the development phase (backloaded project).

The return of a process change can be estimated relatively accurately since savings in the production process are fairly easy to identify. To what extent the risks and effort can be projected depends on several factors: *Factors for the risk assessment of process changes*

- What data exist supporting the intended change? It is possible that during development the data had already been collected that can now help to reduce the effort?

- Is the process in this step sensitive to changes? A process step is sensitive to changes if small changes have a large impact on product quality. This can, for example, be the case if baseline separation depends on the gradient of the mixed elution buffer. Another example could be the change of solubility properties of a lyophilized protein after replacement of a formulation component. An example for process robustness could be if the protein is expressed equally in different fermenter sizes.

- What influence could the change have on the final product quality? In principle, changes made further downstream, and therefore closer to

Process and product understanding

the patient, are to be treated more carefully than others. This is mainly due to the fact that the effects of these changes cannot be detected further downstream and are not controlled by additional process steps. Exceptions to this rule are changes to the fermentation process since metabolic processes of the cell can be affected that can have a major impact on the byproduct and impurity profile.

Specifications

- How tight are the specifications of the individual process steps and of the final product? If the specifications are very narrow it is more probable that the 'new' product is not comparable to the 'old' product. For example, if a product concentration of $90 \pm 3\%$ is the acceptance criterion for further processing of a chromatography fraction, a process change yielding a concentration of 94% would lead to rejection of this fraction, whereas the fraction would pass if the specification would have been $90 \pm 5\%$. However, this example should not suggest that specifications are the only comparability criteria; in fact, a comprehensive characterization of the protein is necessary. In this example it should additionally be considered whether the process step is robust enough if the result ends up at the upper limit of the specification.

Ability to characterize

- How well can the product be characterized? How well known are the glycosylation forms and byproducts, and can they be detected qualitatively and quantitatively? For a start it is very helpful to see how the product changes with the process. Under this aspect it is good to be able to fully characterize the product. Problems occur in the second step – what clinical consequences do these changes have? Seeing more does not mean knowing more. Since biotechnological drugs are mixtures with complex physiological mechanisms of action, it is very difficult to predict the clinical effect of changes. The better the measurements are, the higher the probability of detecting a change. It has to be assessed very carefully whether this change is relevant when compared to the current marketed product.

Clinical effect of impurities

- What experiences are there for the clinical effects of glycosylation forms or byproducts? This question is closely connected to the one in the preceding point (i.e. how well can detected changes be interpreted).

The situation for optimization is ideal if:

- Data for the intended change exist.
- The process is robust.
- The change is made after fermentation as far as possible upstream of purification.
- Specifications are broad.
- The product can be well characterized and the clinical consequences of changes are known.

Almost all changes to the direct manufacturing process of biotechnological products have to be submitted to authorities prior to implementation and need regulatory approval. This does not only generate waiting times, but also fees and additional risks as the changes are subject to approval by third parties.

Optimizations after launch are done very frequently. However, it is wise to use a process in phase III which is able to cope with market demand for at least 3–4 years after launch. This time can be used for optimization or – if that fails – for the implementation of a second identical production line. The work that supports the production ramp-up should be continued and not be stopped after regulatory submission of the drug dossier.

Due to the difficulties in showing comparability for biotechnological products it would be risky to rely on the potential of post-launch process optimization only, since this potential may be achieved only at high costs and effort.

12.6
Supply-Chain Management

The path of the product to the customer leads over several stations that all together are denoted as supply chain. The organization of the supply chains should ensure market supply, while both economic optimization as well as the special ethical responsibility of the drug manufacturer has to be considered. The drug-specific responsibility means, for example, that the patients always have access to the drug, that the drug is supplied with a suitable shelf-life and that the drug is delivered under consideration of relevant transport conditions.

Optimization of the supply chain coherently gears to the aims of service quality (no supply shortfalls, flexibility) and economic goals. The costs of the supply chain that should be minimized include: *Costs of the supply chain*

- Storage (refrigeration, facilities).
- Insurance for stored goods.
- Capital bound in circulating assets (inventories).
- High idle costs generated through poor usage of manufacturing facilities.
- Transportation routes and conditions.

Figure 12.4 shows the components of the supply chain for a typical biotechnological product. Several options for economic optimization appear along this chain.

Inventory management plays a central role in supply-chain management. The pharmaceutical company gets the orders from wholesalers and the company should be able to satisfy these orders without having to *Inventory management*

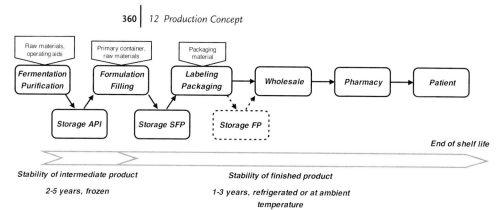

Figure 12.4 Supply chain for the market supply of a drug and typical values for the shelf-lives at the holding points. SFP = semi-finished product; FP = finished product.

start a production campaign for each order. Since the orders do not always arrive on a regular basis, it is advisable to build up a stock for buffering demand peaks. Since biotechnological products usually have limited shelf-life (2–3 years), this stock cannot be too large as the shelf-life of goods sold to the market could fall below an acceptable level.

The supply chain offers some possibilities to stagger this storage so that manufacturing can be essentially decoupled from orders and therefore planned most optimally. Usually, there are three holding points in the supply chain:

Holding points in the process

- Between product isolation and purification. At this step the product often can only be stored a short time (days to weeks) refrigerated or a longer time frozen (months).

- After purification, before formulation (API stage). At this stage the product can often be stored for years in the frozen state.

- After filling, before labeling and packaging (semi-finished goods, labeled vials). The clock for the final shelf-life starts ticking from the timepoint of filling into the primary container. Therefore, it should be made as close to delivery as possible. This step if followed by labeling and packaging, which should be the only steps in which the final product gets its country-specific features. Consequently, a country-specific stock would have to be built here.

It has to be considered that the clock starts running again at each step. An intermediate on the API stage can, for example, be stored frozen for 4 years and after filling be stable for a further 2 years, so that the overall shelf-life adds up to 6 years starting from purification.

The possibility to decouple fermentation and purification is limited, while the gap between purification and filling can be significant.

When designing a production concept, aspects of supply safety play a major role. Safety can be enhanced by diversification or redundancy of supply units. Ideally, storage and manufacturing should be balanced in a way that market supply can be kept up any time – if the manufacturer fails, the goods should come from stock until the manufacturer can deliver again or an alternative is found. Mitigating a significant risk like the loss of a manufacturing facility by fire or other circumstances is only possible by qualifying a second independent source. Inventory keeping for such a long timeframe is not economically feasible and often it is simply not possible to provide the capacity for building up the stock. The qualification of a second supplier is expensive and challenging from an organizational and contractual standpoint; thus second suppliers are usually only implemented for strategically important products. Further measures to ensure supply safety are to distribute the stock to several storage facilities and to improve reliability (redundant installations) of the manufacturing facility. *Supply safety*

When managing shortfall situations, all the inventories in the supply chain have to be considered. This requires the distributing site to have an overview over the inventories in the individual supplied warehouses. Due to this overview, which is enhanced by the central position of the manufacturer, more and more suppliers act as the coordinating dispatch function (vendor-managed inventory). A typical case for a stock build-up to bridge capacity gaps consists of launch phases for which manufacturing starts before market launch in order to be able to react quickly to the rising market demand. *Second supplier*

A supply chain typically consists of several components that transform into the final product along the value chain. Figure 12.4 indicates that the contemplation should start with the raw materials. Due to the fact that the suppliers need to be qualified, raw materials can become scarce without being easily resupplied. In particular, critical raw materials like affinity gels for chromatography columns or substances of animal or human origin (HSA, FBS) should be taken into consideration. Also, packaging materials need consideration since they are country-specific and contain a great deal of information with legal relevance. The primary container (vial, ampoule, pre-filled syringe) is always supplied from outside; some product presentations also contain a more or less complex application device (syringe, pen, injector). *Vendor-managed inventory*

Critical raw materials

The bottlenecks and risks of this supply chain should be analyzed and, if deemed necessary, appropriate corrective actions implemented to secure supply.

References

Arvinte, T. (2003) Therapeutical activity predicted and explained by *in vitro* analytical techniques, in *Proceedings of the Conference on Comparability and Immunogenicity of Biologicals*, Brussels.

Braganz, A. (2001) Seminarunterlagen zu 'Claims Management', Düsseldorf.

Brauer, J.-P. (1996) *DIN EN ISO 9000–9004 Gestaltungshilfen zum Aufbau Ihres Qualitätsmanagementsystems*, Hanser Verlag, Munich.

Chmiel, H. (1991) *Einführung in die Bioverfahrenstechnik*, Gustav Fischer Verlag, Stuttgart.

Hartmann, K. (2003) Arzneimittelsicherheit, in Jaehde, U., Radziwill, R., Mühlebach, S. and Schunack, W. *Lehrbuch der Klinischen Pharmazie*, Wissenschaftliche Verlagsgesellschaft mbH, Stuttgart, pp. 165–183.

ISPE (2001) *Baseline Pharmaceutical Engineering Guides for New and Renovated Facilities, Vol. 5 – Commissioning and Qualification*. ISPE, Tampa, FL.

Jaehde, U., Radziwill, R., Mühlebach, S. and Schunack, W. (2003) *Lehrbuch der Klinischen Pharmazie*, Wissenschaftliche Verlagsgesellschaft mbH, Stuttgart.

Lottspeich, F. and Zorbas, H. (1998) *Bioanalytik*, Spektrum Akademischer Verlag, Heidelberg.

Madigan, M.T., Martinko, J.M., Parker, J., Brock, T.D. and Goebel, W. (eds) (2000) *Brock Mikrobiologie*, Spektrum Akademischer Verlag, Heidelberg.

Olfert, K. (2003) *Investition*, Kiehl, Ludwigshafen.

Plinke, W. and Rese, M. (2002) *Industrielle Kostenrechnung*, Springer, Berlin.

Rautenbach, R. (1997) *Membranverfahren, Grundlagen der Modul- und Anlagenauslegung*, Springer, Berlin.

Schneppe, T. and Müller, R.H. (2003) *Qualitätsmanagement und Validierung in der pharmazeutischen Praxis*, Editio Cantor, Aulendorf.

Stadler, E. (1998) Dual purpose fermenter and bioreactor – a capital quandary? *Pharmaceutical Engineering*, **18**, 74–84.

Wheelwright, S.M. (1991) *Protein Purification Design and Scale-Up of Downstream Processing*, Wiley-Interscience, New York.

Further Reading

Biotechnology General

Borem, A., Santos, F.R. and Bowen, D.E. (2003) *Understanding Biotechnology*, Prentice Hall, Upper Saddle River, NJ.

Dübel, S.(ed.) (2007) *Handbook of Therapeutic Antibodies (3 Volumes)*, Wiley-VCH Verlag GmbH, Weinheim.

Gad, S.C.(ed.) (2007) *Handbook of Pharmaceutical Biotechnology*, John Wiley & Sons, Inc., Hoboken, NJ.

Kahl, G. (2004) *The Dictionary of Gene Technology: Genomics, Transcriptomics, Proteomics*, 3rd edn, Wiley-VCH Verlag GmbH, Weinheim.

Schmid, R.D. and Hammelehle, R. (2003) *Pocket Guide to Biotechnology and Genetic Engineering*, Wiley-VCH Verlag GmbH, Weinheim.

Walsh, G. (2003) *Biopharmaceuticals: Biochemistry and Biotechnology*, 2nd edn, John Wiley & Sons, Ltd, Chichester.

Walsh, G. (2007) *Pharmaceutical Biotechnology: Concepts and Applications*, John Wiley & Sons, Ltd, Chichester.

Whitford, D. (2005) *Proteins: Structure and Function*, John Wiley & Sons, Ltd, Chichester.

Wiley-VCH (eds) (2007) *Ullmann's Biotechnology and Biochemical Engineering (Two Volumes)*, Wiley-VCH Verlag GmbH, Weinheim.

Fermentation

Gelissen, G.(ed.) (2004) *Production of Recombinant Proteins: Novel Microbial and Eukaryotic Expression Systems*, Wiley-VCH Verlag GmbH, Weinheim.

McNeil, B. and Harvey, L.(eds) (2008) *Practical Fermentation Technology*, John Wiley & Sons, Ltd, Chichester.

Purification

Jornitz, M.W. and Meltzer, T.H. (2007) *Filtration and Purification in the Biopharmaceutical Industry*, 2nd edn, Informa Healthcare, London.

Ladisch, M.R. (2001) *Bioseparations Engineering: Principles, Practice, and Economics*, John Wiley & Sons, Inc., Hoboken, NJ.

Shukla, A.A., Etzel, M.R. and Gadam, S. (2006) *Process Scale Bioseparations for the Biopharmaceutical Industry*, CRC Press, Boca Raton, FL.

Sofer, G. and Hagel, L. (1997) *Handbook of Process Chromatography*, Academic Press, New York.

Zydney, A.L. and Zeman, L.J. (1996) *Microfiltration and Ultrafiltration: Principles and Applications*, CRC Press, Boca Raton, FL.

Aseptic Filling and Lyophilization

Banga, A.K. (2005) *Therapeutic Peptides and Proteins: Formulation, Processing, and Delivery Systems*, 2nd edn, CRC Press, Boca Raton, FL.

Bontempo, J.A. (1997) *Development of Biopharmaceutical Parenteral Dosage Forms*, Informa Healthcare, London.

Jennnings, T.A.(ed.) (1999) *Lyophilization: Introduction and Basic Principles*, Informa Healthcare, London.

Rey, L. and May, J.C. (2004) *Freeze-Drying/Lyophilization of Pharmaceutical and Biological Products*, 2nd edn, Informa Healthcare, London.

Bioanalytics

Gault, V. (2009) *Understanding Bioanalytical Chemistry: Principles and Applications*, Wiley-Blackwell, Chichester.
Venn, R.F. (2000) *Principles and Practice of Bioanalysis*, CRC Press, Boca Raton, FL.

Regulatory

Berry, I.R. (2004) *The Pharmaceutical Regulatory Process*, Informa Healthcare, London.
Clark, M.E. (2007) *Pharmaceutical Law: Regulation of Research, Development, and Marketing*, BNA Books, Arlington, VA.
Pisano, D.J. and Mantus, D.S. (2008) *FDA Regulatory Affairs: A Guide for Prescription Drugs, Medical Devices, and Biologics*, 2nd edn, Informa Healthcare, London.
Siegel, E.B.(ed.) (2008) *Development and Approval of Combination Products: A Regulatory Perspective*, John Wiley & Sons, Inc., Hoboken, NJ.

For regulatory guidelines published by agencies, see Chapter 7 and Weblinks.

Pharmacy and Clinical Development

Bohaychuk, W. and Ball, G. (1999) *Conducting GCP-Compliant Clinical Research: A Practical Guide*, John Wiley & Sons, Ltd, Chichester.
Day, S. (2007) *Dictionary for Clinical Trials*, 2nd edn, John Wiley & Sons, Ltd, Chichester.
Gad, S.C.(ed.) (2008) *Preclinical Development Handbook: ADME and Biopharmaceutical Properties*, John Wiley & Sons, Inc., Hoboken, NJ.
Kelly, W.N. (2006) *Pharmacy: What It Is and How It Works*, 2nd edn, CRC Press, Boca Raton, FL.
Lee, C.-J., Lee, L.H., Wu, C.L., Lee, B.R. and Chen, M.-L. (2005) *Clinical Trials of Drugs and Biopharmaceuticals*, CRC Press, Boca Raton, FL.
Mann, R.D. and Andrews, E.B.(eds) (2007) *Pharmacovigilance*, 2nd edn, John Wiley & Sons, Ltd, Chichester.
Meibohm, B.(ed.) (2006) *Pharmacokinetics and Pharmacodynamics of Biotech Drugs: Principles and Case Studies in Drug Development*, Wiley-VCH Verlag GmbH, Weinheim
Piantadosi, S. (2005) *Clinical Trials: A Methodologic Perspective*, 2nd edn, John Wiley & Sons, Inc., Hoboken, NJ.

Quality and Validation

Avis, K.E., Wagner, C.M. and Wu, V.L. (1998) *Biotechnology: Quality Assurance and Validation*, CRC Press, Boca Raton, FL.
Bliesner, D.M. (2006) *Validating Chromatographic Methods: A Practical Guide*, John Wiley & Sons, Inc., Hoboken, NJ.

Ermer, J. and Miller, J.H.McB. (2005) *Method Validation in Pharmaceutical Analysis: A Guide to Best Practice*, Wiley-VCH Verlag GmbH, Weinheim.

Goetsch, D.L. and Davis, S.B. (2001) *Understanding and Implementing ISO 9000 and Other ISO Standards*, Prentice Hall, Upper Saddle River, NJ.

Goetsch, D.L. and Davis, S.B. (2005) *Quality Management*, Prentice Hall, Upper Saddle River, NJ.

ISPE (2001) *Baseline Pharmaceutical Engineering Guides for New and Renovated Facilities, Vol. 5 – Commissioning and Qualification*. ISPE, Tampa, FL.

Nunnally, B.K. and McConnell, J.S. (2007) *Six Sigma in the Pharmaceutical Industry: Understanding, Reducing and Controlling Variation in Pharmaceuticals and Biologics*, CRC Press, Boca Raton, FL.

Prichard, F. and Barwick, V. (2007) *Quality Assurance in Analytical Chemistry*, John Wiley & Sons, Ltd, Chichester.

Sarker, D.K. (2008) *Quality Systems and Controls for Pharmaceuticals* John Wiley & Sons, Ltd, Chichester.

Good Manufacturing Practice

Haider, S.I. (2001) *Pharmaceutical Master Validation Plan: The Ultimate Guide to FDA, GMP, and GLP Compliance*, Informa Healthcare, London.

Miller, J.M. and Crowther, J.B.(eds) (2000) *Analytical Chemistry in a GMP Environment: A Practical Guide*, John Wiley & Sons, Inc., Hoboken, NJ.

Nally, J.(ed.) (2006) *Good Manufacturing Practices for Pharmaceuticals*, Sixth edn, Informa Healthcare, London.

Sharp, J. (2004) *Good Pharmaceutical Manufacturing Practice: Rationale and Compliance*, Informa Healthcare, London.

Facility Design

ISPE (1999) *Baseline Pharmaceutical Engineering Guides for New and Renovated Facilities, Vol. 3 – Sterile Manufacturing Facilities*. ISPE, Tampa, FL

ISPE (2001) *Baseline Pharmaceutical Engineering Guides for New and Renovated Facilities, Vol. 4 – Water and Steam Systems*. ISPE, Tampa, FL

ISPE (2004) *Baseline Pharmaceutical Engineering Guides for New and Renovated Facilities, Vol. 6 – Biopharmaceuticals*. ISPE, Tampa, FL

Odum, J.N. (2004) *Sterile Product Facility Design and Project Management*, 2nd edn, CRC Press, Boca Raton, FL.

Clean Rooms

Dixon, A.M. (2006) *Environmental Monitoring for Cleanrooms and Controlled Environments*, Informa Healthcare, London.

Seiberling, D.A. (2007) *Clean-in-Place for Biopharmaceutical Process*, Informa Healthcare, London.

Whyte, W.(ed.) (1999) *Cleanroom Design*, 2nd edn, John Wiley & Sons, Ltd, Chichester.

Whyte, W. (2001) *Cleanroom Technology: Fundamentals of Design, Testing and Operation*, John Wiley & Sons, Ltd, Chichester.

Project Management

Frame, J.D. (2003) *Managing Projects in Organizations: How to Make the Best Use of Time, Techniques, and People*, 3rd edn, Jossey Bass, San Francisco, CA.

Kerzner, H. (2006) *Project Management: A Systems Approach to Planning, Scheduling, and Controlling*, 9th edn, John Wiley & Sons, Inc., Hoboken, NJ.

Engineering

Clark, F. and Lorenzoni, A.B. (1996) *Applied Cost Engineering*, 3rd edn, CRC Press, Boca Raton, FL.

Kleinfeld, I.H. (1992) *Engineering Economics Analysis for Evaluation of Alternatives*, John Wiley & Sons, Inc., Hoboken, NJ.

Simonsen, C.B. and Shaw, S. (2006) *Essentials of Environmental Law*, Prentice Hall, Upper Saddle River, NJ.

Stewart, R.D., Wyskida, R.M. and Johannes, J.D. (eds) (1995) *Cost Estimator's Reference Manual*, 2nd edn, John Wiley & Sons, Inc., Hoboken, NJ.

Economy

Bozarth, C. and Handfield, R.B. (2007) *Introduction to Operations and Supply Chain Management*, Prentice Hall, Upper Saddle River, NJ.

Epstein, A. (2008) *Contract Law Fundamentals*, Prentice Hall, Upper Saddle River, NJ.

Hartman, J.C. (2006) *Engineering Economy and the Decision-Making Process*, Prentice Hall, Upper Saddle River, NJ.

Kimmel, P.D., Weygandt, J.J. and Kieso, D.E. (2006) *Financial Accounting: Tools for Business Decision Making, Working Papers*, 4th edn, John Wiley & Sons, Inc., Hoboken, NJ.

Weblinks

Bundesinstitut für Arzneimittel und Medizinprodukte (BfArM): www.bfarm.de
Eudralex: http://ec.europa.eu/enterprise/pharmaceuticals/eudralex
European Medicines Agency (EMEA): www.eu.emea.int
European Directorate for the Quality of Medicines: www.pheur.org
Food and Drug Administration: www.fda.gov
International Conference on Harmonization: www.ich.org
International Organization of Standardization: www.iso.org
International Society of Pharmaceutical Engineering: www.ispe.org
Japan Pharmaceutical Manufacturers Organization: www.jpma.or.jp
Parenteral Drug Association: www.pda.org
Paul-Ehrlich-Institut: www.pei.de
Pharmaceutical Inspection Convention and Pharmaceutical Inspection Cooperation Scheme: www.picscheme.org
United States Pharmacopoeia (USP): www.usp.org
World Health Organization: www.who.org
Zentralstelle für Gesundheitsschutz bei Arzneimitteln und Medizinprodukten (ZLG): www.zlg.de

Index of Abbreviations

AA	Amino Acid (=AS)
ADR	Adverse Drug Reaction
AE	Adverse Event
AIEX	Anion Exchanger
AMG	Arzneimittelgesetz
AMWHV	Arzneimittel- und Wirkstoffherstellungsverordnung
AP	Aqua Purificata
API	Active Pharmaceutical Ingredient
APR	Annual Product Review
AR	Adverse Reaction (=ADR)
AR	Annual Report
ATP	Adenosine triphosphate
AUC	Area Under the Curve
AVP	Aqua Valde Purificata
BAS	Building Automation System
BDS	Bulk Drug Substance
BfArM	Bundesinstitut für Arzneimittel und Medizinprodukte
BHK	Baby hamster kidney
BLA	Biological License Application
BOD	Basis of Design
BP	Basen Pair
BR	Batch Record
BRR	Batch Record Review
BSE	Bovine Spongiforme Encephalopathie
CAPA	Corrective Action Preventive Action
CBE30	Changes Being Effected in 30 Days
CBER	Center for Biologics Evaluation and Research
CDER	Center for Drug Evaluation and Research
CDRH	Center for Devices and Radiological Health
CDW	Cell Dry Weight
CE	Capillary electrophoresis

Manufacturing of Pharmaceutical Proteins: From Technology to Economy. Stefan Behme
Copyright © 2009 WILEY-VCH Verlag GmbH & Co. KGaA, Weinheim
ISBN: 978-3-527-32444-6

CFR	Code of Federal Regulations
CFSAN	Center for Food Safety and Applied Nutrition
CFU	Colony Forming Unit
cGMP	Current Good Manufacturing Practice
CHMP	Committee for Medicinal Products for Human Use
CHO	Chinese hamster ovary
CI	Chemical Ionisation
CIEX	Cation Exchanger
CIP	Cleaning in Place
CJD	Creutzfeldt-Jakob Disease
CMC	Chemistry, Manufacturing and Control
CMO	Contract Manufacturing Organisation
CoA	Certificate of Analysis
CoC	Certificate of Compliance
COMP	Committee for Orphan Medicinal Products
COP	Cleaning out of Place
CPG	Gel-permeation chromatography
CPMP	Committee for Proprietary Medicinal Products
CRF	Case Report Form
CTA	Clinical Trials Authorization
CTD	Common Technical Document
CVM	Center for Veterinary Medicines
CVMP	Committee for Medicinal Products for Veterinary Use
DIN	Deutsches Institut für Normung
DNA	Desoxyribonucleic Acid
DQ	Design Qualification
DSC	Differential Scanning Calorimetry
EBA	Expanded-bed adsorption
EBR	Electronic Batch Record
ED	Effective Dose
EDQM	European Directorate for the Quality of Medicines
EIS	Electron Impact Spectroscopy
ELISA	Enzyme Linked Immunosorbent Assay
EMEA	European Medicines Agency
EP	European Pharmacopoeia (=PharmEur)
EPO	Erythropoetin
ESI	Electrospray ionization
FAB	Fast Atom Bombardment
FBS	Fetal Bovine Serum
FCS	Fetal Calf Serum
FDA	Food and Drug Administration
FMEA	Failure Mode and Effect Analysis
FOIA	Freedom of Information Act

FP	Final Product
GAMP	Good Automated Manufacturing Practice
GCP	Good Clinical Practice
G-CSF	Granulocyte Colony Stimulating Factor
GEP	Good Engineering Practice
GFC	Gel Filtration Chromatography
GLP	Good Laboratory Practice
GM-CSF	Granulocyte Macrophage Colony Stimulating Factor
GMO	Genetically Modified Organism
GMP	Good Manufacturing Practice
GPC	Gel Permeation Chromatography
GSP	Good Storage Practice
GSS	Gerstmann-Sträussler Syndrom
GTP	Good Tissue Practice
HACCP	Hazard Analysis and Critical Control Point
HCP	Host Cell Protein
HEPA	High-efficiency particulate air
HIC	Hydrophobic Interaction Chromatography
HIV	Human Immunodeficiency Virus
IIMPC	Herbal Medicinal Products Committee
HPLC	High Pressure Liquid Chromatography (also High Performance LC)
HPMC	Hydroxypropylmethyl- cellulose
HSA	Human Serum-Albumin
HVAC	Heat Ventilation Air Conditioning
ICH	International Conference on Harmonization
IEC	Ion-exchange chromatography
IEF	Isoelectric Focusing
IEX	Ion Exchanger
IF	Interferon
IFN	Interferon
IGG	Immunoglobulin G
IL	Interleukin
IMP	Investigational Medicinal Product
IMPD	Investigational Medicinal Product Dossier
IND	Investigational New Drug
IOM	Investigations Operations Manual
IPC	In-Process Control
IQ	Installation Qualification
IR	Infrared
ISO	International Organization of Standardization
ISPE	International Society for Pharmaceutical Engineering
JP	Japanese Pharmacopoeia

LADME	Liberation, Absorption, Distribution, Metabolism, Excretion
LAL	Limulus Amebocyte Lysate
LD	Lethal Dose
LFH	Laminar Flow Hood
LIMS	Laboratory Information Management System
LOD	Limit of Detection
LOQ	Limit of Quantification
MALDI	Matrix Assisted Laser Desorption Ionisation
MBR	Master Batch Record
MCB	Master Cell Bank
MCO	Molecular Cut Off (= MWCO)
MF	Microfiltration
MHLW	Ministry of Health, Labor and Welfare
MS	Mass spectrometry
MSA	Manufacturing and Supply Agreement
MTD	Maximal Tolerated Dose
MWCO	Molecular Weight Cut Off
NCTR	National Center for Toxicological Research
NDA	New Drug Application
NPV	Net Present Value
OC	Office of the Commissioner
OMCL	Official Medicines Control Laboratories
OOS	Out of Specification (QC context) or Out of Stock (logistical context)
OQ	Operational Qualification
ORA	Office of Regulatory Affairs
PAB	Pharmaceutical Affairs Bureau
PAGE	Polyacrylamid Gel Elektrophoresis
PAS	Prior Approval Supplement
PCR	Polymerase Chain Reaction
PD	Pharmacodynamics
PD	Plasma Desorption
PDA	Parenteral Drug Association
PEG	Polyethylen glycole
PEI	Paul-Ehrlich-Institut
PFBS	Pharmaceutical and Food Safety Bureau
PharmEur	European Pharmacopoeia
PIC/S	Pharmaceutical Inspection Convention/Scheme
PK	Pharmacokinetics
PM	Posttranslational Modification
PMDA	Pharmaceutical and Medical Devices Agency (= KIKO)
PoC	Proof of Concept (= PoP)
PoP	Proof of Principle (= PoC)

PQR	Product Quality Review
QA	Quality Assurance
QAA	Quality Assurance Agreement
QC	Quality Control
QM	Qualitätsmanagement
rFVIII	Rekombinant Factor VIII
RNA	Ribonucleic Acid
ROI	Return on Investment
RPC	Reversed Phase Chromatography
RP-HPLC	Reversed Phase HPLC
RPM	Regulatory Procedures Manual
RT	Reverse transcription
SDS	Sodium dodecylsulfate
SDS–PAGE	Sodium dodecylsulfate–polyacrylamide gel electrophoresis
SEC	Size Exclusion Chromatography
SIP	Sterilization in Place (also Steaming in Place)
SOP	Standard Operating Procedure
SPC	Statistical Process Control
SPC	Supplementary Protection Certificate
TEM	Transmission Electron Microskopy
TFF	Tangential Flow Filtration
TOC	Total Organic Carbon
TOF	Time of Flight
TSE	Transmissible Spongiform Encephalopathie
UF	Ultrafiltration
URS	User Requirements Specification
USP	United States Pharmacopoeia
UV	Ultra Violett
WCB	Working Cell Bank
WFI	Water for Injection
WHO	World Health Organisation
ZLG	Zentralstelle für Gesundheitsschutz bei Arzneimitteln und Medizinprodukten

Index

a

Absorption (of an active substance) 141
Absorption spectrum 122
Accelerated stability study 110
Acceptance (econ.) 292
Acceptance criteria
– analytical 102
– contractual 353
Access control 190, 200
Accuracy 111
Acetic acid 116
Action of a drug 140
Active pharmaceutical ingredient 37
Activity 107
– assay 107
Actual account 305
Addiction 145
Additional services 290
Adenosine-triphosphate 21
Administration 231
Adsorbent 75
Adsorption process 29, 75
Adsorption-desorption mechanism 80
Adverse drug event 160
Adverse drug reaction 160
Aeration **56**, 261
Aerobic respiration 21
Aerosol 95
Affinity chromatography 84, 105, 120
Affinity gel 91
Agarose-dextrane material 91
Aggregation **29**, 106
Air changes 270
Air duct 270
Air filter 270
Air shower 267
Airlock 19, **267**
Alanine 97
Alarm threshold 282
Algae 17
Allocation key 306
Allowance factor 285
Aluminum-oxide 93
Ambient air 266
Ambient temperature 259 pp
Amino acid **26**, 137
– analysis 105, 113
Amphoteric behavior 30
Amplification steps 47
Ampoule 96
Anabolism 21
Analogue data 147
Analysis of primary structure 104
Analytical artifact 176
Analytical chromatography 120
Analytical report 200
Analytical transfer 353
Analyzer 123
Animal experiment 149
Animal husbandry 46, 60
Animal model 127
Anion exchanger 84
Annuity 340
Antibiotics **16**, 71
Antibody **29**, 119
Anti-foaming agent 71, **133**
Antigen 118
Antigen-antibody reaction 119
API production 37
API-plant 232
Apoptosis 23
Appearance 97
Application patent 301

Aqua purificata 255
Aqueous phase 88
Archae bacteria 17
Area under the curve (AUC) 144
Aspergillus niger 43
Assessment matrix 343
Association 30
Asymmetrical membrane 93
Attenuation 12
AUC 144
Audit trail 199
Audit 179
Automation of water systems 259
Availability of oxygen 48

b

Baby-hamster kidney 45
Bacteria **17**, 108, 126
Balance of interests 288
Base 31
Baseline (-separation) 121
Basic engineering 279
Basis of Design 196
Batch adsorption 75
Batch culture 50
Batch documentation 202
Batch mode 50
Batch number 200
Batch record 200
Batch-record review 202
Batch-to-batch cleaning 194
Bed height 92
Beta elimination 30
Betaferon® 45
beta-sheet structure 122
BfArM 218
BHK 45
Bidding process 287
Bid-proposal preparation 313
Binding affinity 72
Binding offer 320
Bioavailability 140
Bioburden 52
Bioburden controlled 70
Bioburden reduction 64
Biogenerics 136
Biological activity 27, **107**
Biological barrier 142
Biological clock 47
Biological effect 107
Biological hazard 223
Biological license application 225
Biological membrane 141
Biological safety class 246

Biological variability 149
Biomass 49
Bioreactor 55
Biosuspension 71
Biotech working party 218
Blinded study 157
Blood cells 55
Blood circulation 142
Blood fractionation 13
Blood serum 61
Blood-brain barrier 142
Boiler 266
Book value 312
Bottleneck analysis 346
Bovine serum 61
Break-even point 300
Bridging study 348
Bridging trial 155
BSE 134
Budget request 285
Buffer exchange 62
Buffer preparation 230
Buffer solution 39
Buffer-preparation area 87
Building automation-system (BAS) 272
Building drawings 280
Building laws 293
Building regulations 292
Building shell 269
Build-or-Buy 332
Bulk filling 39
Butanol 88
Byproduct 48

c

Calculation surcharge 291
Calibration 112
Campaign mode 249
Cap 97
CAPA 203
Capacity usage 345
Capillary electrophoresis 115
Capital commitment 238
Capital costs 304
Capital demand 325
Capital interest 311
Capital investment 238
Capital value 339
Carbohydrate structure 124
Carbohydrates 31
Carbon dioxide 62
Carbon source 61
Carcinogenicity 160

Cardiovascular system   151
Carry-over   189
Cartridge filter   78
Cascade concept   244
Case report form   156
Catabolism   21
Cation exchanger   92
CBER   215
CDER   214
CDRH   215
Ceiling diffusor   270
Ceiling-air inlet   270
Cell
– activity   48
– debris   68
– density   **48**, 41
– dry weight   26
– expansion   47
– growth   48
– growth-rate   345
– lifetime   47
– nucleus   20
– retention system   51
– substrate   38 pp, 47
– viability   129
Cell bank   46
Cell culture   127
Cell division   22
Cell line   43
Cell mass   49
Cell separation   72
Cell therapy   13
Cell-culture medium   61
Cellulose   31
Center effect   156
Central compartment   142
Central nervous system (CNS)   151
Centrifugation   72
Certificate of analysis   176, 202
Certificate of compliance   176, 202
CFSAN   216
Chain-termination method   126
Chamber filter-press   73
Change control   178
Change management
   (engineering)   291
Change management (quality)   203
Change of scope   291
Change request   206
Change requiring regulatory
   approval   205
Change-over investment   328
Charge/mass ratio   115, 123
Chemical compatibility   253

Chemical inactivation   90
Chicken eggs   46
Chinese hamster ovary   45
Chlorination   256
CHMP   217
CHO   45
Chromatogram   **121**, 131
Chromatography   80
– gel   89
– skid   87
Chromosome   21, 25
CI   123
CIP   251, **265**, 259
CIP-return line   63
Circular dichroism   127
Claims management   290
Clarity   109
Classified environment   190
Clean room   265
Clean steam   261
Clean utilities   254
Cleaning agent   56
Cleaning area   243
Cleaning in Place   See CIP
Cleaning procedure   265
Cleaning protocol   265
Cleaning validation   198
Cleanliness class   267 pp
Cleanliness requirements   233
Clean-room cascade   190
Clean-room class   188, 234
– clothing   188
– concept   239
– finishing   267
– monitoring   32
Clearance   143
Client-manufacturer relation   179 pp
Climate chamber   198 pp
Clinical endpoint   150
Clinical trials   102, **149**
– application   225
– directive   184, 207
Closed process   190
Co medication   160
Code of Federal Regulations   216
Codon usage   26
Codon   25
Cold chain   198
Cold processing   42
Cold spot   258
Cold storage   236
Cold withdrawal   259
Colony-forming unit   127
Color   110

Color particle   133
Column-packing   88
– station   88
Combined production   319
Commercial supply   346 pp
Commissioning   281
Common technical document   221
COMP   217
Comparability   206 pp
Comparability protocol   207
Comparability study   155, 353
Comparative trial   154
Comparison of profitability   338
Compartment (body)   142
Compartment (environment)   293
Compendial water quality   255
Complaint   177
Complex media   61
Computer validation   199
Concept engineering   279
Conceptual design   236
Concomitant administration   145
Conductivity measurement   133
Confidentiality   287, 290
Confidentiality agreement   287
Conflict resolution   352
Conformance runs   197, 283
Consideration   288
Construction costs   285
Construction time   284
Consumable   242
Contact inhibition   22, 55
Contact switches   253
Contamination   189 pp
Content   104
Contingency   285
Continuous cell line   22
Continuous mode   51
Contract design   287
Contract manufacturer   349
Contract negotiation   349
Contract partner   180 pp
Contract review   180 pp
Contractor   287
Contractual risk   288
Contraindication   145
Contribution margin   322
Control group   156
Controlled environment   190
Cooling loop   261
Cooling media   261
Cooling tower   261
Coomassy blue   116
Cooperation structure   350

COP   265
Core process   229
Corn   46
Corrective action   175, 179, **203**
Cost
– accounting   305
– allocation   306
– comparison   338
– control   351
– estimate   280, 285
– function   321
– risk   301
Cost center accounting   308
Cost inertia   307
Cost model   313
Cost of piping   285
Cost structure analysis   321
Cost type accounting   309
Cost unit accounting   309
Cost-distribution sheet   319
Costs of labor   286
Costs-by-cause principle   306
Counter ion   84
Creutzfeldt-Jacob disease   134
Critical observations   179
Critical parameter studies   42
Critical path   284
Critical system   282
Cross contamination   189
Crossover study   157
Cryo conservation   46
Cryo protectant   97
Crystallization   88
Cumulation   145
Currency risk   274
Current GMP   183
CVM   216
CVMP   217
Cytoplasm Membrane   32
Cytoplasm   19

d

Data accessibility   200
Data safety   199
Data-archiving system   199
Dead volume   258
Dead-end filtration   72
Deamidation   30
Dear Doctor letter   164
Death kinetics   53
Decanter   72
Dedicated plant   249
Deep-freeze system   254
Degradation product   49, 69, 111

Degree of automation 247
Dehumidification 271
Deionization 61
Deionized water 256
Demand peak 360
Denaturation 105
Department of Health and Human Services 214
Depreciation 311
Depth filtration 77
Desalting 61
Design control 171
Design qualification 196
Design validation 182
Design verification 182
Desorption 75
Desoxy-nucleotide-triphosphate 125
Destruction 177
Detail engineering 279
Development agreement 351
Development report 201
Deviation investigation 200
Deviation management 201
Deviation 200
Deviation report 200 pp
Dextrane 93
Diabetes 14
Diafiltration 39, **71**
Dictiosomes 20
Differential scanning calorimetry 128
DIN ISO 9000 169
DIN 212
Direct capture 76
Direct costs 305
Direct-capture chromatography 76
Direct-impact system 282
Discoloration 110
Displacement flow 267
Disposable bag technology 253
Dissolved oxygen 129
Distillation 257
Distribution 142
Disulfide bridge 27, 105, 116
Dithio-threitol 116
DNA sequencing 125
Document control 202
Document transfer 353
Documentation 200
Documentation for regulatory permits 293
Donor anamnesis 133
Dose recommendation 149
Dose-finding study 154

Dose-response relationship 140
Dossier 225
Double blinding 157
Double peak 121
Doubling time 22, **45**
Downstream 39
Drinking water 61
Drug
– allergy 160
– approval 225
– study 149
Drug interaction 145
Drug product 37
Drug safety 159, 177
6-D rule 258
Dumas 115

e
Echerischia coli (E. coli) 43
ED_{50} 146
Edman degradation 113
EDQM 218
Effect of investment 328
Effective dose 146
Efficacy 140
Electrical power supply 262
Electronic batch record (EBR) 199, 272
Electronic signature 199
Electrophoresis 115
Elimination 142
ELISA 119
Eluate 81
Elution buffer 81
Elution step 81
EMEA 216
Emergency-power generator 262
Endoplasmic reticulum 20
Endotoxin 126
Endotoxin load 69
Energy balance 280
Engineering run 283
Enhanced design review 283
Environmental legislation 293
Environmental monitoring 132
Environmental protection 293
Enzymatic reaction 118
Episomal plasmid DNA 44
Epogen® 45
Equilibration 81
Equipment cleaning 265, 345
Equipment layout 280
Erythropoietin 14

Escape door   234
ESI   123
Ethical responsibility   300
Ethics committee   155
Eudralex   218
Eukaryotic cell   20
European Union   216
Evaluation software   122
Excipient   187
Exclusion criterion   156
Excretion capacity   142
Excretion organ   142
Executed batch record   200
Exhaust-air scrubber   264
Expandability   238
Expanded bed adsorption   76
Expansion areas   250
Expansion investment   328
Exponential growth   49
Expression strength   48
Expression system   43
External service   310
Extinction coefficient   106
Extinction   122
Extracellular expression   44
Extracellular fluid   142
Extraction   88

f

FAB   123
Facade area   251
Factor VIII   14, 52
Factory acceptance-test (FAT)   283
Failure costs   168
Failure prevention   174
– costs   168
Failure root-cause   174
Fast-track review   226
Fast-track status   303
Fatty acids   32
FDA   214
Feasibility study   280
Fed batch mode   52
Federal Register   212
Fee-for-service contract   288
Fermentation   48
– analytics   129
– area   236
– capacity   344
– conditions   36
– course   56
– debris   264
– titer   41
– volume   67

Fetal bovine serum   133
Fields of activity of ISO 9000   169
Filgrastim   149
Filling   95
Filling level   56
Filter
– cake   73
– cassette   73
– cloth   73
– residue   73
– test   62
Filtration   72
Filtration membrane   93
Final container   97
Final formulation   256
Final purification   79
Final rinse   198, 265
Final treatment step   256
Financial planning   325
Finished product   99
Fire code   293
Fixed costs   306
Fixed investment   328
Fixed unit costs   308
Flexibility   246
Flexible hose   251
Flexible piping   251
Flocculation   71
Floor plan   236
Flow resistance   91
Flow visualization, smoke study   271
Flowrate   91
Fluidization   76
Fluidized bed reactor   59
FMEA-analysis   177
Foam control   57
Foaming   56
Folding   **27**, 104
Folding box   96
Follow-up investment   249, **328**
Follow-up product   110
Follow-up stability study   110
Form of administration   95
Formulation buffer   96
Formulation   **95**, 131
Fouling   77
Fractal plant   250
Fraction collection tank   81
Fraction   81
Fractionation   123
Freedom of information act   216
Functional description   196

g

Gastrointestinal tract   148
G-CSF   149
Gel electrophoresis   115
Gel staining   116
Gel-filtration chromatography   84
Gel-permeation chromatography   71
Gene expression   19
Gene sequence   105
Gene therapy   13
Gene vector   24
General contractor   287
Generic drug   135
Genetic code   25
Genetic instability   47
Genetic stability   23
Genetically modified organisms   263
Genome   44, 126
Genotype characterization   18
Gerstmann-Straeussler syndrome   134
Glass ampoule   96
Glycine   97
Glyco analytics   124
Glycogen   31
Glycolipid   32
Glycolysis   21
Glycoprotein   **31**, 124
Glycosylation   **27**, 14
Glycosilation pattern   124
GM-CSF   14
GMO-free   263
GMP flows/diagram   239
GMP review   182
Go/no-go criterion   330
Gold standard   303
Golgi-Apparatus   20
Good automated manufacturing practice   168
Good clinical practice   168
Good engineering practice   168
Good laboratory practice   168
Good manufacturing practice   183
Good storage practice   168
Good tissue practice   168
Gowning   188
Gowning area   278
Gradient elution   87
Gram staining   18, **127**
Grant of a patent   151
Gross investment   328
Growth factor   61
Growth limitation   49

h

Half-life   144
Hansenula Polymorpha   43
Hard piping   251
Head of manufacturing   188
Head of quality control   188
Healthy volunteer, test person   153
Heat sterilization   53
Heat supply   261
Heating jacket   56
Heating-circuit (-media, -steam)   261 pp
Heat-transfer system   261
Heavy metals   255
Helical structure   105
HEPA-Filter   270
High-pressure homogenization   74
HIV   23
HMPC   217
Holding point (in the process)   39
Homogenizer   74
Horizontal design   236
Hormone   61
Horse-shoe crab   126
Host cell   43
Host cell protein   68 pp
Host cell protein ELISA   119
Hot storage   258
HPLC chromatography   52 pp
HPMC   97
Human growth-hormone   14
Human insulin   28
Human serum albumin   133
Humidity   269
Humulin®   45
Hurricane   275
HVAC installation   269
Hybridization   125
Hydrogen bonding   116
Hydrolysis   30, 113
Hydrophobic interaction chromatography   85
Hydrophobicity   85
Hygiene policy   188
Hypersensitivity reaction   160

i

ICH   221
Identity   104
– test   104
Idiosyncratic side effect   145
Immune system   152
Immuno-electro blot   118
Immunologicals Working Party   218

Improvement process   187
Impurity   68
Impurity profile   106
Imunogenicity   27, 134
Inclusion body   **19**, 44
Inclusion criterion   156
IND   225
Indirect impact system   282
Infectivity   127
Informed Consent   156
Infrared spectroscopy   122
Initial application   145
Injection site   148, 152
Inoculation   **47**, 232, 244
Inorganic salts   69
In-process control (IPC)   128
Insect cell   45
Installation qualification   196
Instrument air   261
Insulin   28
Integrated plant   250
Integrator   121
Interest rate for calculation   336
Interface management   349
Interferon   14
Interleukine   14
Intermediate   187
Intermediate precision   111
Intermediate release   104
Internal rate of return   340
Intracellular expression   44
Intravenous injection   141
Inventory management   359
Investigational drug   207
Investigational medicinal product dossier   225
Investigational new drug application   225
Investigations operations manual   225
Investment
– appraisal   336
– calculation   336
– costs   285
– decision   329
– planning   334
In-vivo stability   149
Ion concentration   69
Ion exchanger   84
Ion source   123
Ion-exchange chromatography   84
Ion-exchanger gel   90 pp
Ionization   123
IP safety   275
IPC   202
IPC analytics   230
IPC lab   129
Irrelevant costs   308
Ishikawa diagram   177
ISO   221
Isoelectric point   116
Isoelectrical focusing   116
Isolation   72
Isolator technology   269
Isometric drawing   280
ISPE   223

j
Japan Pharmaceutical Manufacturers Organization   220

k
Kidney   142
KIKO   220
Kjeldahl   115
Know-how safety   275
Kogenate®   45

l
Labeling   40 pp
Labor legislation   293
Laboratory information management system (LIMS)   199
LADME   141
Lag phase   49
LAL-test   125
Laminar air flow   267
Lantus®   149
Large-scale process   353
Launch   300 pp, 346 pp
LD_{50} 148
Leachable   69
Leaching   92
Lead time   284
Lepirudin   14
Lethal dose   148
Letter of intent   288
Leukine®   45
Liability   289
Liberation   141
License payment (royalty)   304
Life cycle   300
Life cycle of investment   327
Ligand   84, 91
Limit of detection   112
Limit of quantification   112
LIMS   199
Linearity   112

Lipid   107
Lipo-polysaccharide   **32**, 44, 69, 106
Liquid nitrogen   261
Liquidation value   335
Liver   142
Living cell mass   49
Loading capacity   345
Loading step   81
Local tolerance   152
Long-term side effect   160
Lower price-limit   305
Lump-sum contract   291
Lymphatic system   142
Lyophilisate   131
Lyophilization   97 pp
– cake   **97**, 149

m

Machine hour-rate accounting   319
Mad cow disease   134
Main drying   97
Main equipment list   286
Main peak   121
Main service   289
Maintenance costs   310
Major change   204
Major observation   179
Major variation   204
Make-or-Buy   355
Make-up   41
MALDI   123
Mammalian cell   45
Mammalian cell-culture   54
Manufacturing and supply agreement   351
Manufacturing costs   41, 155
Manufacturing documentation   200
Manufacturing license   218 pp, 224 pp
Marginal costing   318
Marker   119
Market exclusivity   302
Market release   201
Market saturation   300
Market share   299
Marketing authorization   225, 352
Mass spectrometry   123
Master batch record   201
Master cell bank   46
Material airlock   191
Material balance   230
Material costs   310
Material flow   234
Material of construction   69, 93, 267

Maturation phase   300
Maturity of application   247
Maxam-Gilbert method   126
Maximum tolerated dose   146
MBR   201
MCB   46
Mechanical completion   279
Media
– component   60
– hold tank   61 pp
– prep tank   63
– preparation   232
– supply   62
Medical investigator   156
Medication error   161
Medium (fermentation)   60
Membrane bioreactor   59
Membrane module   59, 178
Mercaptoethanol   116
Metabolic system   35
Metabolism   **21**, 142
Metazoan systems   45
Method validation   111
MHLW (ksei-roudou-sho)   220
Microbial contamination   126, 243, 256
Microbial count   53
Microfiltration   77
Microorganisms   **17**, 254, 276
Microparticle   148
Microtiter plate   119
Milk   46
Mill   74
Minimal therapeutic concentration   144
Minimal toxic concentration   144
Ministry of Health Labor and Welfare   220
Minor change   204
Minor observation   180
Minor variation   204
Mitochondria   20
Mobile CIP-skid   252
Mock run   119
Mode of action   134
Model process   249
Model virus   90
Moderate change   204
Modular construction   251
Molecular weight   105
Molecular weight cut-off   92
Monocellular layer   55
Monocentric   150
Monoclonal antibody   13, **28**, 45

Mortality rate   150
Moss cell   46
Mouse-myeloma   45
Movable tank   252
MTD   146
Multicentric trial   156
Multiple application   145
Multi-purpose plant   193 pp
Municipal sewer system   264
Municipal water   257
Mutagenicity   152
Mutation   **23**, 47
MWCO   79
Mycoplasma   107

n

Nasal administration   148
Nasal mucosa   148
Natural disaster   275
NCTR   216
Net cash flow   339
Net charge   116
Net costs   304
Net investment   328
Net present value   339
Net profit   322
Neulasta®   149
Neutralizing antibodies   152
New drug application   225
Nitrocellulose matrix   118
Noise protection   293
Nominal accounting   305
Nominal capacity   307
Non-binding orientation bid   320
Northern blot   118
Note for Guidance   213
Novo Seven®   45
Novolin®   45
NPH-Insulin   149
NPV   337
NS0   45
Nuclear magnetic resonance   105
Nucleic acids   31
Nucleotide building-block   125
Nutrient medium   60

o

Obligation to co-operate   290
Occupational health   293
Office of the commissioner   216
Official Journal of the European
  Communities   212
Off-label use   303
Oil-free compressor   261

Olfactory nuisance   264
OMCL   218
Once-through cleaning   265
OOS Result   203
Open process   190
Open study   157
Operating assets   312
Operating costs   246
Operating time   345 pp
Operational qualification   196
Operational readiness   307
ORA   215
Oral mucosa   148
Organelle   20
Organic phase   88
Organic solvent   42, 88
Organics   253, 264
Organization schedule   283
Orphan drug   226
Orphan drug status   302
Outer appearance   104
Overall yield   66
Overfill   345
Overhead calculation   305
Overhead costs   304
Oxygen supply   62
Ozonization   256

p

Package units   272
Packaging   **99**, 131
Packing homogeneity   88
Paper-based documentation   271
Parallel study   157
Pareto-analysis   177
Parking-lot effect   159
Particle   131, 271
Particle filtration   77, 256
Past pollution   293
Patent   301
– protection   301
Patient compliance   159
Patient-information leaflet   96, 146
Paul-Ehrlich Institut   218
Payback time (static/dynamic)
  339 pp
PCR   123
PDA   223
Peak area   121
Peak height   121
Peak profit   300
Peak sales   300
Peak symmetry   121
PEG   148

Peptide Mapping   114
Peptide pattern   114
Peptide   26
Perforated ceiling   270
Performance qualification   197
Periplasmic space   44
Permit procedure   293
Permit situation   274
Personnel airlock   268
Personnel flow   240
Personnel hygiene   243
Personnel investment   328
pH-adjustment   56, 256
Phages   108
Pharmaceutical affairs bureau   220
Pharmaceutical and food safety bureau   220
Pharmaceutical and medical devices agency   220
Pharmaceutical legislation   211
Pharmaceutical manufacturing   38
Pharmacodynamics   144
Pharmacokinetics   141
Pharmacopoeia   212, 218, 221
Pharmacopoeial quality   61
Pharmacovigilance   162
Pharmacovigilance agreement   351
Phase IV   162
Phase of clinical trial   153
Phenomenon of acquired tolerance   151
PhEur   218
pH-gradient   116
Phosphoric acid   265 pp
Photosensitivity   122
Photosynthesis   59
pH-swing   90
Phylogenetic tree   18
PIC/S   223
Pichia Pastoris   43
Pilot plant   346 pp
Piping and instrumentation diagram (P&ID)   280
Pivotal study   154
PK-profile   140
Placebo   156
Placebo-controlled study   156
Placenta   142
Plant maintenance   234
Plant usage   307
Plaque test   127
Plasma   13
Plasma concentration   143
Plasma desorption   123

Plasmid   19
Plasmid DNA   44
Plastic tube   253
Platform technology   238, 249, 349
Point-of-use heat exchanger   258
Points to consider   213
Polarity   72
Pollen dispersal   60
Polyethylene glycol   88
Polymer membrane   93
Polymerase   125
Polysaccharides   31
Pooling   346
Pore size   77, 92
Posttranslational modification   **27**, 104
Potatoe   46
Precipitation   29, 88
Precision   111
Preclinical study   151
Pre-fermenter   62 pp
Prefilled syringe   96
Pregnancy   152
Preparation area   242
Preparative chromatography   80
Present value   339
Pressure
– drop   91
– gradient   243 pp
– test   62
– vessel   56
Pressurized-air system   254
Prevalidation study   353
Price regulation   212
Pricing   303
Primary cell   16
Primary container   95, 109
Primary structure   **27**, 113
Primer   125
Prion   134
Priority review   226
Procedural segregation   233
Process
– air   261
– analytics   128
– change   203
– consistency   271
– control   199, 272
– control system   272
– description   201, **230**, 282
– documentation   271
– flow-diagram   235
– optimization   357
– patent   301

– piping 252
– support 230
– validation 197
Processing time 62, 90
Process-related impurity 102
Procurement 170
Product
– change-over 249
– comparability 133
– concentration 67
– flow 240
– instability 110
– life-cycle 302
– loss 66
– patent 301
– release 201
– safety 233
– stability 42
– testing 103
– yield 66
Production facility 229
Production fermenter 55
Product-related impurity 68 pp
Product-related substances 68 pp
Profit 300
Profit and loss account 306
Profit comparison analysis 336
Profit margin 350
Programmable logic control 271
Project organization 286
Project time-plan 284
Prokaryotic cell 19
Proleukin® 45
Promoter 19
Promotion 300
Proof of concept 347
Proof of principle (concept) 154
Prospective validation 199
Protease inactivation 75
Protease inhibitor 75
Protease 107
Protein aggregate 69, 74
Protein expression 43
Protein structure **26**, 105
Protein-A chromatography 85
Proteins 26
Protein-sequence analysis 113
Proteolytic degradation 49
Protropin 45
Psychogenic effects 156
Pulmonary administration 148
Purchasing documents 278
Purification 39
Purification Area 236
Purified water 61
Purity 106
Pyrogen 126

q

QA 202 pp
QC 202 pp
Quadrupol MS 123
Qualification 182, 195 pp, 260 pp, 279 pp
Qualification protocol 195
Qualification report 195
Qualified Person 188
Qualitative method 111
Quality assurance 202 pp
Quality assurance agreement 180, **351**
Quality
– control 202 pp
– deficiency 176
– management system 171
– of life 150
– policy 179
– system 169
Quality control lab (QC-lab) 273
Quality working party 218
Quantitative method 111
Quartiary structure 27

r

Rapid-alert system 164
Rapporteur 217
Rationalization investment 329
Raw material 60, 187
Raw-material testing 133
Recall 164, **177**
Receptor 17
Recombinant protein 10
Recombinate® 45
Recomittal rate 150
Reduction factor 90
Re-examination 168
Reference project 349
Reference sample 122
Reference standard 108
Reference substance 198
Refludan® 45
Refolding 73
Regulatory affairs 201 pp
Regulatory approval 348
Regulatory dossier 181
Regulatory inspection 224
Regulatory procedures manual 225
Release behavior 148

Release certificate   200
Release specification   102
Release status   231
Relevant costs   308
Reliability   230
Repeatability   111
Replacement problem   331
Reprocessing   176
Reproducibility   111
Reservation fee   354
Residence time   141
Residual moisture   107
Residual solvent   107
Resolution   90
Respiratory tract   148
Restriction enzyme   125
Retention capacity   93
Retention time   121
Retrospective validation   199
Retrovirus   70
Reverse osmosis   256
Reverse transcriptase   152
Reversed phase chromatography   84 pp
Review phase   330
Rhizosecretion   60
Ribosome   19
Right of continuance   249
Right to use   330
Risk
– analysis   136
– assessment   263
– communication   177
– control   177
– identification   177
– of flooding   275
Risk-based approach   177
Risk-benefit assessment   139
Risk-management process   177
Robustness of the manufacturing process   42
Rodent   152
Roller bottle   59
Room air   267 pp
Room book   280
Room-in-room concept   269
Root-cause analysis   203
Route of administration   95
Routine audit   179
Routine operation   271

S

Saccharomyces cerevisiae   43
Saccharose   97
Safety interlock   197
Safety stock   43
Safety working party   218
Salt content   81
Sample port   129
Sample-size calculation   156
Sampling   259 pp
Sandwich ELISA   120
Sanger-Coulson method   126
Sanitization   56
Saturation phase   300
Scalability   253
Scale-up   353
Scale-up step   38
Scale-up study   353
Scaling method   334
Scientific advice   226
Scope of work   289 pp
SDS PAGE   116
Search phase   330
Second Supplier   361
Secondary drying   97
Secondary packaging   97
Secondary structure   27
Sector field   123
Seed train   47
Segregation steps   233
Selection problem   331
Self sanitizing   258
Self-draining system   258
Semi-finished goods   99
Sensitivity analysis   336
Separation principles   64
Separation zone   82
Sequence ladder   125
Serum   61 pp, 133
Shake flask   38
Shear stress   55
Side effect   145
Sight glass   129
Silica gel   91
Silver-nitrate solution   116
Singular event   177
SIP   265
Site acceptance-test (SAT)   283
Site location factor   273
Size-exclusion chromatography   86
Small-scale model   353
Smoke study   271
Sodium dodecylsulfate   114
Sodium hydroxide
Solid waste   264
Solid-liquid extraction
Solubility of proteins   28 pp

Solubilization   73
Solubilizing agent   74
Solvency   326
Source of contamination
Southern Blot   118
Special permit
Specifications
– analytical   102
– devices   42
Specificity   111
Spectrophotometer   122
Spectroscopy   122
Spectrum   122
Spectrum of byproducts   36 pp
Spine, backbone   250
Sponsor   155
Spray ball   56
Sprinkler system   234
Stability   110
Stability data   348
Stability of the expression system   47
Stability profile   110
Stability test   239
Stabilization   75
Stainless steel technology   253
Standard costing   318
Standard operating procedure   172
Standard technology   348
Standard therapy   157
Starting culture
Starting material   187
State of the art   213
Statistical methods   171, 178
Statistical relevance   156
Steam
– condensate   261
– distribution   261
– sterilizability   258
– sterilization   53
– trap   266
Steering committee   287
Stem cells   16
Step yield   66
Sterile filter   63
Sterile filtration   89, 191
Sterile gases   261
Sterile manufacturing   245
Sterile technology   52
Sterility   52
Sterilization in place   56
Sterilization method   52
Sterilization time   52
Stirred bioreactor   55
Stirrer rotational speed   129

Stopper   97 pp
Storage buffer   96
Storage conditions   111
Strategic investment   329
Street clothing   242
Structural data   105
Study design   150
Study protocol   155
Subcontractor   287
Subcutaneous injection   148
Subloop   258
Success of cleaning   265
Sugar chain   105
Sugar molecule   124
Supernatant   38
Supplementary protection
   certificate   301
Supplier qualification   180
Supply chain   359
Supply safety   361
Supply-chain management   359
Surface filtration   76
Suspension culture   55
Symmetrical membrane   93
Symmetry test   88
Synthetic medium   61
Syringe   96
System wall   269
Systemic application   143

t

Tactical investment   329
Tangential flow filtration   76
Target location   140
Target profile   150, 303
Tax incentives   273
Technical complexity   247
Technical risks   288
Technical specification   278
Technology transfer   352
TEM   127
Temperature logger   198
Tender vetting   305
Teratogenicity   152
Term sheet   288
Terminal sterile filtration   243
Termination fee   354
Termination of a contract   290
Tertiary structure   27
Testing costs   168
Testing status   170
Therapeutic dose   146
Therapeutic window   144
Thermal bridge   54

Thermal inactivation 75, 90
Throughput 90
Time of flight 124
Time point of investment 346 pp
Time-concentration curve 144
Time-point of market approval 110
Time-to-market 110
Tissue culture 58
Tissue engineering 16
Tissue-plasminogen activator 14
Titer 41
Tobacco plant 46
TOC method 133
TOF 123
Tolerable impurity 134
Tolerance formation 145
Tooth-brush effect 159
Topical application 142
Total body water 142
Total organic carbon (TOC) 133
Total protein content 106
Total spectrum of activity 146
Toxicity of metabolites 48
Toxicity 151
Toxicology 151
Traceability 99
Training 178, 188
Transcription 25
Transfection 25
Transfer panel 252
Transformed cell line 16
Transgenic systems 46
Translation 25
Trans-membrane flow 77 pp
Trans-membrane pressure 131
Transmission electron microscopy 127
Transport conditions 111
Transport protein 26
Transport validation 198
Trehalose 97
Trend analysis 178
Triplet 25 pp
TSE 134
Turbulent flow 270
Two phase extraction 88
Type A, B, C, D - reactions 160

u

Ultra/diafiltration 77
Unidirectional personnel flow 242
Unit costs 308
Unit profit 322
United States Pharmacopoeia 221

Upfront investment 238
Upstream 39
Use point 259
Useful economic life 335
User requirements 277
User requirements specification 196
Utilities 254 pp
UV/VIS spectroscopy 122
UV-detection 81

v

Vaccination 11
Vaccines 11
Validation 197
– batch 348
– master plan 199
– protocol 199
– report 199
– run 283
Value engineering 286
Value-benefit analysis 332
Variable costs 306
Vendor-managed inventory 361
Verum 156
Vial 96 pp
Virus 23
– contamination 23
– inactivation 90
– load 89
– production 11
– reduction 89
– testing 127
– validation 90
Visual inspection 131
V-model 195

w

Wage level 274
Warehouse-management system 199
Warning letter 216
Warranty 290
Washing step 81
Waste flow 243
Waste incineration 243
Waste management 263
Waste treatment 263
Water 254 pp
Water solubility 72
WCB 47
Western blot 117
WFI 61
WHO 221
Wholesaler 158
Wipe test 133

Working cell bank   47
World Health Organization   221

y
Yeast extract   61
Yeast   43

Yield loss   67
Yield   **41**, 66, 344

z
ZLG   219
Zoning concept   243